Reviews of Human Factors and Ergonomics

VOLUME 2 Edited by
Robert C. Williges

PUBLISHED BY THE HUMAN FACTORS AND ERGONOMICS SOCIETY

Reviews of Human Factors and Ergonomics

VOLUME 2 Edited by
Robert C. Williges

PUBLISHED BY THE HUMAN FACTORS AND ERGONOMICS SOCIETY

Human Factors and Ergonomics Society
P.O. Box 1369
Santa Monica, CA 90406-1369 USA
310/394-1811, Fax 310/394-2410
http://hfes.org, info@hfes.org

Copyright 2006 by Human Factors and Ergonomics Society

Individual readers of this book and nonprofit libraries acting for them are permitted to make fair use of the material in it, such as to make a copy for use in teaching or research. Permission is granted to quote excerpts from the book in scientific works with the customary acknowledgment of the source, including the author's name and book title, year, and page(s). Permission to reproduce any chapter or substantial portion (more than 500 words) thereof, or any figure or table, must come from the Human Factors and Ergonomics Society (HFES). Republication or systematic or multiple reproduction of any material in this book is permitted only under license from HFES. Address inquiries and notices to Communications Director, Human Factors and Ergonomics Society, P.O. Box 1369, Santa Monica, CA 90406-1369 USA; 310/394-1811, fax 310/394-2410.

Additional copies of this book may be purchased from HFES at the address above; $80 for HFES members, $95 for nonmembers, plus shipping/handling and sales tax if shipped to a California address. Multiple-copy discounts apply for orders of five copies or more. To order, visit the Publications page of the HFES Web site, http://hfes.org, or call or write to HFES at the address above.

ISBN 0-945289-27-8
ISSN 1557-234X

REVIEWS OF HUMAN FACTORS AND ERGONOMICS
Series Editor: Raymond S. Nickerson
Volume 1, 2006, Edited by Raymond S. Nickerson
Volume 2, 2006, Edited by Robert C. Williges
Upcoming in 2007: Volume 3, Edited by Deborah A. Boehm-Davis

Library of Congress Cataloging-in-Publication Data

Reviews of human factors and ergonomics / edited by Raymond S. Nickerson.
    p. cm.
  Includes bibliographical references and index.
  ISBN 0-945289-25-1 (alk. paper)
  1. Human engineering. I. Nickerson, Raymond S. II. Human Factors and Ergonomics Society.
  T59.7R49 2005
  620.8'2—dc22
                                                    2005020989

CONTENTS

Preface ix

Chapter 1. Situation Awareness Catches On: What? So What? Now What? 1
Yvette J. Tenney & Richard W. Pew

What? Defining and Measuring SA 2
Measuring SA 9
So What? 14
Now What? 28
Acknowledgments 30
References 30

Chapter 2. Crew Resource Management Training Research, Practice, and Lessons Learned 35
Eduardo Salas, Katherine A. Wilson, C. Shawn Burke, Dennis C. Wightman, & William R. Howse

Why CRM and CRM Training? 35
What Are CRM and CRM Training? 36
Theoretical Drivers of CRM Research and Training 37
What Have We Learned from CRM Research? 41
Observations About the Research and Practice of CRM 60
Where Do We Go from Here? 66
Conclusion 66
Acknowledgments 67
References 67

Chapter 3. Representation Aiding to Support Performance on Problem-Solving Tasks 74
Philip J. Smith, Kevin B. Bennett, & R. Brian Stone

Leveraging Human Cognitive and Perceptual Processes 75
Mappings: The Agent, the Domain, and the Interface 75
Representations to Support Anticipated Scenarios 77
Representations to Support Problem Solving in Unanticipated Scenarios 81
Interaction Design: Crafting the Details 92
Representation Aiding: Empirical Evaluations 94

Conclusions and Future Research 104
References... 106

Chapter 4. Usability Assessment Methods 109
Joseph S. Dumas & Marilyn C. Salzman

Usability Testing... 111
Inspection Methods... 118
Surveys, Interviews, and Focus Groups............................. 124
Field Methods.. 127
Discussion... 133
References... 136

Chapter 5. Satisfying Divergent Needs in User-Centered Computing: Accounting for Varied Levels of Visual Function 141
Julie A. Jacko & V. Kathlene Leonard

Divergent User Needs... 142
An Organized Approach to Accessibility........................... 143
A Framework of Interaction Thresholds from Functional Abilities .. 152
Deriving Performance Thresholds on GUI Interaction for Visual Impairments ... 155
Conclusions and Future Directions 160
Acknowledgments.. 162
References... 162

Chapter 6. Multidimensional Aspects of Slips and Falls 165
Krystyna Gielo-Perczak, Wayne S. Maynard, & Angela DiDomenico

System Components of Slips and Falls 166
Connectivity Considerations of the System Components 175
Considerations in Reducing Slips and Falls....................... 177
Future Directions 191
References... 192

Chapter 7. Protection and Enhancement of Hearing in Noise 195
John G. Casali & Samir N. Y. Gerges

What Constitutes Noise? 195
Noise Regulation and Hearing Conservation 202
Noise Influence on Hearing Loss 206
Noise Influence on Performance, Bodily Response, and Annoyance 209
Signal Audibility in Noise 212
Augmented Hearing Protection Technology for Improved Hearing in Noise 221
Conclusions 236
Acknowledgments 237
References 237

Chapter 8. Designing Effective Warnings 241
Kenneth R. Laughery & Michael S. Wogalter

Purpose of Warnings 243
Brief History of Warnings Research and Applications 245
Where Do Warnings Fit In? A Systems Approach 246
Theoretical Approaches 247
Design Factors That Influence Warning Effectiveness 249
Nondesign Factors That Influence Warning Effectiveness 255
Discussion 258
The Future 259
Conclusions 266
References 267

About the Authors 272

Index 281

PREFACE

Robert C. Williges

Volume 2 of *Reviews of Human Factors and Ergonomics* presents eight reviews of current topics of practical significance to human factors/ergonomics (HF/E) researchers and practitioners. Although some of the chapters focus primarily on research and others focus on applications, each chapter attempts to provide both future research and real-world application implications for the reader interested in the design of user-centered devices, systems, and processes.

The first three chapters deal with cognitive, organizational, and information-processing aspects of HF/E common to a variety of transportation, military, and occupational applications. In Chapter 1, Yvette Tenney and Richard Pew review various definitions, measures, applications, and future research directions in assessing an individual's situational awareness in complex systems. In Chapter 2, Eduardo Salas, Katherine Wilson, Shawn Burke, Dennis Wightman, and William Howse review the literature on crew resource management training by listing lessons learned from research and providing observations concerning future research and practice. In Chapter 3, Philip Smith, Kevin Bennett, and Brian Stone review representation aiding strategies in visual displays to support effective human problem-solving performance.

The next two chapters focus on usability considerations in product design. In Chapter 4, Joseph Dumas and Marilyn Salzman review major categories of usability assessment methods and describe new challenges in usability assessment. In Chapter 5, Julie Jacko and Kathlene Leonard extend the concept of usability to address the needs of users with varying sensory, motor, and cognitive capabilities. They demonstrate a framework based on the user's functional abilities for satisfying divergent user needs in user-centered computing for individuals with varied levels of visual function.

The last three chapters in Volume 2 deal with components of user safety. In Chapter 6, Krystyna Gielo-Perczak, Wayne Maynard, and Angela DiDominico provide a review of various biomechanical, tribological, and operational aspects involved in understanding the causes of slips and falls, and they describe how this information can be used in developing prevention strategies. John Casali and Samir Gerges review the role of noise in human hearing impairment in Chapter 7. They summarize a variety of methods for auditory danger signal design as well as hearing protection and enhancement techniques for use in noisy environments. Finally, Kenneth Laughery and Michael Wogalter review key factors in the design of effective visual warning displays in Chapter 8 and discuss future research and design opportunities.

The chapter authors and I gratefully acknowledge the helpful critiques provided by several of our scientific colleagues during the development of these chapters. Their suggestions and recommendations greatly improved the scientific quality, content coverage, and clarity of Volume 2. The individuals who provided peer reviews for Volume 2 include

Armando Barreto, Catherine Burns, Nancy Cooke, Mica Endsley, Rhona Flin, Jean Fox, Raoul Grönqvist, Ellen Haas, Lee Hager, Vicki Hanson, Deborah Hix, Michael Kalsher, Simeon Keates, Brian Kleiner, Mark Lehto, Mary Lesch, James Lewis, Arnold Lund, Kathleen Mosier, Robert Nullmeyer, Mark Redfern, Tonya Smith-Jackson, Randy Tubbs, Kim Vicente, Christopher Wickens, William Yost, and Jaije Zhang.

Series Editor Raymond Nickerson spent considerable time reviewing each submitted chapter and provided valuable comments on each chapter for final revision. Lois Smith at the HFES central office provided expert supervision in guiding the copy editing, production layout, and publication of this volume.

I appreciate the opportunity to serve as editor of Volume 2 and have a small role in establishing the *Reviews of Human Factors and Ergonomics* as a continuing publication series for HFES. My trusted friend, colleague, and mentor, Frederick Muckler, would have been proud to see this new publication finally come to fruition some 20 years after his initial attempt to start the series in 1984. I thoroughly enjoyed working with all my colleagues involved in various aspects of Volume 2 preparation, and their combined expertise and professionalism made my responsibilities quite straightforward.

All of us involved in the preparation of Volume 2 hope the readers will share our goal that the new *Reviews* series will quickly establish itself as the primary reference for an overview summary and status review of central human factors/ergonomics topics—one that is important to students, researchers, practitioners, and the informed public.

Blacksburg, Virginia
April 2006

CHAPTER 1

Situation Awareness Catches On: What? So What? Now What?

By Yvette J. Tenney & Richard W. Pew

> Human factors/ergonomics professionals regularly study the situation awareness (SA) problems of pilots, air traffic controllers, automobile drivers, power plant workers, ambulance dispatchers, urban search and rescue professionals, and unmanned vehicle operators, to mention a few. The challenge has been to define SA operationally and devise measurement strategies that focus on attention and recall on the one hand and relevant actions on the other. Although there have been many successes, challenges remain. This review is organized in three parts. Under "What?" we discuss definitions of SA, their interpretations in terms of human information processing and ecological psychological theories, as well as techniques for measuring SA. Under "So what?" we introduce SA applications—information requirements, technological tools, and training systems—that support the way people work and live. Finally, under "Now what?" we discuss future directions for SA research.

At the time of this writing, 10 years have passed since *Human Factors: The Journal of the Human Factors and Ergonomics Society* published a special issue on situation awareness (SA)—the ability to maintain the "big picture" in a dynamically changing environment—and in that time the topic has caught on (Gilson, 1995). SA originated in the aviation community to explain the perceptual skills responsible for a fighter pilot's success. Its opposite—*loss of SA*—was proposed as a more diagnostic and intuitive explanation for the cause of commercial aviation accidents than the ubiquitous "human error" label.

Originally of interest to aviation psychologists, the topic began drawing the attention of researchers in a number of domains. Some were skeptical at first, both about the need for a new term and the way it was defined. Turf battles ensued. However, the practical necessities of designing systems and training programs to improve performance and minimize accidents compelled the research community to seek common ground. As a result, significant progress has been made in both theory and practice. The questions posed in this review—"What?" "So what?" "Now what?"—were taken from Strater et al.'s (2004) depiction of the three levels of situation awareness in Endsley's (1987) definition.

The review begins with the definition of SA and the methodologies to assess it ("What?"), along with the controversies they spawned and the consensus that was reached. The second part of the review ("So what?") highlights significant findings across diverse domains in four practical areas: SA requirements, system design, training, and selection. The final section ("Now what?") discusses future directions—expected new technology trends and further questions for research.

WHAT? DEFINING AND MEASURING SA

Some Examples

Before reading a formal definition of SA, it may be helpful to consider a few examples. It is hard to pick up a newspaper without coming across a current event that illustrates good or bad SA. For example, in August 2005, we learned that all 309 flight crew and passengers aboard an Air France flight had the foresight to evacuate the aircraft after it overshot the runway in Toronto—in time to avoid the explosion that occurred five minutes after the aircraft landed. Fortunately, the crew was aware of the situation, including the time constraints, and acted accordingly.

Not long ago, a Yankee outfielder caught a fly ball and did nothing until he caught sight of a runner from the opposing team and realized it was not the third out. It is rare, but not unheard of, for a professional player to lose track of the number of outs, especially in a long inning.

In a lighter vein, the 18th-century French playwright Molière delighted his audiences through a literary device that capitalizes on a loss of SA. Two characters talk to each other, each one thinking the other is talking about a different situation than the one the other person has in mind.

It is noteworthy that these examples of SA, or the loss thereof, involve more than one person. The original definitions considered only the individual operator. As we will show, the expansion of the definition to teams is a recent, significant advancement.

Endsley's Definition

A number of definitions of SA can be found in the literature. Some are general; for example, "the congruence between the subjective interpretation of an event and objective measures of the actual event" (Flach, 1996, p. 3). However, it was Endsley's highly specific definition, created in the context of aviation, that caught on and has been adapted across a variety of domains:

> the perception of the elements in the environment within a volume of time and space, the comprehension of their meaning, and the projection of their status in the near future. (Endsley, 1988, p. 97)

Endsley's definition, with its clear, intuitive appeal and neat three-way classification, seemed to fit all domains. One did not have to be a psychologist to understand it and adapt it to one's own needs. For example, Wong and Blanford (2004) applied it to ambulance dispatchers; Scholtz, Young, Drury, and Yanco (2004) to robot operators; and Levin, Tenney, and Henri (2000) to cyber security managers. Endsley's definition appeared to serve everyone well.

Frameworks for Understanding SA

The practical appeal and widespread use of Endsley's definition did not mean that psychologists agreed on how to think and talk about SA. Although none of the perceptual

Situation Awareness Catches On: What? So What? Now What?

theories had addressed the perception of units as large as situations (usually going no further than isolated events), it was inevitable that some of the same arguments, such as those between the information-processing and ecological schools, would resurface regarding this new phenomenon. As one set of reviewers of the topic remarked, "It is interesting that two of the frameworks paramount in basic cognitive research, often existing in tension or as mutually exclusive alternatives, are reflected in this applied area" (Durso & Gronlund, 1999, p. 285).

Table 1.1 summarizes some differences between the information-processing (Neisser, 1967) and ecological psychology (Gibson, 1979) approaches. Psychologists in both camps have contributed solutions to problems of training, selection, and interface design. That said, we find the differences in emphasis and language to be fascinating and worth recounting.

Information-processing psychologists see SA as separable from performance. They like to point out that good performance can accompany poor SA (as, for example, when you do not know why your computer is behaving strangely but, after trying a limited set of possibilities such as rebooting, the problem is fixed) and, conversely, that bad performance can accompany good SA (for instance, when time pressure or other forms of stress produce poor execution). Some see the acquisition of SA as a multistage process in which disparate elements are integrated into a meaningful whole, or product. They distinguish the product of SA (what can be reported at any given moment) from the process of obtaining it, referred to as *situation assessment* (e.g., active search of the environment or team communication patterns).

Although it is possible to derive both kinds of measures, information-processing psychologists have tended to favor product measures (e.g., responses to midscenario probe questions). They like to speculate on where the contents of SA reside—whether in short-term memory, working memory, or, most recently, long-term working memory. They like the notion of long-term working memory because it explains why experts have such remarkable SA (Ericsson & Kintsch, 1995). Chess masters, for example, can recall the

Table 1.1. Two Theoretical Views of Situation Awareness

Aspect	Information Processing	Ecological Psychology
Emphasis	"What's inside your head"	"What your head's inside of" (Mace, 1977)
Key ideas for SA	Mental representation, long-term working memory	Direct perception, affordances of objects, events [and situations]
Goal of research	Diagnostics—what is noticed and remembered; what is overlooked	Adaptation—how does organism learn to distinguish situations and their action implications?
Nature of stimulation	Sensation-based, elemental	Rich, nested structure, invariants over time and space
Meaning and action potential	Derived from memory, knowledge, inference	Perceived directly as affordances of objects and events

Sources: Neisser (1967), Gibson (1979).

placement of every chess piece on the board at any point during a game while showing very poor memory for randomly placed pieces. Rather than memorize the placement of each piece, experts need only take note of the pattern, store a pointer to that pattern in short-term memory, and then retrieve the pattern from long-term memory when needed. The SA of novices, on the other hand, places a heavier burden on short-term memory, a less effective strategy for maintaining awareness (Sohn & Doane, 2004).

Finally, information-processing psychologists distinguish SA from workload, which has to do with competing demands on limited cognitive resources. Situation assessment is thought to require cognitive resources and therefore contributes to workload. They note that it is easier, therefore, to maintain SA when workload from other competing demands is low and that, without extra effort, SA will suffer under high-workload conditions (Wickens, 2002). Even under reasonable workload, however, they are concerned about how easily events—even bizarre and highly striking events—can be missed when they are outside the immediate focus of attention. They term this side effect of focused attention *inattentional blindness* or *change blindness* and see it as a significant safety issue for operators of complex systems (Durlach, 2004; Varakin, Levin, & Fidler, 2004).

Ecological psychologists, on the other hand, are less enamored of the idea that SA and performance can be out of synch. They emphasize the close relationship between perception and action that occurs in the natural environment and the importance of the larger structure, or context, surrounding a situation. To take an example, a doctor may have good SA with respect to the patient's diagnosis but show poor judgment in advising the patient. Rather than dismiss this case as an example of poor performance with good SA, an ecological psychologist would want to understand the SA that would be required in the larger context that included patient behaviors and reactions. This context would certainly include the doctor's need to gain the patient's acceptance for treatment options.

Two points emerge from this way of looking at situations. First, ecological psychologists feel that situations have indeterminate boundaries—they can be conceived of in a narrow context (e.g., the diagnosis) or in a larger context (the patient-doctor relationship), so there is not a single boundary. Second, in the ecological view, SA is tightly bound to actions. Once one becomes fully aware of the situation (how the patient is likely to react under particular circumstances), the action implications are usually clear (for fostering compliance).

What, then, would an ecological psychologist make of the compelling demonstrations of inattentional blindness when, for example, participants who are instructed to focus on one of two superimposed films of basketball teams at practice fail to notice a woman with a red umbrella as she walks across the court, let alone the presence of a gorilla (Neisser & Becklen, 1975; Simons & Chabris, 1999)? To an ecological psychologist, these experiments are interesting but secondary. Ecological psychology argues that illusions (including illusions of normalcy in the inattentional blindness studies), though intriguing, are not the basis on which to form a theory of perception; they are the exception rather than the rule. Ecological psychology is more interested in the fact that participants, even when faced with a three-ring circus, are able to perceive the actions of whichever event they choose to follow. From this perspective, this demonstration says more about "how operators discover meaning within complex work domains" (Flach, 1996, p. 3) than about how they can be fooled.

Situation Awareness Catches On: What? So What? Now What?

Finally, one could ask what an ecological psychologist would say to mode-change blindness—for instance, a tendency for operators not to notice automation-induced mode changes when a function unexpectedly switches from automated to manual control (Sarter & Woods, 1995). Gibson did not live to see the computer age. If he had, it is safe to say that he would have seen the issue as one of poor design. He would say that designs with the right affordances, or features, should invite or demand the proper actions from the operator.

To summarize, the distinction between the two schools with respect to situation awareness can be expressed as follows: Information-processing psychologists emphasize the limitations of attention, whereas ecological psychologists ponder the opposite, action-oriented question—why we so rarely walk into walls or fall off cliffs.

Achieving Common Theoretical Ground

Although theoretical disagreements persist, fortunately, they have not impeded research. The need to understand SA in enough depth to improve performance and prevent accidents has led to a dialogue that has advanced both theory and practice. One attempt at synthesis is the perceptual cycle, which emphasizes the importance of characteristics of situations as well as the nature of awareness.

Perceptual cycle. Neisser's (1978) perceptual cycle has been adapted by a number of theorists concerned with SA (e.g., Adams, Tenney, & Pew, 1995; Durso & Dattel, 2004; Flach, 1995; Klein, Phillips, Rall, & Peluso, in press) and has much in common with Endsley's (1988) model. It is not surprising that Neisser's ideas have had broad appeal, given that he switched camps from information processing (Neisser, 1967) to the ecological view (Neisser, 1978) and that his perceptual cycle merges these approaches by emphasizing both the environment and the operator. The perceptual cycle is a flow model in which knowledge, in the form of schemas (called *products* by some), influences looking and exploratory behavior; these perceptual processes, in turn, influence the content of the schema. The updated schema redirects looking behavior recursively, yielding a continuous cycle. With respect to the environment, the cycle recognizes the potentially unlimited amount of information available either immediately or through further exploration. With respect to the operator, it recognizes that people have expectations that lead them to perceive certain information more readily than other information.

Neisser's cycle has two important implications for SA. First, it supports the rapid, holistic perception of experts, such as the chess players observed by Chase and Simon (1973). Knowledge of the game leads to expectations, which, when the pieces are properly placed, leads to rapid grasp of (and memory for) the entire board. Second, it introduces the possibility of bias, when expectations lead one to overlook subtle differences between situations. Thus, a situation in which the pilot experiences inadequate lift because he has forgotten to set the flaps on takeoff can lead him to attribute the problem to wind shear, especially if it is raining and he has been warned to expect it. The cycle does not stop there, however. Theoretically, at least, the pilot could seek further information to disambiguate the situation, either alone or with help from crew members.

Characteristics of situations. Interestingly, there has been relatively little discussion of situations, even among ecological psychologists who have taken pains to describe the environment to be perceived. Some have explicitly called for such a discussion (Flach, 1995). What distinguishes one situation from another? How can different kinds of situations be classified? Some have proposed treating complex situations, often encountered in work settings, as a set of simpler, concurrent situations, each reflecting a high-level goal and entailing a set of prescribed tasks and procedures (Deutsch, Pew, Rogers, & Tenney, 1994). According to this view, events that are either externally driven or produced through operator actions occur within a situation. Furthermore, each concurrent situation has associated with it a set of possible futures, depending on the operator's actions. Unexpected happenings give rise to new goals with new concurrent situations. SA depends on differentiating a situation from possible similar situations as well as determining its scope (e.g., are we facing a series of coincidental failures or an act of sabotage?). Situations can be characterized as routine, clear, and easily managed at one extreme and dangerous, confusing, and requiring extraordinary expertise at the other.

In the last of his classic works in which he described the world to be perceived, J. J. Gibson (1979) included a brief mention of episodes. (Interestingly, he chose to use the word *episode* rather than *situation*.) He viewed an episode as consisting of nested events:

> The flow of ecological events consists of natural units that are nested within one another—episodes within episodes, subordinate ones and superordinate ones.... If we can understand the nested sequences it may be possible to understand how it could be that in some cases the *outcome* of an event sequence is implicit at the *outset*—how the end is present at the beginning—so that it is possible to *foresee* the end when an observer *sees* the beginning. (Gibson, pp. 101–102)

In this short passage, Gibson tried to do for situations (or episodes) what he did earlier for the stimulus—namely, to point out that they consist of nested units of structure and can be viewed at different levels of granularity. Depending on expertise, goals, and motivation, different participants will attend to different aspects of "the stimulus" or "the situation." Finally, for Gibson, attention to higher-order structure makes anticipation—Endsley's highest level of SA—possible.

The nature of awareness. Why call it *situation awareness* and not *situation perception*, as we do when we speak of object or event perception? Is the notion of awareness important? The kind of awareness required can differentiate among situations. Situations may require awareness that there is a problem (noticing something needs attention), making discoveries (seeing a link between two symptoms), finding explanations (diagnosis), preventing accidents, and preparing for future actions (Klein et al., in press).

Which of these types of situations qualify as pertinent to SA specialists, and where do we draw the admittedly fuzzy line in selecting relevant studies? For this review, we looked for situations that either rely on or are likely to result in the operator's ability to remember and report details. We focused on these studies because SA is often thought of as a product (i.e., something that can be reported). We also included studies in which SA is inferred from appropriate actions to hazards, abnormal states, or other unexpected events. The detection of abnormal states is close to the topic of system diagnosis, for which there already is a large literature (e.g., Rasmussen & Rouse, 1981; Swets & Pickett, 1982). We feel that diagnosis is an important part of SA and have included some recent studies involving medical diagnosis.

Another area in which the boundary between SA and other topics becomes fuzzy concerns guidance systems. Does a system that guides a pilot or a driver to a location through visual, verbal, or even tactile directional prompts enhance the operator's SA or just improve performance? We decided that directional guidance, though it could be considered a case of good performance with poor SA (when one follows the directions without forming a cognitive map), nevertheless does provide an important kind of awareness—namely, that one is on track and on the way to a destination—so we decided to include it.

Team SA

Although SA has long been assessed as part of team training (see Salas, Wilson, Burke, Wightman, & Howse, 2006), it has taken a while for this aspect of SA to catch on. The importance of team training and the recent popularity of multiplayer games no doubt have stimulated further research on enhancing SA in teams.

What is a team? Team members have common goals and interdependent roles (Salas, Dickinson, Converse, & Tannenbaum, 1992). "In a team, the success of each person is dependent on the success of other team members" (Endsley, Bolte, & Jones, 2003, p. 194). Similarly, one can define a "team of teams" in which the individual teams share goals and play interdependent roles. Thus, team SA is "the degree to which every team member possesses the SA required for his or her responsibilities" (Endsley et al., p. 195). It includes awareness of how your work affects what others are doing, and vice versa.

Given this formulation of the needs of team members, what can be done to foster team SA? Endsley et al. (2003) proposed three solutions. The first is use of shared displays and communication devices, as well as the benefits that accrue from being colocated. Being able to observe where the other person is looking provides information about what the other person is attending to and what he or she may have missed. The second consists of establishing shared mental models, or a common understanding of goals, tasks, and procedures. Team members acquire shared mental models by working together or engaging in cross training. The third consists of incorporating practices that make for good teamwork; these include contingency planning, making sure everyone understands and is pursuing a common goal, and fostering an atmosphere that encourages asking questions.

Modeling SA

Another approach to the definition of SA that has received a lot of attention in recent years is through computer modeling of the behavior of the operator whose performance is being analyzed. The goal of these models is to be able to predict performance from an understanding of the task and environmental constraints, but in the process, the model must make some assumptions about how a situation is assessed and what the product of that assessment is. There is nothing quite so illuminating (and challenging) as the requirement to specify in sufficient detail how SA takes place so it can be coded into a model. A number of cognitive architectures and models in the literature describe human performance generally (see Pew & Mavor, 1998; Ritter et al., 2003), but only a few purport to represent SA explicitly. We describe a few that seem to have made the most progress with respect to modeling SA.

Wickens and his colleagues (McCarley, Wickens, Goh, & Horrey, 2002; Wickens et al., 2005) developed a model of situation assessment, grounded in information-processing theory, that builds on the allocation of attention to events and locations (referred to more generally as *channels*). SA is treated in the aggregate as an average over the specific content of the information acquired. SA is assumed to improve as valuable events are attended to and to decay over time, especially when workload is spent on unrelated activities.

These authors define the probability of attending to an event or area of the visual field (which may contain specific displays) as dependent on three variables that are defined in advance, specifically for the domain and tasks under study: (a) the salience of the events; (b) the value of an event, wherein value is defined by its intrinsic contribution to SA for the task at hand; and (c) the likelihood that it will change since last inspected. They also considered a fourth variable, the effort required to move from the currently attended event to the new event (e.g., the size of the eye movement required), but in validation studies, this variable was shown to play only a minor role in the results. The investigators carried out three validation studies in the aviation domain, one applied to predicting taxi behavior on the ground at O'Hare Airport and two applied to evaluating the usefulness of synthetic vision displays for approach and landing (Wickens et al.).

Warwick and his colleagues (Warwick, McIlwaine, Hutton, & McDermott, 2001) described a model whose goal is to describe decision making according to Klein's concepts of *recognition-primed decision making* (Klein et al., in press). The model supports several parallel processes internal to the operator as well as external processes determined by the environment in which the operator works. As does the Wickens model, it incorporates an explicit information-processing theory of SA. The operator's primary cognitive resources are devoted to noticing and remembering cues from the environment. When the cues are noticed, thus updating perception, they are stored in limited-capacity working memory—limited in terms of both how many cues can be remembered and how long any one cue will be remembered. The cue-updating process is repeated as each new cue is noticed, updated, and interpreted until some time, defined independently of the cue-sampling cycles, when a recognition-primed decision is triggered. At that time, SA is made up of the current set of cues combined with data derived from long-term memory. The long-term memory representation is assumed to be a table having one row of data representing each situation the decision maker has experienced, together with the expectancies generated from SA for that situation and an associated appropriate course of action. The model has been used to represent the SA of an automobile driver, air traffic controller, and soldier, but it has been validated only in the sense that it has been shown to be useful in these applications (Warwick et al., 2001).

The third model is the work of Zacharias and colleagues and is called Situation Awareness Model for Person-in-the-Loop Evaluation (SAMPLE; see Hanson, Sullivan, & Harper, 2001; Pew & Mavor, 1998). This model treats SA as a diagnostic reasoning process under uncertainty and employs belief networks. It might be considered to represent Neisser's perceptual cycle. Situations are hypothesized to be "states of the world"—at least, the possible states within the bounds of the task environment under study. These possible states must be enumerated in advance but may include a "catch-all state," anything not enumerated in the remainder of the network. Belief networks are used to represent the inferential process linking raw beliefs to situations. A belief network, which also must be populated in advance, consists of a hierarchical network of nodes, each of which contributes a Bayesian probability of each situation element, given the cumulative probability of the previously aggregated beliefs. Events in the environment are sensed and converted into beliefs about likely situations using fuzzy logic interpretations. Thus, SA is defined as the distribution of the various probabilities of the possible alternative states of the world given the sensory data or events in the environment and their diagnostic value for assessing the states of the world.

The SAMPLE model has been applied to situation assessment for a tactical aviation pilot (Mulgund, Harper, Zacharias, & Menke, 2000), unmanned air vehicle management (Hanson et al., 2001; Hanson & Harper, 2000), distributed decision making for air traffic management (Hanson, Harper, Endsley, & Reszonya, 2002), and military operations in urban terrain (Aykroyd, Harper, Middleton, & Hennon, 2002). It has been shown to have face validity but has not been validated empirically.

It should be evident from these brief summaries that modeling of SA is in its infancy. Although it could be argued that each project represents the germs of a "theory" of SA, until such quantitative models are more extensively validated, we are left with much room for discussion about what really matters in SA.

MEASURING SA

Techniques for measuring SA have proliferated, generating great interest and discussion. The techniques fall into five categories: recall, anticipation (part-task), critical events, subjective ratings, and physiological indicators. Although the research community has been discussing these techniques for at least 10 years, their relative strengths and weaknesses are still subject to debate (Durso & Dattel, 2004; Durso & Gronlund, 1999; Endsley, 2000; Endsley et al., 2003; Jones & Endsley, 2000; Pew, 2000; Snow & Reising, 2000).

Today, however, these techniques are not just being talked about—they are being put to use, separately and in combination. In this section, we describe the techniques and illustrate each with examples. In the next section, we consider more explicitly what can be gained by applying these techniques to solve practical problems.

Recall Techniques

Recall measures of SA typically require the participant to answer questions about a dynamic situation while it is happening. The questions are often asked by means of probes inserted into a scenario. The usual procedure is to freeze the simulation briefly at random points, make the screen go blank, and present the participant with a set of randomly selected probe questions regarding the scenario (Endsley, 2000). The SA score is the number of probes of each type answered correctly.

There are several variations of this procedure, all of which have both face validity and practical utility (Endsley, 2000; Jones & Endsley, 2000). Recently, researchers have begun to investigate and compare them more thoroughly (Durso & Dattel, 2004; Venturino, 1997; Vidulich, 2000). We review this progress briefly, beginning with the first direct measure (SAGAT), the concerns it generated, and the research it spawned.

SAGAT. Working in the area of aviation, Endsley (1987, 2000) introduced a probe method of assessing SA, the SA Global Assessment Technique (SAGAT). To assess SA, the screen is periodically blanked and a set of questions is displayed. Sample questions might be "What is the altitude?" "What is the distance to the next waypoint?" "How close is another aircraft?" When the participant responds or exceeds the time allowance for a response, the simulation resumes. The percentage of probes of each type answered correctly constitutes the measure(s) of SA. (Each probe question is considered a *type*.) For this measure to be valid, care must be taken to create questions that adequately reflect the SA requirements for the scenario.

Some sample probes that have been used to assess SA in aviation and other domains include the following:

- During a combat simulation: "Estimate how much you are currently above or below your commanded airspeed" and "From your current position, what is the best escape route to avoid terrain?" (Snow & Reising, 2000, p. 3.51)
- During an air traffic control simulation: Indicate for all aircraft, "Flight level, Ground speed, Heading, Next sector, Destination, . . . Conflict, Type of conflict, Time to separation violation, Call sign of conflicting a/c." (Hauss & Eyferth, 2003, p. 424)
- At the conclusion of a live gallbladder surgery: "Were there stones in the gall bladder in the preoperative ultrasound?" "How many clips (proximal and distal) were placed on the cystic artery?" and "Did any adverse events, 'near-misses,' or errors occur during this case?" Answers were scored with reference to video and digital recordings taken during the surgery. (Guerlain et al., 2005, p. 33)

Concerns about SAGAT. The practical appeal and use of SAGAT have raised some issues. For example, does SAGAT require operators to keep more in their head than they normally would? Does SAGAT interfere with SA by providing an unnatural interruption? Alternatively, does taking the test enhance SA by judiciously directing attention to variables that might otherwise be overlooked?

Building consensus regarding SAGAT. The widespread use of SAGAT has led to a number of carefully designed experiments that are helping to build consensus regarding the validity of the technique. For the most part, the results have been positive, lending credence to the technique (Endsley, 2000). Certain results continue to suggest that caution be exercised in the design and interpretation of probe studies. Here we provide some illustrative examples of how rigorous experimentation is leading to greater understanding of the strengths and weaknesses of SAGAT-like techniques. It should be noted that the other types of measures have not yet been subjected to the same scrutiny.

Concerning the first issue—memory load—Jones and Endsley (2000) compared performance in a surveillance and targeting simulation task under two probe conditions. The first condition consisted of the conventional use of SAGAT, in which the simulation was frozen periodically, the screen was blanked, and a probe question appeared. In the second condition—a variation of SAGAT called SPAM (Situation Present Assessment Method; see Durso & Dattel, 2004)—the probe appeared while the simulation was still running, so the participant could seek the needed information rather than relying on memory. The SA measure was time to respond. The results showed that the scores on the two tests yielded similar results, suggesting that both were measuring SA and could be used interchangeably.

However, another experiment, extending early work by Yntema and Mueser (1960), suggested that there might be a limit to how many attribute changes can be tracked simultaneously. Demonstration of such a limit would have practical implications for the design of SAGAT tests. In this study (Venturino, 1997), participants were asked to play the role of a fire station dispatcher. The task consisted of keeping track of the current values of attributes for one or more fire engines. Performance was nearly perfect when participants had to track the value of two different attributes, one for each of two engines (e.g., the number of gallons of water for a tanker engine and the location of a pumper engine). Performance was still high, at 80%, when participants were asked to keep track of a different attribute for each of four types of fire engine. However, performance dropped radically when, instead of keeping track of a different variable for each engine, the dispatchers had to keep track of the same set of attributes each time. It made no difference whether the attributes were associated with one object or multiple objects; in each case, tracking multiple instances of the same attribute proved difficult.

In discussing implications of the results for SAGAT, Venturino noted, "The robustness of memory for dynamically changing variables must be questioned, particularly when those variables occur within an information stream consisting of other similar changing variables" (p. 538). These findings do not invalidate SAGAT but, rather, suggest caution in the construction of tests to avoid imposing unnatural memory requirements.

The second and third issues concern the effect of the test itself. The test could affect SA for at least two reasons. First, it might require effort to reorient to the situation following the interruption posed by the probe. Second, the content of the probes could affect SA by directing attention to aspects of the scenario that otherwise might have been ignored. Both these influences were examined in an experiment involving a simulated driving task (McGowan & Banbury, 2004).

Participants in this study were asked to watch video footage of the road ahead, taken from the vantage point of a driver, and click on the location of any hazard to which they would need to react if they were driving. Time to anticipate the hazard was the measure of SA. Embedded in the video were 70 events that experts had marked as hazards. Experts also indicated the earliest and latest time that each hazard could be detected. A 10-second interruption preceded half the hazards. There were four kinds of interruptions, only two of which required a response: (a) SA query (e.g., "Where is the pedestrian?"), (b) orienting interruption (e.g., "Pay attention to pedestrians."), (c) irrelevant query (unrelated to hazards), and (d) pause interruption (blank screen for 10 seconds).

An interruption was introduced by freezing the scenario at a random point during the interval in which the hazard was visible. The scenario resumed 10 seconds after the beginning of each interruption. Participants had to respond to each hazard whether or not an interruption preceded it, as soon as they detected it. They were scored on how quickly they responded within the time window defined by the experts. Each participant was assigned to one of the four aforementioned interruption conditions.

The results in McGowan and Banbury's study showed that the type of interruption affected how soon the hazard was detected. The fastest detection occurred with the SA query, followed by the orienting interruption. Both interruptions led to faster detection than did the pause interruption, suggesting that SAGAT-like tests could overestimate SA by directing attention to the important variables, especially if the queries were repeated. One way of mitigating this effect is to include many different queries so participants will not be expecting any particular one. The results also showed an increase in time to detect hazards following an interruption regardless of the type of interruption. Thus SAGAT-like tests may underestimate SA simply by interrupting processing of the event.

These results do not invalidate SAGAT. In fact, Endsley has reported a number of experiments that show no effect of interruptions (Endsley, 2000). Nonetheless, these concerns need to be kept in mind in designing SA tests.

Recall Techniques Applied to Teams

One way of measuring SA in teams is to extend the probe method to include questions not only about a participant's own memory for particular parameters but also about what a participant knows about a teammate's understanding of those parameters and the allocation of tasks.

Andersen, Pedersen, and Andersen (2001) applied this technique to a study of anesthesiologists in a medical simulation. Each time the scenario was frozen and the screen blanked, participants were asked for their own estimate of parameters, their estimate of their colleague's knowledge of parameters, their own knowledge of task assignments (roles and responsibilities), and their estimate of their colleague's knowledge of task assignments. The answers to these questions, scored both objectively and with respect to the answers provided by the appropriate team member, constituted the measure of team SA.

Similarly, Entin and Entin (2000) measured team SA in a command and control simulation by presenting a series of questions at the conclusion of the scenario. Participants were asked to remember, with respect to each of a series of events in the scenario, what task each of their teammates was performing when the event occurred. They also had to indicate what task they themselves had performed. Team SA was scored as the congruence between the team's judgments about their teammates' activities and what participants themselves said they were doing. The congruence score was validated by showing that it correlated positively with a subjective measure: observer ratings of teamwork.

The studies just described were predicated on the assumption that good teamwork requires an awareness of how tasks are delegated and what other members are doing and thinking. Lack of such awareness can affect team performance, as the accident literature attests.

Anticipation Techniques (Part-Task)

SA can also be measured with part-tasks that involve detecting hazardous conditions at strategic moments—hazards that, if not properly addressed, could lead to errors or accidents. For example, air traffic controllers had to detect conflicts in 90-second scenarios that involved altitude restrictions versus free flight (Remington, Johnston, Ruthruff, Gold, & Romera, 2000); soccer players had to judge the direction that a dribble/pass would take, given video footage excerpted from an actual game and paused at a key moment (Ward & Williams, 2003); and anesthesiologists had to detect and explain the cause of events depicted via readings taken from an operating room (Michels, Ing, Gravenstein, & Westenskow, 1997).

The advantage of using part-tasks is that trials are short enough that multiple trials can be administered. The disadvantage is that the short trials may not be representative of the workload demands posed by a longer scenario.

Critical Event Techniques

Another popular technique involves planting a critical event—often involving an anomaly or problem of some sort—into a scenario and observing when and how it is handled. A favorite technique for creating scenarios with the potential for error is the so-called garden path. A garden path scenario is one in which the operator is led to make certain assumptions and predictions about a scenario that turn out not to be true. SA requires that the person recognize that he or she is on a false path before making an error.

For example, cadets in a live military exercise were told to expect to encounter guerilla forces at a certain location. The test of SA was whether they were able to ascertain that the prior intelligence they had received was incorrect and that the people were innocent civilians (Strater et al., 2004).

Subjective Techniques

Self-ratings. Several scales have been created for self-reports of SA (see Pritchett & Hansman, 2000). Because they are subject to possible biases on the part of participants, they are frequently used in combination with other measures. The most frequently used scale is the SA Rating Scale, or SART (Taylor, 1990; Taylor & Selcon, 1994), which consists of three subscales. Some researchers have questioned the usefulness of the scale because it seems to confound SA with workload demands (Endsley et al., 2003).

Expert judgments. Expert rating scales are frequently used to assess team SA. Often the ratings are tied to critical incidents planted in the scenario (see Salas et al., 2006). For example, in a recent study, experts rated two-pilot aviation teams on their handling of 21 critical events using a three-point scale. The scale, taken directly from Endsley's definition of SA, was 1 = *notice,* 2 = *understand,* and 3 = *anticipate* (Hormann, Banbury, Dudfield, Lodge, & Soll, 2004).

Physiological Techniques

Although the field of neurophysiology has made strides in measuring physiological functions that are indicative of awareness, thus far only eye movement recording has become popular with SA researchers. It is assumed that the direction of eye gaze indicates the aspect of the environment to which the observer is attending. This approach has face validity, but caution must be exercised because it is entirely possible that the observer is staring at one object and thinking about something different. As with the subjective measures discussed earlier, the use of eye movements is frequently accompanied by other measures. Discussion of eye movements and other physiological techniques can be found in Kramer and McCarley (2003).

To summarize, this section has focused on the basics—the "what" of defining and measuring SA. Although comparisons of the different measures abound, including excellent tables summarizing their strengths and weaknesses (e.g., Endsley et al., 2003; Pritchett & Hansman, 2000), our take-home message is simple: *There is no single right way to select or create measures.* The goal, as in other areas of research, is to get meaningful, clean results. If one measure is showing a ceiling or floor effect, try another measure. Multiple measures are always a good idea and increase the chances that one will work. If two measures show the same trend but the absolute values are different, don't worry. However, if the type of measure produces different trends for different conditions, then you have an interpretation problem and will need to experiment further.

With respect to SAGAT, it is nice when different probe questions yield comparable results. Interpretation can be tricky when probe questions yield different results. The differences could be diagnostic (e.g., one display really is better than the other with respect to a particular type of information) or spurious (e.g., the assumption, based on task analysis, that both types of information are equally important in the scenario could be mistaken). When participants consistently succeed or fail at answering a probe, interpretation is easy. When they fall in between, interpretation is more difficult because there is little basis for predicting performance. What are the implications for performance when a variable in a probe is answered correctly 64% of the time—is that good or bad?

Finally, although some believe that different schools of psychology prescribe different measurement techniques, we do not believe that is the case. Gibson, for example, relied on verbal reports as well as performance measures in many of his studies of perception.

In the next section, we consider actual problems involving SA and how they have been addressed through training, selection, and system design. Only through a consideration of actual results and accomplishments can one understand why SA has caught on. It is time to ask the practical question, "So what?"

SO WHAT?

In this section, we consider the practical implications of studying SA by discussing studies specifically aimed at ameliorating SA, when it is weak or incomplete, across a variety of domains (see Table 1.2). In these domains, safe operation hinges on the operator's ability to make accurate and timely assessments of the flow of events, often under ambiguous circumstances and high workload. Mitigation strategies begin with an understanding of SA requirements and range from better design to better training and selection.

Table 1.2. Studies by Application Area

Application Area	SA Requirements	Technology Impact (+)	Technology Impact (-)	Training	Selection
Air traffic control and flight/aviation		Farley et al., 2000; Jennings et al., 2004; Olmos et al., 2000	Endsley et al., 1997; Remington et al., 2000	Hormann et al., 2004	Carretta et al., 1996; Gopher, 1982; O'Hare, 1997; Roscoe, 1997
Driving		Reagan et al., 2004	Drews et al., 2004; McCarley & Vais, 2004; Strayer et al., 2004	Fisher et al., 2004	
Infantry				Strater et al., 2004	
Groupware	Gutwin & Greenberg, 2002				
Medical and medical dispatch	Gorman et al., 2000; Wong & Blandford, 2004	Michels et al., 1997		Guerlain et al., 2005	
Robotics	Casper & Murphy, 2003; Scholtz et al., 2004				
Sports					Ward & Williams, 2003

SA Requirements

The first step in any SA amelioration project or experiment is to determine the information requirements. This process involves applying cognitive engineering techniques to understand the tasks, roles, environmental conditions, and work practices particular to a domain (e.g., Cooke, Stout, & Salas, 2001). Here we discuss some interesting cases in which requirements analysis proved to be especially valuable.

Groupware. As groupware applications become part of the Internet, the idea of distributed, synchronous work is becoming a reality. The notion of two or more people collaborating on a work project using the same application at the same time but from different locations—for example, to produce a drawing, page layout, or outline—introduces new SA requirements that are not supported on the Internet (Gutwin & Greenberg, 2002). This research suggests the following SA requirements for groupware collaborators:

- *Who?* Collaborators need to know who their partners are and who is at work on the task. Possible solution: include participant list with status information.
- *What?* Collaborators need to know what the other person is doing. This requirement becomes a problem when people scroll to different parts of a window or navigate to different windows and cannot see one another's actions. Possible solution: Include a secondary display showing individual viewpoints.
- *When?* Collaborators need to determine the best time to share findings or offer assistance. They need to understand enough of what the other person is doing to anticipate that person's needs and to avoid interrupting at inopportune times. Possible solution: Include individual viewpoints.

As this list shows, adding only one additional person to an application creates a host of awareness requirements that will require new Web features. Add a whole team, and the problems multiply.

Medicine—intensive care. The medical community has long been interested in problems of maintaining SA in settings such as operating rooms and intensive care units. This interest has grown as efforts have intensified to reduce the number of medical errors.

Gorman et al. (2000) observed nurses and other personnel in intensive care and cardiac intensive care units. Interestingly, the nurses saw their job as one of "getting the story straight, thinking the problem through" (p. 22). They formulated the following requirements: observing whether patient has been stabilized; determining the effects of medications; keeping track of informal thoughts, questions, and suggestions concerning care of a patient; and ensuring shift continuity.

The researchers noted that these requirements are now being served by low-tech artifacts: the Cardiac Unit Flow Sheet, a timeline with columns indicating time progression and separate rows for each vital sign and each medication, and Kardex, a temporary, erasable sheet that stays with the patient and then is destroyed. Gorman et al. pointed out that when introducing new automation, it is important to take into account the function of these artifacts in promoting team SA.

Situation Awareness Catches On: What? So What? Now What?

Medicine—emergency dispatch centers. Most recently, attention has turned to the SA requirements involved in dispatching emergency medical technicians. Wong and Blandford (2004) observed a European center with 400 ambulances that received 3,500 calls a day requesting help. Their report suggests the following SA requirements and strategies:

Maintaining an up-to-the-minute picture of available resources. Dispatchers try to maintain the big picture of how their resources are allocated. Interestingly, they say they do not try to keep track of the exact location of ambulances en route to a scene. However, when they receive a radio call indicating that an ambulance is returning, they track it precisely so they can reroute it, if necessary, before it gets back to the center.

Anticipating further need for resources. Dispatchers report using visualization strategies to fill in the details of what they hear on the phone. They try to imagine possible outcomes so they can reserve resources in case the situation worsens.

Recognizing false alarms. Dispatchers have to be alert to the possibility that the call might be a hoax. On the other hand, they do not want to deny true requests.

In short, trainers and system designers interested in enhancing the SA of medical dispatchers need to be mindful of the mental effort involved in listening to, visualizing, and anticipating the needs of emergency victims. It seems likely that maintaining an atmosphere free of visual distractions would be crucial for these kinds of activities.

Robotics. The use of robots to perform dangerous rescue operations is now a reality. Controlling a robot requires that the operator maintain SA of the status and location of both the robot and the victim. A Robot Rescue Competition takes place annually (e.g., Scholtz, Young, Drury, & Yanco, 2004). Robots have also been observed in real rescue operations, such as at the World Trade Center following the September 11, 2001, terrorist attack (Casper & Murphy, 2003). These studies point to the following SA requirements:

Determining the state of a victim remotely. Once a body is recognized, the state of the victim must be determined. In the case of a live victim, rescuers might dispatch a robot to bring medical supplies. Robots were not employed in this way at the World Trade Center.

Avoiding hazards. A major difficulty in using robots is that they are likely to become stuck, entangled, or destabilized. They also are prone to sliding on slopes. Operators need to anticipate, as much as possible, when a robot is likely to get into trouble.

Diagnosing robot problems. A robot may malfunction for a number of reasons. It may be trapped, or there could be a break in communication. When a robot ceases to function, the controller must determine the cause and whether it can be fixed remotely.

Keeping track of location. Operators need to be aware of the robot's path to avoid missing areas or covering the same ground. Limited camera views, lack of information on camera direction, and poor lighting can result in disorientation on the part of the operator. At the World Trade Center, the homogeneity of the environment (i.e., rubble containing only glass, metal, and sand) contributed to the difficulty of keeping track of the robot's position.

The use of a team of robots, whether paired one-to-one or many-to-one with human operators, poses additional requirements. Understanding the needs of robot teams will be critical for making robot technology the preferred and trusted emergency rescue measure.

Having concentrated on SA requirements, a prerequisite for ameliorating SA, we turn now to studies that evaluate new technologies designed to enhance SA.

Technology to Enhance SA

In this section, we review a growing body of literature that uses the assessment techniques discussed earlier to demonstrate that a particular new technology is superior to traditional technology not only in improving performance but also in fostering SA. The studies to be described are representative of a large number of investigations conducted in an increasing number of domains. They highlight exciting new developments in visual displays (e.g., for air traffic control, aviation piloting, driving, and medicine) and tactile displays (e.g., for an aircraft landing on a carrier). The improved displays either offer better data integration, in the form of emergent properties (i.e., by use of color and shape to depict abstract relationships), or introduce more naturalistic features that make the tasks more similar to everyday activities. We end this section with a caveat concerning technologies that could impede SA.

Air traffic control. The advent of a digital datalink between air and ground means that controllers, who in the past have had to rely on pilots for weather reports, and pilots, who have had to rely on controllers for traffic status, can now have access to both kinds of information. Changing the distribution of information can affect team performance. Although pilots have typically deferred to controllers on issues of traffic, and controllers have deferred to pilots on issues of weather, the duplication of information through datalink could change these dynamics, possibly even introducing contention.

Farley, Hansman, Amonlirdviman, and Endsley (2000) investigated these issues in a study of datalink. Six controllers were paired with six pilots in a part-task simulator that enabled the members of each pair to interact with each other from a different room, as well as with simulated pilots. Each participant completed scenarios under two conditions: with and without a datalink connection. The SA measures consisted of responses by both pilots and controllers to the critical traffic and weather events.

The major finding was that the weather information helped the controllers to become aware of the deteriorating weather in time to be proactive. With advance weather information, they made fewer errors in handling weather-related requests. The investigators found that, contrary to concerns, datalink did not induce contention between pilot and controller.

Aviation—flight. The aviation world continues to focus, to its advantage, on issues of SA. We chose the two studies to be discussed, involving visual and tactile aviation displays, from among many others because they suggest clearly defined and highly effective alternatives to existing technologies, as well as a rigorous approach to display design.

Olmos, Wickens, and Chudy (2000) carried out two experiments designed to investigate display requirements of air combat pilots. In Experiment 1, they identified strengths and weaknesses of three canonical display types with respect to spatial awareness:

   a. A conventional display (2-D coplanar) consisting of two separate views: a 90-degree look-down perspective and a profile view (orthogonal to the aircraft's moment-to-moment location).

Situation Awareness Catches On: What? So What? Now What?

A disadvantage is that acquiring this information involves mentally integrating the two views, which is a demanding task because the aircraft is facing in different directions in the two views (see Figure 1.1a).

b. A tethered, over-the-shoulder (exocentric) view that combines vertical and lateral information. A disadvantage is that the depiction of altitude is ambiguous (see Figure 1.1b).

c. A 3-D split screen consisting of two separate views: an immersed (egocentric) view showing the scene as it would appear to the pilot through the windscreen (see Figure 1.1c) and a small-scale global (top-down) view, which is necessary for threat detection in the rear (not shown). No disadvantage was apparent at the outset.

Eight air combat pilots flew missions in a flight simulator with each of the three display types. The task was to fly to waypoints, avoid hazards, and report threats. Unbeknownst to the participants, critical events were planted in the scenarios consisting of pop-up threats and sudden weather hazards that required action on the part of the pilot. The measure of SA consisted of performance measures that were indicative of SA or its lack: time to complete each leg, time spent in contact with terrain or hazards, and correct and timely threat reports.

The results confirmed the hypothesized shortcomings of each display, as described earlier. Compared with the other displays, the traditional coplanar display fared worst in helping pilots to avoid hazards. The 3-D split-screen display, which provided the best hazard avoidance, surprisingly proved least helpful for detecting moving targets. Debriefing of participants revealed that the immersed view was so compelling that they forgot to check the global view for targets.

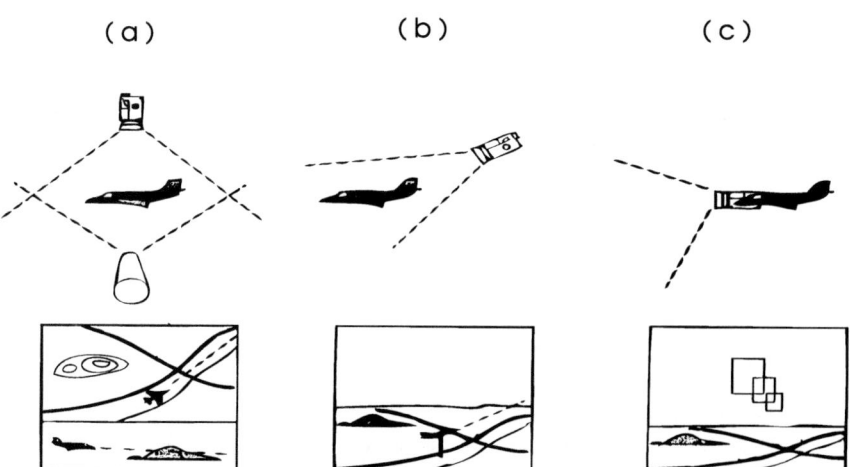

Figure 1.1. Three canonical viewpoints for aviation displays and the resulting display depiction are (a) a 2-D-coplanar, (b) exocentric or tethered, or (c) egocentric or immersed display. From "Tactical Displays for Combat Awareness: An Examination of Dimensionality and Frame of Reference Concepts and the Application of Cognitive Engineering," by O. Olmos, C. D. Wickens, and A. Chudy, 2000, The International Journal of Aviation Psychology, 10, p. 249. Copyright 2000 by Lawrence Erlbaum Associates. Adapted with permission.

In Experiment 2, the investigators sought to remedy the problems they had identified with each display and see if they could eliminate display differences.

For the 2-D coplanar display, they predicted that rotating the side-view 90 degrees so that the aircraft would be facing in the same direction in both views would facilitate mental integration. For the exocentric (over-the-shoulder) display, they predicted that adding color coding to indicate the altitude of threats and terrain with respect to ownship would remove the ambiguity associated with altitude depiction. For the 3-D split-screen display, they predicted that adding a warning tone would draw attention to the global view when a threat appeared.

The results with 27 additional pilots using the same scenarios showed that the task analysis and cognitive engineering paid off in attenuating the hypothesized weaknesses of each of the display types. This study illustrates how a rigorous approach to display design based on task analysis and cognitive engineering can be used to enhance SA.

Jennings, Craig, Cheung, Rupert, and Schultz (2004) investigated ways of helping pilots maintain spatial orientation while landing a helicopter on the deck of a ship, a difficult task in rough seas and darkness. The investigators developed and tested an aid that makes use of an unusual modality—touch. The system consists of a special vest worn by the pilot, which translates path errors into tactile feedback using a gyro-sensor, a computer, and a pulse generator. The experience, from the pilot's point of view, is akin to bumping up against a wall. If the aircraft is forward of its desired position, the front of the vest is activated, sending tactile feedback. Wearing the vest puts pilots into a hypothetical sphere in which touching a wall indicates the need to back away.

This system, called Tactile SA System (TSAS), was evaluated in a flight simulator in which two variables could be manipulated: sea-state (either 3 = *moderate breeze* or 5 = *strong breeze* on a scale used by the U.S. National Weather Service) and the visual environment (either good or degraded). Eleven pilots were tested on simulated landings under all four conditions of vision and sea-state, both wearing and not wearing the TSAS vest. The direct measure of SA was their self-report on the China Lake SA scale (Gawron, 2000). The results showed that wearing the vest increased SA (as measured by self-reports) in the hardest condition, with high sea-state and degraded vision. Performance (as measured by low RMS error) was also significantly enhanced by wearing the vest. The results showed that drawing on the so-called natural ways people behave (i.e., with respect to the affordances of walls) can enhance SA.

Driving. The introduction of navigation systems in automobiles can be a safety feature as well as a convenience. When drivers are lost or are unsure of their route, the added uncertainty can lead to dangerous maneuvers, such as changing lanes without checking traffic. A well-designed navigation system enables drivers to concentrate on the task of driving by enhancing traffic awareness and removing the need to maintain awareness of the route. Although the system makes the driver aware of being on target—an important aspect of SA—it does little to foster awareness of the specific route being traversed. You may have had the experience of arriving at your destination under the guidance of a navigation system without a clue as to how you got there.

To document and remedy this possible shortcoming of navigation systems, Reagan, Baldwin, and Carpenter (2004) tested three versions of a hypothetical new navigation system: "In two blocks turn left onto 5th Avenue" (traditional version), "Turn left onto 5th Avenue heading north" (cardinal direction version), and "Turn left onto 5th Avenue after the police station" (landmark version).

Twenty-four college students drove three routes in a driving simulator, using a different version of the navigation system for each route. After completing a run with the first version, participants were tested for their memory of the route without the system. This procedure was repeated with each of the other two versions.

Three measures of SA were scored for each navigation condition: number of trials to criterion, total number of errors during the runs, and number of requests for help from the experimenter. The assumption was that systems that fostered greater route awareness would lead to faster route learning. These measures all proved to be sensitive to the type of navigation system.

Participants reached criterion sooner and made fewer errors with the landmark version compared with either the cardinal or the traditional version. There were no significant differences between the cardinal and the traditional versions, suggesting that compass directions were not helpful. The results showed that relatively simple modifications to the databases in navigation systems could have a huge impact on driver SA with respect to route knowledge. Of course, one could argue that such knowledge will no longer be needed once all drivers come to rely on such systems. The counterargument, true of most technology, is that drivers still need to stay in the loop in case the equipment fails.

Medicine. In the field of medicine, it is perhaps not surprising that anesthesiologists have been among the first to embrace the concept of SA. The number of independent displays they must monitor during an operation creates conditions of high workload, reminiscent of those faced by aircraft pilots in the past. Recently, researchers have developed and tested integrated medical displays that could be part of future expert systems aimed at enhancing SA.

Michels et al. (1997) compared the ability of anesthesiologists to detect life-threatening events with integrated versus traditional displays. The traditional display consisted of a series of waveforms presented in an arbitrary order. The integrated display showed boxes that filled or emptied as a particular variable departed from a steady state. It is important to note that the boxes were arranged by function, so variables pertaining to respiratory processes, cardiac functioning, or fluid management could be seen at a glance.

Ten anesthesiologists were tested on their ability to detect critical events with the two kinds of displays. Half the participants viewed the traditional display and half the integrated display. SA was measured by showing participants four simulations, each containing a critical event, and asking them to respond as quickly as possible once they detected a problem. Events consisted of blood loss, inadequate paralysis, cuff leak, and depletion of soda lime. After indicating there was a problem, the participant had to diagnose what had happened by choosing from a set of multiple-choice options. The results showed that the integrated display led to significantly faster detection times for all four problems. For more information about creating effective displays, see Smith, Bennett, and Stone (2006).

Caveat: Technology With the Potential to Impede SA

Air traffic control. The idea of free flight, in which aircraft no longer have to adhere to route and altitude restrictions, raises the specter of increased SA problems for controllers. Can separation be adequately maintained under free-flight conditions? The results thus far have been mixed, in part because different assumptions have been made about how free flight will operate.

A study by Remington et al. (2000) showed mostly positive results for the free-flight conditions that were examined. SA was assessed in a part-task simulation under four conditions: fixed path (with and without altitude restrictions) and no fixed path (with and without altitude restrictions). Four retired air traffic controllers viewed a series of 90-second scenarios, each containing one conflict. Each participant experienced all four conditions. Participants were instructed to hit a key as soon as they detected a conflict. The response stopped the scenario, allowing the participant time to click on the two aircraft in conflict. The direct test of SA was how rapidly participants detected the conflict in each scenario.

The results showed that there were no significant differences in time to detect the conflict among any of the conditions (although there was one marginally significant difference that needs further investigation), suggesting that neither path nor altitude restrictions were a problem. Other variables, on the other hand, such as traffic load and conflict geometry, were highly significant, ruling out the possibility that the measure was not sensitive. This study, it should be pointed out, was based on a small number of participants and a part-task paradigm.

Endsley, Mogford, Stein, and Hughes (1997) envisioned free flight in a slightly different way and tested the impact on controllers during 45-minute simulator sessions. The participant played the role of controller and interacted with simulated pilots. A baseline condition, which consisted of current practices (assigned routes and no deviation without permission), was compared with three experimental conditions. In all three, the simulated pilots filed their intended routes with the controller in advance. The three conditions differed, however, in the free-flight rules: In the first condition, the simulated pilots were not allowed to deviate from the filed route. In the second, they were required to notify the controller of their intention before they deviated. In the third, they were allowed to deviate without notifying the controller. Ten controllers from an en-route center completed two scenarios in each condition. SA measures consisted of responses to SAGAT probes and performance ratings by subject matter experts. In addition, participants completed the NASA Task Load Index (TLX) workload rating.

The results showed a decline in controller SA as flight conditions became less restricted. In particular, the controller's ability to report, when probed, aircraft location and impact of weather declined significantly as restrictions were eased, suggesting that the controller's SA had been compromised by the unpredictability of the pilots' actions. Together, these two studies provide important first steps, as well as some cautions, regarding the impact of free flight on controller SA.

Driving. It has long been known that visual and motor distractions impair driving. Therefore, when cell phones appeared to contribute to accidents, it was natural to assume the problem was related to the distraction of manipulating the phone. The design of hands-free phones was intended to solve the problem. However, recent research suggests that hands-free may not be the answer.

In a revealing experiment using recognition of objects along the route as a measure of SA, Strayer, Cooper, and Drews (2004) demonstrated a significant loss of SA with hands-free cell phone use. Undergraduate licensed drivers drove two different routes in a simulator, one with and one without a cell phone. In the cell phone condition, a research assistant engaged the participant in a cell phone conversation on a topic predetermined to be of interest. Eye movements were recorded. At the end of each of the two routes, the participant was tested for recognition of objects that had been positioned along the route. A forced-choice recognition paradigm was used in which 30 pairs of objects were presented, one of which had appeared and one of which had not. In each case, participants had to say which of the two objects had appeared.

Recognition for objects along the route was significantly lower after the dual task of driving and engaging in a hands-free cell phone conversation than after simply driving. Interestingly, the eye movement data showed no difference between conditions in the number of times the participants looked at the objects. Even when only those objects that had been fixated were considered, drivers still had lower recognition scores in the cell phone than in the no-phone condition, suggesting that while conversing, participants were not deeply processing the objects they were looking at.

Further analysis indicated that the cell phone users were not simply tuning out unimportant aspects of the scene, because their recognition in the cell phone condition was just as poor for objects rated (by the participants themselves) as highly relevant as for those objects rated irrelevant. By contrast, recognition in the no-phone condition was greater for objects rated relevant than for those rated irrelevant. Measures of driving performance were not reported.

This study is one of a growing number to demonstrate the "negative" benefit of hands-free cell phones by showing a clear loss of SA in drivers who talk on the phone while driving. Other studies support these findings (e.g., Drews, Pasupathi, & Strayer, 2004; McCarley & Vais, 2004). The fact that drivers looked at but did not remember objects suggests that the limitation imposed by cell phones is more cognitive than physical.

The driving public needs to take seriously the growing body of research on the hazards of cell phone use while driving. Until technology manages to simplify the driving task, or makes the person on the other end of the phone aware of the traffic situation, drivers are advised to curtail their phone conversations.

In summary, research over the past 10 years suggests that technology can increase the SA of pilots, drivers, and medical personnel, along with others, through relatively simple means: vests that provide tactile indicators of orientation, displays that simulate the external view, navigation systems that point out landmarks, and displays that depict disease states through emergent graphic features. The research studies reviewed in this section show that technologies needed to enhance SA are easily within reach and, in many cases, could be incorporated into actual products. The research also shows that other innovations, such as free-flight air traffic control procedures, which have shown mixed results in studies of SA, need more investigation before they can safely become the new norm. In this vein, research strongly suggests that drivers limit their use of cell phones while in motion (for more on driver safety, see Lee, 2005).

The next section addresses the question of how SA can be trained.

Training Programs to Enhance SA

The SA training studies to be discussed in this section draw on the latest in training technologies. These innovations include live exercises with after-action reviews (infantry, medical), simulation and game-based team training (infantry, medical), individual PC trainers (driving, infantry), and formal training programs that combine many of the aforementioned technologies (aviation). Each of the studies is unique in its focus on training SA, as distinct from performance. The studies are reported by domain.

Aviation—flight. Popular flight training programs include the Line Oriented Flight Training (LOFT) simulator for teaching flying skills, and crew resource management training for teaching safety-related, interpersonal team skills, including how and when to question authority. Recently, several European airlines adopted a new program, Enhanced Safety through SA Integration (ESSAI), in response to a need for training specifically geared toward SA skills. The program consists of several parts (Hormann et al., 2004):

 a. *An interactive, computer-based introduction.* This 50-minute DVD developed by the British Airlines Human Factors Department provides an introduction to the skills needed to maintain SA during unexpected events.
 b. *A low-tech tactical decision game for two pilots.* Pilots are assigned roles in a scenario that calls for "communication, planning, anticipation, management of ambiguous information, workload management, and updating one's mental picture and strategy under moderate time-pressure" (p. 220). The instructor conducts a debrief at the completion of the game.
 c. *Simulator scenarios.* Three scenarios were developed for the LOFT simulator. These scenarios present the trainees with both expected and unexpected threats. Expected threats include missing items from the minimum equipment list, low visibility, and high terrain near the airport. Unexpected threats include equipment failure, weather deterioration, air traffic control delays, air traffic control errors, and distracting cabin events.

Hormann et al. tested the effectiveness of the ESSAI program by comparing two groups of pilots: a trained group that underwent one day of ESSAI training (DVD introduction, low-tech decision game, and LOFT simulator exercises) and a control group that completed one day of regular LOFT simulator training. Both groups consisted of eight two-person teams and completed a pretest and posttest on additional LOFT scenarios. Trained observers evaluated performance on the approximately 20 critical events embedded in each of the tests (pretest and posttest). For the handling of each incident, a participant received SA ratings corresponding to Endsley's three levels of awareness (1 "notice," 2 "understand," 3 "anticipate") and a performance score (1 "ineffective," 2 "partially effective," and 3 "effective").

The results showed significant improvement in SA and performance from pretest to posttest in the trained group but not in the control group. The trained group improved from "understand" on the pretest to "think ahead" on the posttest, whereas the control group began and ended at "understand." Ratings were particularly high for the quality of briefings in the trained group, suggesting that members of that group had learned to engage in contingency planning.

The study is impressive in showing gains in SA with only one day of training. Other airlines will undoubtedly adopt the ESSAI program to complement their use of the popular LOFT and crew resource management programs. ESSAI could be expanded to include SA training for equipment malfunctions. Giving crews the opportunity to experience problems in a forgiving, simulated environment and talk about it afterward increases the likelihood that they will be prepared in an actual emergency.

Driving. One way of increasing driver SA is to teach drivers to recognize risky situations (i.e., events that require careful monitoring). Examples include the following: obscured obstacle following a warning (e.g., a stop sign ahead is indicated but is obscured until the last minute by a bush), obscured pedestrian (e.g., a pedestrian who is visible is briefly obscured by a fence as the driver approaches the intersection), and obscured traffic (e.g., a truck in the opposing left lane blocks the driver's view of cars to the truck's right—important if the driver is turning left).

Fisher, Narayanaan, Pradhan, and Pollatsek (2004) developed and tested a PC-based risk awareness program to train drivers to recognize risky situations, such as those described in the earlier examples. They recruited 48 novice drivers with six months or less of driving experience and divided them into a training group and a control group. The training group worked on 10 PC-based scenarios. For each scenario, trainees viewed a static, top-down view of a driving situation. The trainee's task was to drag a yellow marker over any person or object that was not visible to the driver but that would constitute an obstacle shortly, and a red marker over any area that would require continuous monitoring. After making a response, the trainee saw the correct answers with an explanation.

The researchers assessed the effectiveness of the training by testing, in a full driving simulator, both the group that had received PC-based training and a control group that had not received training. They used eye-movement recordings to determine whether participants detected the risk in scenarios similar to as well as different from those used in training. Based on the eye-movement data, the researchers found that the trained group was significantly more likely than the control group to detect risks on scenarios similar to those in training and slightly more likely to detect risks in novel scenarios. These results suggest that SA training for drivers does not require a full simulator. Lessons learned about risk detection on a PC using static, top-down views transferred to a more realistic driving situation.

Infantry. In a different domain, researchers have utilized both PC-based and real-life exercises to train SA skills. Strater et al. (2004) developed a program to help infantry platoon leaders learn to assess and deal with unusual situations. Training consisted of a CD that used video clips, photos, maps, auditory soldier reports, and civilian queries to depict events during a mission. For each scenario, the trainee's task was to evaluate the information presented and select an action to take from choices on a menu. The system scored the responses and provided feedback. As part of the training, the scenarios were periodically interrupted for a SAGAT probe question.

The researchers evaluated the CD-based training by comparing, in a live exercise in which all participated, the performance of platoon leaders who had received training with that of those who had not. The test of SA was whether the participant was led down the garden path of thinking that refugees in a camp were terrorists when, in fact, they were civilians. Although the number of participants was small, the results were suggestive. Of the eight squads that participated, only two correctly recognized that the alleged enemy was friendly, and both of those leaders had been trained. Trained participants also rated the mental difficulty of the exercise as higher than did untrained participants. Evidently the training made them aware of the need to expend effort to "predict what will occur next" and "decide how best to achieve mission goals" (Strater et al., 2004, p. 668).

Medicine. Working within the context of high-risk operating rooms, Guerlain et al. (2005) developed a state-of-the-art tool for facilitating after-action reviews with the following features: audio capture from eight locations (capturing different personnel); video capture from four feeds (laparoscope, anesthesia data, table camera, and room camera), time-stamping and annotation facility—observers can note important events (critical events, errors made) that should be brought out during the after-action review; and data retrieval. "All data are compressed, saved, synchronized, selectable, indexable, and protected" (p. 31).

This tool is intended to be used with an SA quiz that is administered to each member of the team at the end of surgery. The quiz is scored manually against the "ground truth" contained in the video and audio capture. Although this after-action-review tool still needs to be evaluated, it seems to have the right components for fostering situation awareness.

The success of the various training programs described rests on principles for developing differentiated perceptual strategies and mental models (Endsley & Robertson, 2000). These include pre-exercise briefings to direct attention, varied practice in realistic situations, diagnostic measures of performance, and structured feedback. We turn now to the question of how to identify and select those individuals who will be best able to maintain SA under demanding conditions.

Techniques for Selecting for SA Abilities

Are there tests that can be administered to determine if someone will exhibit good SA in a particular domain? This selection question has found a tentative answer in the domain of aviation, where two tests have been developed that correlate with job performance. Some have questioned whether these tests, which assess performance in dynamic, complex, gamelike environments, are measuring SA apart from performance. This criticism may prove to have little practical import because selecting for performance may turn out to be the same as selecting for SA.

Aviation—Flight. In the 1980s, researchers demonstrated that performance on a simple dichotic listening test was sufficient to predict the success of candidates for flight training (Gopher, 1982). Based on these findings, a multitasking test was developed by Roscoe and Corl (1987) named the WOMBAT Situational Awareness and Stress Tolerance Test. The test measures an individual's ability to "scan multiple information sources, evaluate alternatives, establish priorities, and select and work on the task that has the highest priority at the moment" (Roscoe, 1997, quoted in O'Hare, 1997, p. 541). The task is an interactive game similar to Space Fortress (Mané & Donchin, 1989).

O'Hare (1997) examined the utility of the WOMBAT test for predicting SA in two experiments. The first experiment examined whether more basic abilities contributed to scores on the WOMBAT. Because WOMBAT itself is gamelike, O'Hare first checked whether game experience would correlate with performance on WOMBAT. He found that computer game experience predicted initial WOMBAT performance ($r = .64$) but not performance after 60 minutes. WOMBAT seemed to be measuring something more than game experience.

The second experiment compared WOMBAT performance for three types of participants: 8 elite pilots attending a soaring competition in New Zealand, 6 experienced pilots, and 12 nonpilot controls. O'Hare hypothesized that pilot status (elite, regular, or control) would correlate with WOMBAT performance. The results showed a significant difference in WOMBAT scores between the pilot groups, as expected: Elite pilots scored highest and nonpilots scored lowest. The findings supported the use of WOMBAT as a selection tool for SA by showing that elite pilots, whose extraordinary skills depend on the ability to maintain awareness, excelled on the WOMBAT test. It does not, however, prove that SA is the underlying factor or that the test has predictive value prior to training.

Carretta, Perry, and Ree (1996) looked for predictors of pilot SA by administering a battery of cognitive and personality tests to F-15 pilots and seeing which subskills correlated with peer and supervisor SA ratings. Once flying hours were controlled for, they found that the best predictors were working memory, spatial reasoning, and divided attention. Furthermore, the cognitive test items were better predictors than the perceptual or motor test items, as would be expected if the test were measuring SA skills and not other aspects of piloting. It should be noted, however, that some of the items on the expert/peer rating scale (e.g., "discipline" and "communication") seem to be only remotely related to SA.

Sports. Ward and Williams (2003) developed and validated SA tests to predict skill in soccer. Interestingly, they were unable to predict skill purely on the basis of visual factors, such as acuity and peripheral vision. The following SA paradigms were developed for the study:

a. *Memory for structured versus unstructured material.* The test required participants to recall the position of players shown in either a structured setting (during a game) or an unstructured setting (break time on the field). Their score was the difference in recall between the two conditions.
b. *Anticipation task.* The test consisted of brief excerpts of video footage from an actual game. In each case, the excerpt ended right before the player threw or dribbled the ball to another player. The participant had to indicate the direction that the ball would take after being released by the player. Responses to this question were scored objectively against the complete video footage. The participant was also asked to indicate which players were in a position to receive the ball and to rate the importance of players on the other team in terms of their attack potential. These questions were scored against the judgments of experts.

To validate these tests as predictors of soccer ability, the investigators administered them to groups known to differ in soccer ability: players in different age groups (9, 11, 13, 15, and 17 years) and players with different status (elite vs. nonelite team members). A total of 137 participants were tested in groups that combined the two factors (i.e., elite and nonelite team members in the different age groups). The results showed that scores on both the memory and anticipation tasks increased with age and elite status. There were no interactions between the two factors. This test measures SA apart from performance but requires prior knowledge of the domain.

To answer the question "So what?" we have shown how the application of SA can anticipate and help solve problems of usability in system design, train and evaluate professionals, and predict skills.

NOW WHAT?

Much progress has been made in the approximately 17 years since the first NATO Advisory Group for Aerospace Research and Development (AGARD) conference was held in Copenhagen to exchange information among existing and potential SA researchers. As we have shown in previous sections, there is growing consensus regarding the definitions of SA, team SA, SA requirements, and even what constitutes a situation. There is now a reasonably wide variety of methods, both objective and subjective, for assessing SA, and many investigators are successfully demonstrating its usefulness. So what is left to do?

With respect to "What?" we identify one promising area for further development and one measurement concern worthy of attention by those wishing to apply the concept. The promising area is that of modeling SA. More advanced models are of interest in their own right, but there is also great interest in what have been called integrated models of human performance, or, as is used in the military, *human behavior representation*. In a 2003 report, Morrison cited 21 such models at development stages ranging from "still in research" to "demonstrated to be usable for particular applications in the hands of expert users." Most, if not all of these 21 models "cheat" when it comes to representing sensory and perceptual processes. Many assume that when a stimulus object in the environment is fixated by an eye movement, it is magically perceived and identified at some level and transferred to some form of working memory. Other models do not represent explicit eye movements but have some kind of attentional process that "absorbs" data from the environment, and instantly they are represented in a memory store. These gaps are exactly what SA is about. Not only would improved models of SA advance opportunities to predict the effects of design developments and training on SA in applied settings, but introducing them into integrated human performance models would upgrade the quality of the integrated models.

Our concern is that those who are actively trying to make use of SA measurement methodologies are relying too heavily on subjective methods such as self-ratings. As many investigators have pointed out, subjective assessments of SA are not the same as objective measures; therefore, it's beneficial to use a combination of subjective and objective measures (Jennings et al., 2004; Snow & Reising, 2000). Operators can judge their awareness only of elements that they know to be relevant. They may have no way of knowing what they have missed. No one has done the relevant experiment, but it is likely that, as reported in other domains, more that 50% of operators would report that they have above-average SA.

Furthermore, there are interesting and potentially confounding relationships between SA assessment and mental workload (Vidulich, 2000). It takes active effort to seek the information needed to support SA. This effort represents an increase in mental workload. It is easy to confuse effort spent to assess SA with the success in acquiring it. In such a case, subjective estimates of SA may be biased by the effort spent obtaining it. So, again, the admonition to investigators is to seek one or more of the objective methods at least in addition to, if not instead of, subjective assessments.

With respect to "So what?" we heartily commend the investigators—including many whose work we could not include because of space limitations—who have taken seriously the need for systematically assessing SA. Their work has already paid off in improved system design and training and perhaps also in personnel selection. They have demonstrated, sometimes ingeniously, that SA can be different and more diagnostic than simply assessing overall performance. There is plenty of room for more of this kind of work.

In addition, the reader will note that many of the references are to technical reports and conference proceedings. The challenge now is to move this work into diversified professional journals, textbooks, and other secondary sources to make it more accessible to students and practitioners so they can apply it to their own work and find the encouragement to create their own applications. Endsley's design guidebook (Endsley et al., 2003) and toolkit (Jones, Endsley & Bolstad, 2004) address this need.

So, where do we envision SA research going in the future, or "Now what?" Although we do not have a crystal ball, we believe that greater challenges to SA in certain domains will arise in the future. One such area is robotics. With the potential for robots to provide household help and even companionship; support military missions in the air, on land, and under water; and provide homeland security and rescue services, there will be a growing need to understand how to maintain SA in human-robot teams.

A second area ripe for more work in SA is driving. We are convinced that driving does not need to be so difficult and fraught with fears of collisions, breakdowns, weather-related issues, and getting lost. Technology holds the promise to make driving safer and easier, perhaps just in time to simplify the lives of an aging population. This promise, however, depends crucially on the ability of the technology to enhance, rather than interfere with, driver SA.

We project greater emphasis in the future on issues of cyber awareness. As more functions come to depend on the Internet, including the telephone, the Internet will become more vulnerable to sabotage. The SA of operators responsible for detecting attacks that could result in denial of service will become critical.

Finally, we expect to see experts in team SA contribute to an understanding of how to foster the situation awareness needed to plan for and deal with large-scale disasters. Hurricane Katrina, which devastated Louisiana in 2005, provides a telling lesson of what can happen, for example, when more than one million people try to head for the roads at once, not aware that gridlock and gas shortages await them; when evacuation efforts fail to take into account the reluctance of people to leave pets behind; or when state and local officials mistakenly assume, as a result of having taken part in practice exercises, that generators stored in locations throughout the country will be available for patients who need oxygen and dialysis. Although there has been at least one study of ambulance dispatchers (Wong & Blandford, 2004), investigations of large-scale relief efforts remain an uncharted territory for SA research.

In short, it is easy to see why SA, a concept about "catching on," has itself caught on. In fact, it is helping to make the field of human factors/ergonomics more cognitive and the field of cognitive psychology more applied. It is also bringing together researchers in a number of domains. We believe that these trends are a healthy and necessary step toward achieving the interdisciplinary understanding of SA that will be necessary for the safe and efficient use of a broad range of critical new technologies in the future.

ACKNOWLEDGMENTS

We are grateful for support from the BBN Science Development Program and first-rate assistance from Jennie Connolly, Penny Steele, and Amy Chitwood of the BBN Library. We thank our reviewers—Chris Wickens, Nancy Cooke, and Mica Endsley—as well as Elizabeth Tenney, for insightful comments on an earlier draft.

REFERENCES

Adams, M. J., Tenney, Y. J., & Pew, R. W. (1995). Situation awareness and the cognitive management of complex systems. *Human Factors, 37*, 85–104.

Andersen, H. B., Pedersen, C. R., & Andersen, H. H. K. (2001). Using eye tracking data to indicate team situation awareness. *Proceedings of the 9th Annual Meeting of the HCI International Conference* (pp. 1318–1322). Mahwah, NJ: Erlbaum.

Aykroyd, P., Harper, K. A., Middleton, V., & Hennon, C. G. (2002). Cognitive modeling of individual combatant and small unit decision-making within the integrated unit simulation system. *Proceedings of 11th Conference on Computer-Generated Forces and Behavior Representation (CGF-BR)* (pp. 597–604). Orlando, FL: Simulation Interoperability Standards Organization.

Carretta, T. R., Perry, D. C., Jr., & Ree, M. J. (1996). Prediction of situational awareness in F-15 pilots. *International Journal of Aviation Psychology, 6*, 21–41.

Casper, J., & Murphy, R. R. (2003). Human-robot interactions during the robot-assisted urban search and rescue response at the World Trade Center. *IEEE Transactions on Systems, Man & Cybernetics, Part B, 33*, 367–385.

Chase, W. G., & Simon, H. A. (1973). Perception in chess. *Cognitive Psychology, 4*, 55–81.

Cooke, N. J., Stout, R. J., & Salas, E. (2001). A knowledge elicitation approach to the measurement of team situation awareness. In M. McNeese, M. Endsley, & E. Salas (Eds.), *New trends in cooperative activities: System dynamics in complex settings* (pp. 114–139). Santa Monica, CA: Human Factors and Ergonomics Society.

Deutsch, S. E., Pew, R. W., Rogers, W. H., & Tenney, Y. (1994). *Toward a methodology for defining situation awareness requirements: A progress report* (Report No. 7983). Cambridge, MA: BBN Technologies.

Drews, F. A., Pasupathi, M., & Strayer, D. L. (2004). Passenger and cell-phone conversations in simulated driving. *Proceedings of the Human Factors and Ergonomics Society 48th Annual Meeting* (pp. 2210–2212). Santa Monica, CA: Human Factors and Ergonomics Society.

Durlach, P. J. (2004). Change blindness and its implications for complex monitoring and control systems design and operator training. *Human-Computer Interaction, 19*, 423–451.

Durso, F. T., & Dattel, A. R. (2004). SPAM: The real-time assessment of SA. In S. Banbury & S. Tremblay (Eds.), *A cognitive approach to situation awareness: Theory and application* (pp. 137–155). Hampshire, England: Ashgate.

Durso, F. T., & Gronlund, S. D. (1999). Situation awareness. In F. T. Durso, R. S. Nickerson, R. W. Schvaneveldt, S. T. Dumais, D. S. Lindsay, & M. T. H. Chi (Eds.), *Handbook of applied cognition* (pp. 283–314). London: Wiley.

Endsley, M. R. (1987). *SAGAT: A methodology for the measurement of situation awareness* (NOR DOC 87-83). Hawthorne, CA: Northrop.

Endsley, M. R. (1988). Design and evaluation for situation awareness enhancement. *Proceedings of the Human Factors Society 32nd Annual Meeting* (pp. 97–101) Santa Monica, CA: Human Factors and Ergonomics Society.

Endsley, M. R. (2000). Theoretical underpinnings of situation awareness: A critical review. In M. R. Endsley & D. J. Garland (Eds.), *Situation awareness: Analysis and measurement* (pp. 3–47). Mahwah, NJ: Erlbaum.

Endsley, M. R., Bolte, B., & Jones, D. G. (2003). *Designing for situation awareness: An approach to user-centered design*. London: Taylor & Francis.

Endsley, M. R., Mogford, R. H., Stein, E., & Hughes, W. J. (1997). Controller situation awareness in free flight. *Proceedings of the Human Factors and Ergonomics Society 41st Annual Meeting* (pp. 4–8). Santa Monica, CA: Human Factors and Ergonomics Society.

Endsley, M. R., & Robertson, M. M. (2000). Training for situation awareness in individuals and teams. In M. R. Endsley & D. J. Garland (Eds.), *Situation awareness: Analysis and measurement* (pp. 349–365). Mahwah, NJ: Erlbaum.

Entin, E. B., & Entin, E. E. (2000). Assessing team situation awareness in simulated military missions. *Proceedings of the XIVth Triennial Congress of the International Ergonomics Association and 44th Annual Meeting of the Human Factors and Ergonomics Society* (pp. 1.73–1.76). Santa Monica, CA: Human Factors and Ergonomics Society.

Ericsson, K. A., & Kintsch, W. (1995). Long-term working memory. *Psychological Review, 102*, 211–245.

Farley, T. C., Hansman, R. J., Amonlirdviman, K., & Endsley, M. R. (2000). Shared information between pilots and controllers in tactical air traffic control. *Journal of Guidance, Control and Dynamics, 23*, 826–836.

Fisher, D. L., Narayanaan, V., Pradhan, A., & Pollatsek, A. (2004). Using eye movements in driving simulators to evaluate effects of PC-based risk awareness training. *Proceedings of the Human Factors and Ergonomics Society 48th Annual Meeting* (pp. 2266–2270). Santa Monica, CA: Human Factors and Ergonomics Society.

Flach, J. M. (1995). Situation awareness: Proceed with caution. *Human Factors, 37*, 149–157.

Flach, J. M. (1996). Situation awareness: In search of meaning. *CSERIAC Gateway, 6*(6), 1–4.

Gawron, V. J. (2000). *Human performance measures handbook*. Mahwah, NJ: Erlbaum.

Gibson, J. J. (1979). *The ecological approach to visual perception*. Mahwah, NJ: Erlbaum.

Gilson, R. D. (Ed.). (1995). Situation awareness [Special issue]. *Human Factors, 37*(1).

Gopher, D. A. (1982). Selective attention test as a predictor of success in flight training. *Human Factors, 24*, 173–183.

Gorman, P., Ash, J., Lavelle, M., Lyman, J., Delcambre, L., Maier, D.,Weaver, M., & Bowers, S. (2000). Bundles in the wild: Managing information to solve problems and maintain situation awareness. *Library Trends, 49*, 266–289.

Guerlain, S., Adams, R. B., Turrentine, F. B., Shin, T. Guo, H., Collins, S. R., & Calland, J. F. (2005). Assessing team performance in the operating room: Development and use of a "black-box" recorder and other tools for the intraoperative environment. *Journal of the American College of Surgeons, 200*, 29–37.

Gutwin, C., & Greenberg, S. (2002). A descriptive framework of workspace awareness for real-time groupware. *Computer Supported Cooperative Work, 11*, 411–446.

Hanson, M. L., & Harper, K. A. (2000). An intelligent agent for supervisory control of teams of uninhabited combat air vehicles (UCAVs). *Proceedings of Unmanned Systems 2000*. Arlington, VA: Association for Unmanned Vehicle Systems International.

Hanson, M. L., Harper, K. A., Endsley, M., & Reszonya, L. (2002). Developing cognitively congruent HBR models via SAMPLE: A case study in airline operations modeling. *Proceedings of the Conference on Computer Generated Forces*. Orlando, FL: Simulation Interoperability Standards Organization.

Hanson, M. L., Sullivan, O., & Harper, K. (2001). *On-line situation assessment for unmanned air vehicles. Florida Artificial Intelligence Research Society (FLAIRS-01) Conference* (pp. 44–48). Menlo Park, CA: AAAI Press.

Hauss, Y., & Eyferth, K. (2003). Securing future ATM-concepts' safety by measuring situation awareness in ATC. *Aerospace Science and Technology, 7*, 417–427.

Hormann, H., Banbury, S. P., Dudfield, H. J., Lodge, M., & Soll, H. (2004). Effects of situation awareness training on flight crew performance. In S. Banbury & S. Tremblay (Eds.), *A cognitive approach to situation awareness: Theory and application* (pp. 213–232). Hampshire, England: Ashgate.

Jennings, S., Craig, G., Cheung, B., Rupert, A., & Schultz, K. (2004). Flight-test of a tactile situational awareness system in a land-based deck landing task. *Proceedings of the Human Factors and Ergonomics Society 48th Annual Meeting* (pp. 142–146). Santa Monica, CA: Human Factors and Ergonomics Society.

Jones, D. G., & Endsley, M. (2000). Examining the validity of real-time probes as a metric of situation awareness. *Proceedings of the XIVth Triennial Congress of the International Ergonomics Association and 44th Annual Meeting of the Human Factors and Ergonomics Society* (p. 1.278). Santa Monica, CA: Human Factors and Ergonomics Society.

Jones, D. G., Endsley, M. R., & Bolstad, C. M. (2004). The designer's situation awareness toolkit: Support for user-centered design. *Proceedings of the Human Factors and Ergonomics Society 48th Annual Meeting* (pp. 653–657). Santa Monica, CA: Human Factors and Ergonomics Society.

Kramer, A., & McCarley, J. S. (2003). Oculomotor behaviour as a reflection of attention and memory processes: Neural mechanisms and applications to human factors. *Theoretical Issues in Ergonomics Science, 4*, 21–55.

Klein, G., Phillips, J. K., Rall, E. L., & Peluso, D. A. (in press). A data/frame theory of sensemaking. In R. R. Hoffman (Ed.), *Expertise out of context: Proceedings of the 6th International Conference on Naturalistic Decision Making*.

Lee, J. D. (2005). Driving safety. In R. S. Nickerson (Ed.), *Reviews of human factors and ergonomics, volume 1* (pp. 172–218). Santa Monica, CA: Human Factors and Ergonomics Society.

Levin, D., Tenney, Y. J., & Henri, H. (2000). Issues in human interaction for cyber command and control. *Proceedings of the DARPA Information Survivability Conference & Exposition (DISCEX II;* pp. 0141–0151) Washington, DC: IEEE Computer Society.

Mace, W. (1977). James J. Gibson's strategy for perceiving: Ask not what's inside your head, but what your head's inside of. In R. Shaw & J. Bransford (Eds.), *Perceiving, acting, and knowing: Toward an ecological psychology* (pp. 43–65). Mahwah, NJ: Erlbaum.

Mané, A., & Donchin, E. (1989). The Space Fortress game. *Acta Psychologica, 71*, 17–22.

McCarley, J.S., & Vais, M.J. (2004). Conversation disrupts change detection in complex traffic scenes. *Human Factors, 46*, 424–436.

McCarley, J. S., Wickens, C. D., Goh, J., & Horrey, W. J. (2002) A computational model of attention/situation awareness. *Proceedings of the Human Factors and Ergonomics Society 46th Annual Meeting* (pp. 1669–1673). Santa Monica, CA: Human Factors and Ergonomics Society.

McGowan, A., & Banbury, S. (2004). Interruption and reorientation effects of a situation awareness probe on driving hazard anticipation. *Proceedings of the Human Factors and Ergonomics Society 48th Annual Meeting* (pp. 290–294). Santa Monica, CA: Human Factors and Ergonomics Society.

Michels, P., Ing, D., Gravenstein, D., & Westenskow, D. R. (1997). An integrated graphic data display improves detection and identification of critical events during anesthesia. *Journal of Clinical Monitoring, 13*, 249–259.

Morrison, J. E., (2003). *A review of computer based human behavior representations and their relation to military* (IDA paper P-3845). Washington, DC: Institute for Defense Analyses.

Mulgund, S. S., Harper, K. A., Zacharias, G. L., & Menke, T. (2000). SAMPLE: Situation Awareness Model for Pilot-in-the-Loop Evaluation. *Proceedings of 9th Conference on Computer Generated Forces and Behavioral Representation*. Orlando, FL: Simulation Interoperability Standards Organization.

Neisser, U. (1967). *Cognitive psychology*. New York: Appleton-Century-Crofts.

Neisser, U. (1978). *Cognition and reality: Principles and implications of cognitive psychology*. San Francisco: W. H. Freeman.

Neisser, U., & Becklen, R. (1975). Selective looking: Attending to visually specified events. *Cognitive Psychology, 7*, 480–494.

O'Hare, D. (1997). Cognitive ability determinants of elite pilot performance. *Human Factors, 39*, 540–552.

Olmos, O., Wickens, C. D., & Chudy, A. (2000). Tactical displays for combat awareness: An examination of dimensionality and frame of reference concepts and the application of cognitive engineering. *International Journal of Aviation Psychology, 10*, 247–271.

Pew, R. W. (2000). The state of situation awareness measurement: Heading toward the next century. In M. R. Endsley & D. J. Garland (Eds.), *Situation awareness: Analysis and measurement* (pp. 33–47). Mahwah, NJ: Erlbaum.

Pew, R. W., & Mavor, A. S. (Eds.). (1998). *Modeling human and organizational behavior: Applications to military simulations*. Washington, DC: National Academy Press.

Pritchett, A. R., & Hansman, R. J. (2000). Use of testable responses for performance-based measurement of situation awareness. In M. R. Endsley & D. J. Garland (Eds.), *Situation awareness: Analysis and measurement* (pp. 189–209). Mahwah, NJ: Erlbaum.

Rasmussen, J., & Rouse, W. B. (1981). *Human detection and diagnosis of system failure*. NY: Plenum Press.

Reagan, I., Baldwin, C. L., & Carpenter, E. M. (2004). Auditory in-vehicle routing and navigation systems and facilitating cognitive map development. *Proceedings of the Human Factors and Ergonomics Society 48th Annual Meeting* (pp. 2276–2279). Santa Monica, CA: Human Factors and Ergonomics Society.

Remington, R. W., Johnston, J. C., Ruthruff, E., Gold, M., & Romera, M. (2000). Visual search in complex displays: Factors affecting conflict detection by air traffic controllers. *Human Factors, 42*, 349–366.

Ritter, F. E., Shadbolt, N. R., Elliman, D., Young, R. M., Gobet, F., & Baxter, G. D. (2003). *Techniques for modeling human performance in synthetic environments: A supplementary review* (HSIAC SOAR -2003-01). Wright-Patterson Air Force Base, OH, Human Systems Information Analysis Center.

Roscoe, S. N. (1997). Predicting and enhancing flight deck performance. In R. Telfer & P. Moore (Eds.), *Aviation training: Pilot, instructor, and organisation* (pp. 195–208). Aldershot, England: Avebury.

Roscoe, S. N., & Corl, L. (1987). Wondrous original method for basic airmanship testing. In R. S. Jensen (Ed.), *Proceedings of the Fourth International Symposium on Aviation Psychology* (pp. 493–499). Columbus: Ohio State University, Department of Aviation.

Salas, E., Dickinson, T. L., Converse, S. A., & Tannenbaum, S. I. (1992). Toward an understanding of team performance and training. In R. W. Swezey & E. Salas (Eds.), *Teams: Their training and performance* (pp. 3–29). Norwood, NJ: Ablex.

Salas, E., Wilson, K. A., Burke, C. S., Wightman, D. C., & Howse, W. R. (2006). Crew resource management training research, practice and lessons learned. In R. C. Williges (Ed.), *Reviews of Human Factors and Ergonomics, Volume 2* (pp. 35–73). Santa Monica, CA: Human Factors and Ergonomics Society.

Sarter, N. B., & Woods, D. D. (1995). How in the world did we ever get into that mode? Mode error and awareness in supervisory control. *Human Factors, 37*, 5–19.

Scholtz, J., Young, J., Drury, J. L., & Yanco, H. A. (2004). Evaluation of human-robot interaction awareness in search and rescue. *Proceedings of IEEE International Conference on Robotics and Automation* (pp. 2327–2332). Washington, DC: IEEE Computer Society.

Simons, D. J., & Chabris, C. F. (1999). Gorillas in our midst: Sustained inattentional blindness for dynamic events. *Perception, 28*, 1059–1074.

Smith, P. J., Bennett, K. B., & Stone, R. B. (2006). Representation aiding to support performance on problem-solving tasks. In R. C. Williges (Ed.), *Reviews of Human Factors and Ergonomics, Volume 2* (pp. 74–108). Santa Monica, CA: Human Factors and Ergonomics Society.

Snow, M. P., & Reising, J. M. (2000). Comparison of two situation awareness metrics: SAGAT and SA-SWORD. *Proceedings of the XIVth Triennial Congress of the International Ergonomics Association and 44th Annual Meeting of the Human Factors and Ergonomics Society* (pp. 3.49–3.52). Santa Monica, CA: Human Factors and Ergonomics Society.

Sohn, Y. W., & Doane, S. M. (2004). Memory processes of flight situation awareness: Interactive roles of working memory capacity, long-term working memory, and expertise. *Human Factors, 46*, 461–475.

Strater, L. D., Reynolds, J. P., Faulkner, L. A., Birch, D. K., Hyatt, J., Swetnam, S., & Endsley, M. R. (2004). PC-based tools to improve infantry situation awareness. *Proceedings of the Human Factors and Ergonomics Society 48th Annual Meeting* (pp. 668–672). Santa Monica, CA: Human Factors and Ergonomics Society.

Strayer, D. L., Cooper, J. M., & Drews, F. A. (2004). What do drivers fail to see when conversing on a cell phone? *Proceedings of the Human Factors and Ergonomics Society 48th Annual Meeting* (pp. 2213–2217). Santa Monica, CA: Human Factors and Ergonomics Society.

Swets, J. A., & Pickett, R. M. (1982). *Evaluation of diagnostic systems: Methods from signal detection theory*. New York: Academic Press.

Taylor, R. M. (1990). Situational awareness rating technique (SART): The development of a tool for aircrew systems design. *Situational Awareness in Aerospace Operations* (AGARD-CP-478, p. 3/1–3/17). Neuilly Sur Seine, France: NATO-AGARD.

Taylor, R. M., & Selcon, S. J. (1994). Situation in mind: Theory, application and measurement of situational awareness. In R. D. Gilson, D. J. Garland, & J. M. Koonce (Eds.), *Situational awareness in complex settings* (pp. 69–78). Daytona Beach, FL: Embry-Riddle Aeronautical University Press.

Varakin, D. A., Levin, D. T., & Fidler, R. (2004). Unseen and unaware: Implication of recent research on failures of visual awareness for human-computer interface design. *Human-Computer Interaction, 19*, 389–422.

Venturino, M. (1997). Interference and information organization in keeping track of continually changing information. *Human Factors, 39*, 532–539.

Vidulich, M. A. (2000). The relationship between mental workload and situation awareness. *Proceedings of the XIVth Triennial Congress of the International Ergonomics Association and 44th Annual Meeting of the Human Factors and Ergonomics Society* (pp. 3.460–3.463). Santa Monica, CA: Human Factors and Ergonomics Society.

Ward, P., & Williams, A. M. (2003). Perceptual and cognitive skill development in soccer: The multidimensional nature of expert performance. *Journal of Sport & Exercise Psychology, 25*, 93–111.

Warwick, W., McIlwaine, S., Hutton, R. J. B., & McDermott, P. (2001). Developing computational models of recognition-primed decision making. *Proceedings of the Tenth Conference on Computer Generated Forces*. Orlando, FL: Simulation Interoperability Standards Organization.

Wickens, C. D. (2002). Situation awareness and workload in aviation. *Current Directions in Psychological Science, 11*, 128–133.

Wickens, C. D., McCarley, J. S., Alexander, A. L., Thomas, L. C., Ambinder, M., & Zeng, S., (2005). *Attention-situation awareness (A-SA) model of pilot error* (Report No. AHFD-04-15/NASA-04-5). Savoy, IL: Institute of Aviation, University of Illinois at Urbana-Champaign.

Wong, B. L. W., & Blandford, A. (2004). Describing situation awareness at an emergency medical dispatch center. *Proceedings of the Human Factors and Ergonomics Society 48th Annual Meeting* (pp. 285–289). Santa Monica, CA: Human Factors and Ergonomics Society.

Yntema, D. B., & Mueser, G. E. (1960). Remembering the present states of a number of variables. *Journal of Experimental Psychology, 60*, 18–22.

CHAPTER 2

Crew Resource Management Training Research, Practice, and Lessons Learned

By Eduardo Salas, Katherine A. Wilson, C. Shawn Burke, Dennis C. Wightman, & William R. Howse

> Crew resource management (CRM) training was introduced to the aviation community in 1979. Since then it has evolved and matured and is now being applied in a number of domains, including health care and offshore oil production. There is abundant literature resulting from research in the area, but there is no recent comprehensive review of the origins, current state, and future direction of CRM. The purpose of this chapter is to perform that review and provide the reader with an understanding of the research, practice, and training associated with CRM. We also provide a number of lessons learned based on the literature and our observations, as well as future needs of the community.

There is an adage in aviation, "Takeoffs are voluntary, landings are mandatory." Unfortunately, the number of takeoffs does not equal the number of *safe* landings. Accidents happen not only in aviation but in health care, the oil industry, and general transportation as well. These are activities in which complex functions are performed by multiple individuals acting in consort and in which errors can have immediate and costly consequences. It is no surprise that in these high-consequence domains, human error contributes to 60% of all accidents (e.g., Nullmeyer, Stella, Montijo, & Harden, 2005). We also know that some incidence of human error in these domains is unavoidable. As a means to minimize human error, organizations and agencies deploy teams in part to insert redundancy into the systems. However, despite the incorporation of teams, errors continue to occur in organizations, some with catastrophic consequences (e.g., loss of life, property damage).

WHY CRM AND CRM TRAINING?

Accidents in commercial aviation and the military are highly visible when they occur because of their infrequent occurrence and the loss of life incurred. In the 1950s and 1960s, mechanical failure predominated over human error as the principal contributing cause of commercial aviation accidents. As aircraft design and construction improved, the proportion of accidents attributed to mechanical failure declined. In addition, understanding of and vigilance for human error increased, and as a result, the proportion of accidents attributed to human error increased.

In the 1970s, the National Transportation Safety Board (NTSB) began citing breakdowns in teamwork as a contributing factor in commercial aviation accidents (e.g., Eastern Air Lines Flight 401). In 1978, United Airlines Flight 173 crashed into a wooded area near Portland, Oregon, because the crew failed to properly monitor the aircraft fuel state while they attended to a landing gear malfunction (www.avweb.com). The accident was attributed to crew error. Among the 189 passengers and crew on board, only 10 deaths occurred, but this accident could be considered the turning point and prompted the U.S. commercial aviation industry to take action.

Similar findings were evident beyond the United States, as was observed when a KLM airliner crashed into a Pan Am airliner in Tenerife, Canary Islands, in 1977, killing 583 passengers and crew. This accident was considered the deadliest commercial aviation accident in history (Weick, 1990). As a result, cockpit resource management (as it was known at that time) was born.

WHAT ARE CRM AND CRM TRAINING?

The term *cockpit resource management* was coined by John Lauber in 1977, although it languished as a largely ignored buzzword for some time. Lauber defined CRM as "using all available resources—information, equipment, and people—to achieve safe and efficient flight operations" (Lauber, 1984, p. 20). In 1979, the commercial aviation industry was formally introduced to the concept of CRM at a NASA-sponsored workshop, "Resource Management on the Flightdeck" (Cooper, White, & Lauber, 1980). In line with Lauber's definition, CRM training was introduced as a way to teach air crews to use all available resources by communicating and coordinating as a team. Some airlines took to this new concept almost immediately, and by the early 1980s, United Airlines and KLM Royal Dutch Airlines were the first commercial airlines to provide CRM training to their flight crews (Hawkins, 1987).

The U.S. military first implemented CRM training in 1986 (Alkov, 1989). Salas and colleagues (Salas, Fowlkes, Stout, Milanovich, & Prince, 1999) modified the definition of CRM from an operational to a training perspective, as one form of team training that seeks "to improve *teamwork in the cockpit* by applying well-tested training *tools* (e.g., simulators, lectures, videos) targeted at specific *content* (i.e., teamwork knowledge, skills, and attitudes)" (p. 163). We and others argue that the emergence of CRM training is one of the greatest successes of aviation and the human factors/ergonomics field (e.g., Helmreich & Foushee, 1993; Salas, Bowers, & Edens, 2001).

We should note that, unfortunately in our opinion, the term *CRM* has taken on a life of its own. It is invoked by many in the different industries that apply it as if it were a psychological or performance-oriented construct. It is not. CRM is merely an umbrella term for characterizing the knowledge, skills, and attitudes needed to succeed as a team in the flight deck, the operating room, the shop floor, the boardroom, or the incident command center, to name a few. The term *CRM* seems to have been abused, misused, and overused by the community creating (sometimes) the belief that all that needs to be done to improve safety or manage error is CRM—when the issue is much more complex and deeper than that. This will be discussed further later.

The generally accepted meaning of the abbreviation CRM has shifted from *cockpit* to *crew* resource management, reflecting a greater concentration on the personnel than the location. Although the history of CRM is relatively short, it has been researched and implemented in commercial and military aviation for nearly two decades, during which its training emphasis has evolved through five generations (see Helmreich, Merritt, & Wilhelm, 1999). This evolutionary process has resulted from knowledge gained regarding human error, team performance, and training. The most current focus of CRM training is on threat and error management. Helmreich and colleagues (Helmreich, Wilhelm, Klinect, & Merritt, 2001) developed a model depicting four levels of threat or risk: external threats, internal threats (i.e., crew errors), crew actions, and outcomes. When risk is present, the crew should employ CRM behaviors to assess and manage this risk. Critical to the success of threat and error management are situation awareness and planning.

The success of CRM in aviation has led to the implementation of CRM training (or some variation; e.g., crisis resource management training, aircrew coordination training, bridge resource management) in other team performance areas in aviation (e.g., maintenance, cabin crew) and domains beyond (e.g., health care, maritime, power companies, fire services). The generations through which CRM training has evolved in commercial aviation have served as a model for programs developed elsewhere.

With the expansion of CRM training to other domains, it seems appropriate to look at CRM in these domains to determine the current state of the community. Several reviews have been conducted with a concentration on various aspects of CRM (e.g., Maurino, 1999; O'Connor, Flin, & Fletcher, 2002; Wiener, Kanki, & Helmreich, 1993), but none has provided a comprehensive assessment of its origins and future direction. In this chapter, first we identify a number of theoretical drivers fundamental to CRM research. Second, we turn to the literature to determine what we have learned (so far) from CRM research. Third, we extract from the literature, and from our understanding of those implementing CRM training, how CRM is being implemented (e.g., what skills are trained, how it is being evaluated) and whether it works—in other words, lessons learned. Fourth, we conclude with some observations and needs that we have identified in the CRM community. Throughout the chapter, we provide a selection of lessons learned based on the literature and our experience.

THEORETICAL DRIVERS OF CRM RESEARCH AND TRAINING

Teams have been used as a strategy for achieving organizational goals and improving safety for a very long time. Since the 1980s, we have witnessed a number of advances, as documented in the literature on teamwork. These advances have strengthened our understanding of factors related to team performance and team effectiveness (e.g., Ilgen, Hollenbeck, Johnson, & Jundt, 2005; Kozlowski & Bell, 2003). In turn, these advances in theory have driven the design and delivery of CRM training (see MacLeod, 2005; Salas, Wilson, Burke, Wightman & Howse, 2006, for a deeper discussion on designing and developing CRM training). These advances can be sorted into several areas: team effectiveness, team competencies (i.e., knowledge, skills, and attitudes), shared mental models, and social-psychological constructs. These areas are briefly discussed in the following sections.

Team Effectiveness

Foushee (1984) was one of the first researchers to examine teamwork in the cockpit and discuss the input (e.g., leadership profiles), process (e.g., communication), and outcome (e.g., performance) variables affecting pilots. Since his study, a number of general conceptual and empirically based models have been developed that differentiate between taskwork and teamwork and that depict the relationships among input (e.g., team characteristics, individual characteristics), process (e.g., communication, coordination, decision making, backup behavior, compatible cognitive structures, compensatory behavior, and leadership), and outcome variables (e.g., increased productivity, increased safety, increased job satisfaction; see Ilgen et al., 2005; Salas, Stagl, Burke, & Goodwin, in press, for a representative sample). These models, as well as empirical research, have illustrated not only the dynamic and multidimensional nature of teamwork but also the importance of process variables in determining the relationship between team input and team output (Swezey & Salas, 1992; Wiener et al., 1993). For example, it has been argued that skills such as adaptability, communication, coordination, decision making, team leadership, situation awareness, workload management, and backup behavior are important process (i.e., teamwork) variables. These efforts have formed the basis for describing the skills needed for effective teamwork in the cockpit.

In addition, teams are dynamic and evolve over time (e.g., Gersick, 1988; Marks, Mathieu, & Zaccaro, 2001). Furthermore, there are two tracks of skills that must be mastered within teams: task work and teamwork (McIntyre & Salas, 1995). Task work skills are those that members must understand and acquire for actual task performance (i.e., technical expertise), and teamwork skills are the behavioral and attitudinal responses that members need to function effectively as part of an interdependent team (Morgan, Glickman, Woodard, Blaiwes, & Salas, 1986). Task work skills are necessary but not sufficient for effective team outcomes; teamwork skills are required as well. It has been argued that the relationship between task work skills and team effectiveness is mediated by teamwork skills (e.g., Hackman & Morris, 1975; Bass, 1990).

Teamwork Competencies

As the theory behind teamwork advanced, researchers began to further delineate teamwork. More specifically, they began to model teamwork as composed of a core set of competencies that involve knowledge (what team members think), skills (what team members do), and attitudes (what team members feel; see Cannon-Bowers, Tannenbaum, Salas, & Volpe, 1995). Cannon-Bowers et al. (1995) identified major teamwork competencies (i.e., knowledge, skills, and attitudes) that are transportable across situations.

Recognition of teamwork competencies represented an important development in the theory behind teams and teamwork. Prior to that, training focused primarily on the behavioral (i.e., observable) side of teamwork, neglecting the cognitive and affective components. Three independent reviews were conducted in the aviation community, each of which uncovered similar critical competencies. Prince and Salas (1993) suggested seven key team process variables necessary for CRM: situation awareness, decision making, leadership, communication, adaptability, mission analysis/planning, and assertiveness. Helmreich and Foushee (1993) suggested similar skills, including situation awareness, leadership, planning, communication, assertions, and decision tasks. Finally, the U.S. Army Aviation Center suggested six basic qualities needed by rotary wing air crews: team leadership, premission planning, situation awareness, communication, decision-making techniques, and assertion (Department of the Army, 1992). Recent work found in the team literature has focused on the examination of team cognition and exactly how cognitive processes relate to teamwork (e.g., shared mental models; see Cannon-Bowers, Salas, & Converse, 1993), metacognition (Klein, 2000), and shared situation assessment (Burke, Fiore, & Salas, 2003).

It has also been recognized that there are many types of teams (Sundstrom, DeMeuse, & Futrell, 1990). Differences in function, composition, and interdependencies within various teams determine the relative importance of specific teamwork competencies. Although researchers have argued for a set of generic, transportable competencies, the weight of each one may differ with the situation. Furthermore, other teamwork competencies have been identified as dependent on team characteristics such as whether (a) team membership is stable or unstable or (b) the team focuses on the same tasks or completes several different types of tasks.

Shared Mental Models

A third theoretical driver for crew resource management originates with what is known about the impact of shared mental models on teamwork and team performance. *Shared mental models* (SMM) refer to team members possessing compatible or similar knowledge structures with regard to the equipment, task, and team (e.g., Rouse & Morris, 1986). Accurate shared cognitive structures allow teams to better combat human error, as well as catch system errors so teams can generate descriptions of systems and make predictions about future system states (e.g., Orasanu, 1994; Rouse et al., 1992). When teams are operating in dynamic, complex environments, explicit coordination and/or communication may not be possible. SMMs help team members share an understanding of the situation (i.e., shared situation awareness and the anticipation of individual needs; see Tenney & Pew, 2006, Chapter 1 in this volume, for an in-depth discussion of situation awareness). Researchers have begun to suggest that shared mental models enable effective teams to maintain consistent performance in complex, stressful environments (Cannon-Bowers et al., 1993; Orasanu, 1990) by facilitating implicit coordination (i.e., without explicit communication).

Foushee and others demonstrated the importance of accurate shared mental models in commercial aviation; for example, flight crews that had SMMs performed better (e.g., fewer operational errors, better coordination) than teams without SMMs, even under conditions of fatigue (Foushee, Lauber, Baetge, & Acomb, 1986). Others have suggested that flight crews with SMMs communicate and perform more effectively (e.g., Mohammed, Klimoski, & Rentsch, 2000) and are more willing to work together in the future (Rentsch & Klimoski, 2001).

Finally, during periods of high workload, SMMs lead to improved and more efficient communication strategies (Stout, Cannon-Bowers, Salas, & Milanovich, 1999). This research suggests the need for flight crews to have accurate and shared mental models so routine and nonroutine situations can be handled efficiently and effectively.

Social-Psychological Constructs

Early research examining CRM training focused on the role of communication and sociopsychological and personality factors (Foushee, 1982). Helmreich and Foushee (1993) presented a model in which multiple factors determine the effectiveness of crews. They argued that when the input and process factors are optimized, CRM training will be a success. The authors provided some suggestions for organizations to better design their CRM training programs.

First, individual factors such as personality can be used as a technique for selecting pilots who are predisposed to team activities and coordination (Helmreich & Foushee). CRM training that focused on personality factors was indicative of early programs in aviation, although those programs were not very successful. Personality traits are considered to be deeply rooted and stable predispositions that influence how an individual responds to situations. However, some have argued that training focused on personality is not likely to result in a desired change in behavior (Chidester, Helmreich, Gregorich, & Geis, 1991) and also that personality is not a reliable predictor of pilot performance (see Hedge et al., 2000). Conversely, Chidester and colleagues (1991) argued that selecting individuals to optimize the fit between personality characteristics and desired outcomes (e.g., performance) may be very beneficial.

Some have argued that organizational input factors also influence crew effectiveness (Helmreich & Foushee, 1993). For example, managers must support the development and implementation of training. When managers show a commitment to and a belief in the importance of training, trainees are more likely to believe in training and respond more positively to it (e.g., Zohar, 2000). Additionally, Helmreich and Foushee argued that training content must be consistent with the culture and practices of the organization. They noted three steps that organizations can take to ensure this consistency: (a) Take a crew rather than an individual approach to training, (b) modify checklists and other cockpit documents for consistency with training concepts, and (c) enhance communication between the cockpit and team members outside the cockpit (e.g., cabin crews, dispatchers, maintenance personnel). Instructors, check airmen, and chief pilots (in other words, those in role-modeling positions) must "walk the walk" and "talk the talk." It is important that these individuals practice and reinforce the concepts trained. Finally, it is argued that selecting individuals with both interpersonal skills and technical expertise will ensure that good CRM behaviors are put into practice (Helmreich & Foushee, 1993).

The third input factor that is argued to drive CRM training is regulatory factors (Helmreich & Foushee, 1993). Although CRM training was once a recommended training program in commercial and military aviation, it has since become mandatory for all air crews. Both commercial and military aviation in the United States and beyond have regulatory organizations (e.g., FAA, ICAO, JAA/ESA) indicating who must receive

CRM training and when. In addition, within the health care community, the Joint Commission on the Accreditation of Healthcare Organizations (JCAHO) is working on developing guidelines for implementing CRM training (see Baker, Gustafson, Beaubien, Salas, & Barach, 2005).

Helmreich and Foushee (1993) also emphasized the need for improvement in process factors to increase the effectiveness of CRM training. They suggested that it is best for trainees to experience the process factors firsthand through practice. Specifically, they discussed the use of Line Oriented Flight Training (LOFT) simulations and skilled instructors, as opposed to just lectures, to guide crew members to self-realization. CRM processes should then be practiced and reinforced through future LOFT simulations, line operations, and recurrent CRM training.

Summary

As can be seen, there are several theoretical drivers for CRM research. It is not likely that the research is guided by only one of the suggested drivers. Rather, as CRM has evolved, it has taken components from each of these theoretical frameworks and developed into a core around which aviation research is driven.

Therefore, the next step in understanding CRM is to take a closer look at several areas in which CRM research is being conducted. We draw our first lessons from the brief overview of the theories driving CRM research:

Lesson 1: Relevant and practical theories abound in CRM research.
Lesson 2: CRM research is solidly guided by meaningful theoretical drivers.

WHAT HAVE WE LEARNED FROM CRM RESEARCH?

A number of research areas were identified in which CRM has been examined. In this section, we discuss research findings regarding individual differences, simulations, and understanding of team process and measurement. We selected these areas because they represent areas of research that are important and relevant but have not been adequately reviewed elsewhere.

Individual Differences

In this section, the focus is on two pockets of research related to CRM: personnel selection and cultural differences.

Selecting Pilots for CRM

Individuals undergo stringent training programs to become rated pilots in commercial and military aviation. But it takes more than just technical flying knowledge and skills for a pilot to be a successful crew member; nontechnical skills are important as well (e.g., O'Connor et al., 2002). Research has demonstrated that it is possible to derive useful predictions of pilot performance using cognitive, psychomotor, personality, and biodata instruments. These measurement domains have been used extensively for pilot selection, but little research has been conducted on how to select pilots for their CRM-related skills or potential for developing them.

Hedge et al. (2000) set out to develop and validate a test of CRM skills. They based their test on the Situational Judgment Test (SJT), which presents respondents with a set of situations related to the job and asks them to select the most and least effective solutions or actions. Based on the limited research that had been conducted on SJTs, the authors developed the Situation Test of Aircrew Response Styles (STARS), targeting critical CRM skills for Air Force flight crews. STARS measures six critical skills: problem solving, decision making, knowledge of how to respond to challenging situations, communication, aircrew management, and interpersonal effectiveness. The original 60-item test was reduced to a subset of those items (13 items) to make the test more feasible for respondents.

The results of the research indicated three statistically significant predictor-criterion correlations (i.e., overall ratings, loadmaster, and radio operator). The largest correlation with CRM skills was associated with the loadmaster position, indicating that subordinates may be good candidates for assessing the leadership of the captain. The researchers noted the need for additional research that takes a superordinate perspective (e.g., performance data gathered by check airmen, instructors). Additionally, Hedge et al. suggested a need to determine the incremental validity of STARS over other psychomotor and cognitive measures alone (i.e., does STARS uniquely contribute to crew performance prediction?).

Damitz, Manzey, Kleinman, and Severin (2003) evaluated the validity of an assessment center for pilot selection. The assessment center method attempts to measure teamwork-related behaviors directly in an interactive situation. Damitz et al. examined nine behavioral dimensions: cooperation, conflict management, empathy, commitment, self-assessment, stress resistance, reliability and discipline, flexibility, and decision making. Their results indicate, at least initially, that the assessment center method for assessing interpersonal and performance-related skills has construct and criterion validity. The research also examined moderating effects of the type of assessor used (i.e., psychologist vs. job expert). Psychologists' ratings were found to be more valid for assessing interpersonal behaviors; however, for performance-related behaviors, ratings of job experts were just as valid as those of psychologists.

Lesson 3: Assessment centers and SJTs are strategies for selecting team members with CRM-based skills. However, more validation is needed.

National Culture

Much research has been conducted regarding cultural factors in training and education, but little has been specific to the interface of CRM and culture. Some studies have looked at the impact of culture in the cockpit, particularly the impact of national culture on pilots' attitudes and behaviors. The results of these studies are briefly discussed.

Attitudes toward CRM. Throughout the 1990s, Helmreich and Merritt investigated the effects of national culture on pilots' attitudes toward CRM. The Flight Management Attitudes Questionnaire (FMAQ) has been used to measure attitudes toward leadership and command, work values, team behaviors, crew interaction and communication, stress, and cockpit automation. This research suggests that many cultures agree on issues such as the importance of crew communication and coordination and the importance of preflight briefings (Helmreich & Merritt, 1996). However, many cultures differ on issues regarding junior crewmember assertiveness, the influence of personal problems on performance, use of automation, and the likelihood of making errors in judgment during an emergency situation (Helmreich & Merritt, 1996, 1998; Merritt & Helmreich, 1995; Sherman, Helmreich, & Merritt, 1997).

For example, Helmreich and Merritt (1998) and Merritt and Helmreich (1995) found that Anglo pilots (e.g., American, Irish, Australian) identify more closely with a flattened command structure with greater two-way communication, prefer that captains treat junior crewmembers as equals and include them in the decision-making process, encourage assertiveness by any crewmember in the interest of flight safety, and disagree that written procedures are required for all in-flight situations. In contrast, non-Anglo pilots (e.g., Brazilian, Philippine) indicate a preference for a more hierarchical command structure, expect that the captain is the ultimate authority in the cockpit, are less likely to question the actions or thinking of senior crewmembers, and believe written procedures for all in-flight situations are required. To preserve these relationships and to maintain harmony among the team, communications tend to be indirect. These differences may have serious consequences on the safety of flight operations. For example, a Korean Airlines flight crashed when a Canadian captain and a Korean copilot disagreed on how to manage the landing (Westrum & Adamski, 1999). Because the copilot's English was not very good, he had a difficult time expressing his concern and instead grabbed the controls from the captain. No fatalities were reported after the plane overran the runway and ignited.

The research conducted by Helmreich and Merritt (1998) indicates that culture influences crew members' attitudes both positively and negatively. The results of their research suggest that although some teamwork skills are generally accepted or seen as important in the aviation community worldwide—specifically, communication, coordination, and shared mental models—there are definitely some that are seen differently: assertiveness by junior crewmembers and the effects of stress on pilot performance. It should be pointed out, however, that pilots' expressions of these values do not necessarily mean that the values determine their behavior.

We next review research examining the influence of culture on behaviors in the cockpit.

Behaviors in the cockpit. We found several studies that examined the impact of culture—specifically, national culture—on CRM-related behaviors in the cockpit. Kuang and Davis (2001) reviewed 22 accident and incident reports obtained from the National Transportation Safety Board aviation database involving foreign (i.e., non-U.S.) flight crews. The reports were analyzed and coded based on variables associated with CRM and team performance: teamwork (e.g., communication, coordination, shared mental models), task performance (e.g., use/disuse of checklists, engagement/disengagement of automation, flight control), and team errors (e.g., communication errors, response to errors, operational decision errors). The authors categorized the accidents by the cultural values of the flight crews: individualism/collectivism (i.e., individual- vs. team-oriented) and high/low power distance (i.e., degree of vertical hierarchy).

The accident and incident data suggest that individualist crews exhibited more positive teamwork behaviors, more positive task performance behaviors, and fewer error behaviors than collectivist crews. Although Kuang and Davis were surprised at this trend, in that collectivist groups should be more team oriented, the nature of collectivist groups to want to maintain group harmony may negatively affect performance in emergency situations. This was evident when an Avianca flight crashed because of the collectivist nature of the crew members, who did not feel comfortable declaring an emergency situation when their fuel situation became critical because this would have placed their aircraft in front of others waiting to land (Helmreich, 1994).

Next, Kuang and Davis (2001) found that low power distance crews exhibited more positive teamwork behaviors, more positive task performance behaviors, and fewer error behaviors than did high power distance crews. This trend can be explained by better communication flow that results from the flattened command structure of low power distance cockpits. Results also suggest that crews high in uncertainty avoidance (i.e., degree of tolerance of ambiguity) exhibited more positive teamwork behaviors and fewer error behaviors than did crews low in uncertainty avoidance. However, crews low in uncertainty avoidance exhibited more task performance behaviors. The authors were surprised by these findings because they did not expect uncertainty avoidance to influence teamwork behaviors. They offered one explanation: Because the crew members would want to avoid uncertainty, they would be more likely to work as a team.

Communication is critical when it comes to flight safety, and it is a key component of CRM. It is through communication that pilots convey intentions, provide and request information, encourage participation, and direct crewmember actions (Kanki & Palmer, 1993). Communication is difficult enough for homogeneous cockpits; adding mixed cultures is an additional complication. Orasanu, Fischer, and Davison (1997) explored the challenges of communications in the cockpit and found that national culture has a significant impact. They gathered reports from the Aviation Safety Reporting System (ASRS) and sorted them into categories based on communication error. A predominant number of incidents fell into the following five categories (in order of frequency, high to low): language/accent (i.e., difficulty understanding clearances), partial readback (i.e., incomplete repetition of clearance transactions leading to confusion), dual language switching (i.e., air traffic control outside the United States switching between English for foreign crews and the native language for local crews), unfamiliar terminology (i.e., use of local jargon rather than ICAO standardized terminology), and speech acts (i.e., address, request, instruction; one speech act misinterpreted as another). In short, the findings could be summarized as problems in the transmission of information, problems concerning the content of communication, and problems arising from social interaction style. CRM training may be one strategy to help mitigate these challenges.

Using simulated flight crews, Davis and colleagues (Davis, Bryant, Liu, Tedrow, & Say, 2003) investigated the impact of national culture on CRM behaviors and error management. The researchers compared homogeneous American and Chinese teams (all team members from the same culture) and multicultural teams (team members from different cultures). Their study identified differences between cultures in terms of team behaviors. Comparisons of American, Chinese, and multicultural teams suggested that homogeneous crews communicate better and more frequently than did multicultural teams. American teams had higher situation awareness, made more decisions, displayed more communication, and made fewer communication errors than did Chinese and multicultural teams. It is not clear, however, if the participants received any specific CRM training prior to flying the scenarios.

Research examining the impact of national culture on CRM-related behaviors has been scarce, but the few studies that have been done indicate that there may be a relationship. However, a recent article suggests that national culture may not contribute to accidents and incidents as often as once thought (Hutchins, Holder, & Pérez, 2002). The authors argued that there is no unambiguous evidence that national culture affects flight safety and that professional and organizational cultures may override national culture. They stated that people function well in a second culture and that changes in safety may actually be attributed to changes in professional and organizational culture, not in national culture (i.e., accidents have led pilots to consider more ways to be safe). For example, consider countries in which economical hardships result in cost cutting and minimized training. Changes in safety could result from changes in organizational culture rather than being a factor of national culture. This area of research warrants further review, but we offer the following lessons learned:

Lesson 4: Culture (organizational and national) matters to CRM-related performance.

Culture research is complex, so clearly more, better, and more robust research is needed. Research in this area is still in its infancy, and clear and precise principles of how culture influences CRM performance outcomes are yet to be developed. However, we do know that culture serves as a filter for actions, behaviors, and processes in team dynamics.

Simulations

Typically, the practice of CRM skills has taken place in high-fidelity simulators or in live training environments. High-fidelity simulation has been shown to be a valuable training delivery tool that can provide a realistic yet safe environment in which to practice CRM skills (Jentsch & Bowers, 1998). However, with the advancement of technology and computer games, low-fidelity simulations have become a viable alternative for the practice and feedback of team behaviors (Bowers & Jentsch, 2001). Low-fidelity simulations are a low-cost alternative that may offer increased experimental control of independent variables (Bowers, Salas, Prince, & Brannick, 1992).

Role-play exercises and critical incidents are other low-cost alternatives for training CRM knowledge, skills, and attitudes. We present findings in the literature regarding the use of low-fidelity simulations—specifically, computer-based systems and Rapidly Reconfigurable Event-Set Based Line-Oriented Evaluations (RRLOE).

Personal computer–based systems. The first area of research identified looks at the application of low-fidelity, PC-based simulations for training CRM competencies. PC-based simulations have been successful in the transfer of flight skills to the cockpit. For example, Dennis and Harris (1998) demonstrated that some technical flight skills could be trained using a tabletop computer system. Additionally, Koonce and Bramble (1998) described several studies in which computer-based training devices were effective for training pilots in complex skills and attaining an instrument rating, as well as for reducing flight training costs. Gopher, Weil, and Bareket (1994) studied the effects of training specific and general skills using a computer game to improve coping with high information-processing and response demands (often required by flight tasks). Their results suggest that training specific and general skills, although related to a computer game, transferred to the cockpit and improved trainees' performance. Therefore, the success of computer-based systems at improving flight skills makes the transition to studying the transfer of CRM competencies a natural one.

We found several articles that describe research conducted in this area. In 1990, Stout, Cannon-Bowers, Morgan, and Salas examined CRM behaviors in low-fidelity simulations. The results of their research indicated that CRM behaviors are elicited when using a PC-based simulation, providing preliminary support for the use of low-fidelity, PC-based simulations for training CRM competencies.

Others have evaluated different approaches (i.e., knowledge based and skills based) for training communication skills using a low-fidelity simulation (Lassiter, Vaughn, Smaltz, Morgan, & Salas, 1990). Participants were asked to complete a two-person helicopter mission following training using the computer game Gunship. Overall team communication ratings were significantly correlated with team mission performance. Additionally, performance data indicated that trainees who received skills-based training performed significantly better than those who received knowledge-based or no training. These results suggest that low-fidelity simulations elicit CRM behaviors, and skills-based training leads to improved performance over knowledge-based training. We would expect these results to generalize not only to similar flight-related tasks but also to other tasks requiring CRM behaviors.

Baker and colleagues (Baker, Prince, Shrestha, Oser, & Salas, 1993) used aviation computer games for training CRM skills. They collected reaction data from participants and found that 90% of aviator participants agreed that a computer-based system as used in their study could be used for training CRM skills. A majority of participants felt that it was a good way to learn CRM and that the system demonstrated the importance of CRM.

Most recently, Brannick, Prince, and Salas (2005) demonstrated that skills learned using a PC-based training system (i.e., Microsoft Flight Simulator) do in fact transfer to a high-fidelity simulator. Overall, the research that employed PC-based systems suggests that the use of aviation computer games is a viable way to learn, practice, and provide feedback regarding CRM skills.

Rapidly Reconfigurable Event-Set Based Line-Oriented Evaluations (RRLOE) generator. As a part of the Advanced Qualification Program in commercial aviation, line-oriented evaluations (LOEs) are used to determine pilot performance levels and proficiency. To use LOEs, organizations must gain approval by the Federal Aviation

Administration (FAA; *http://pegasus.cc.ucf.edu/~rrloe/About_RRLOE.html*). As an alternative, with RRLOE it is proposed that instead of using a small number of LOEs, which creates the possibility of a reduction in validity of the training or the need to obtain FAA approval of a new LOE, a basic LOE structure be developed and approved. Specific LOEs could be created from that structure without the need for further approval. For this system to be practical, researchers have worked to overcome obstacles in four key areas: skill-based training, event-based assessment, applied research issues, and human factors aspects of the software.

Because practice and feedback mechanisms must be present for successful performance, one of the primary initiatives of RRLOE has been to provide appropriate opportunities for practice and feedback within the training scenarios. Not only does RRLOE focus on skill-based training, but it also has built-in mechanisms for event-based assessment to increase its practicality. In other words, performance measures are linked to embedded events within the scripted training scenarios.

The RRLOE system has several characteristics that have demonstrated its validity and realism. First, to track the realism of the assessment system, RRLOE adopted the "Domino Method" (see Bowers, Jentsch, Baker, Prince, & Salas, 1997), which has allowed for better tracking of flight-relevant parameters than could be done by a human observer. RRLOE also mimics the operational environment, in that scenarios can be scripted around various weather conditions to increase the difficulty of the scenarios. To ensure that the scenarios created within RRLOE are within the range of difficulty desired by the airlines, a mathematical model was created that is executed for each LOE. Finally, the RRLOE software was developed based on human factors/ergonomics principles of human-computer interaction and is directly related to the operational environment of the commercial aviation community.

Lesson 5: Simulations are a viable alternative to train teams in a safe yet realistic learning environment.

Lesson 6: Low-fidelity simulations offer a cost-effective, viable alternative to large-scale, high-fidelity simulations for training CRM-related skills.

Team Process and Performance Measurement

Since the 1980s, there has been an explosion in research associated with examining team processes and behavior measurement. The dynamic nature of teams and their tasks makes it difficult to understand and measure team processes and behaviors in the cockpit and other complex environments (Prince & Salas, 1999). The fact that team performance consists of both individual and team member actions makes measurement using valid and reliable tools even more challenging.

In the early years of CRM's inception, self-report measures were often preferred over objective measures—specifically, the Cockpit Management Attitudes Questionnaire (CMAQ; Helmreich, 1984) and the Flight Management Attitudes Questionnaire (FMAQ; Helmreich, Merritt, Sherman, Gregorich, & Wiener, 1993). Over time, the aviation community developed more behaviorally based tools, such as Line/LOS worksheet, Team Dimensional Training, and TARGETs (Targeted Acceptable Responses to Generated Events or Tasks). Even so, more was needed. Measurement research throughout the 1990s and 2000s has focused on the development of more and better objective measures of team process. Old tools have been adapted and new tools developed (e.g., NOn-TECHnical Skills, or NOTECHS). Here we discuss several tools being used to measure team performance in aviation and beyond, specifically behavioral-based and cognitive-based instruments.

Behavioral-Based Instruments

TARGETs. The Targeted Acceptable Responses to Generated Events or Tasks (TARGETs) is a measurement tool that facilitates the observation of team performance behaviors in a training setting by embedding learning scenarios (i.e., events), which are defined a priori (Fowlkes, Lane, Salas, Franz, & Oser, 1994). This tool serves as a checklist that lists each event and the acceptable response to that event. The observer then uses the checklist to mark whether the behavior was present or absent.

Three key methodological issues make this an appropriate, tested approach for observing CRM-related performance outcomes. First, TARGETs is event based, in that the scenarios provide trainees with an opportunity to practice and exhibit targeted team competencies. Second, learning is controlled through TARGETs in that opportunities are provided that elicit the behaviors that one would want to observe. Finally, TARGETs behaviors are clearly observable to all observers regardless of experience. Salas and colleagues (Salas, Fowlkes, Stout, Milanovich, & Prince, 1999) successfully used the TARGETs methodology to observe CRM behaviors in the naval aviation community. Their studies showed that pilots who had received CRM training hit more of the TARGETs than did untrained pilots, especially in high-workload periods.

Line/LOS Checklist. Similar to TARGETs, the Line/LOS Checklist (LLC) is used to rate key team performance behaviors (Helmreich, Wilhelm, Kello, Taggart, & Butler, 1991). The LLC was developed for the aviation community to observe crew behaviors during four critical segments of flight: predeparture, takeoff and climb, cruise, and decent and landing. By observing all segments of flight, observers hope to be able to delineate the chain of events should an error occur (Beaubien & Baker, 2002). Pilots are rated on each behavior on a scale of 1 (*poor*) to 4 (*outstanding*).

The latest version of the LLC (LLC4) observes performance on six dimensions: team management and crew communication, situational awareness and decision making, automation management, special situations, technical proficiency, and overall crew performance. The LLC4 also gathers information related to demographics, crews' experience flying together, and observer comments. It has been used to show differences between domestic and international operations at one airline (Law & Wilhelm, 1995).

Line Operations Safety Audit (LOSA). The LOSA technique was developed for the aviation community to assess the impact of CRM training in the cockpit (Croft, 2001). Trained judges observe pilots and crew in flight and rate their responses to threats and coordination of their efforts. The crew's performance is rated on a scale of 1 = *poor* (observed performance had safety implications), 2 = *marginal* (observed performance was barely adequate), 3 = *good* (observed performance was effective), and 4 = *outstanding* (observed performance was truly noteworthy).

At the completion of these observations, the judges interview crews to gather additional information regarding the individual and team processes at work in the cockpit. This information is used to determine weaknesses within the training program. The end result is that practical information is given to the organizations to help with (a) identifying threats in the airline's operating environment, (b) identifying threats from within the airline's operation, (c) assessing the degree of transference of training to the airline, (d) checking the quality and usability of procedures, (e) identifying design problems in the human-machine interface, (f) understanding pilots' shortcuts and workarounds, (g) assessment of safety margins, (h) providing a baseline for organizational change, and (i) providing a rationale for allocation of resources (Helmreich, Klinect, Wilhelm, & Sexton, 2001).

Helmreich, Klinect, and Wilhelm (1999) used LOSA to observe 3,500 flights to detect errors and threats to safety. The purpose of these observations was to provide the commercial aviation community with a representation of normal flight operations so that the amount of risk associated with a certain movement or a particular environment could be evaluated. Helmreich et al. argued that many of the types of errors committed by flight crews can be linked to decrements in CRM behaviors. For example, procedural errors indicate a failure in monitoring and cross-checking, whereas communication errors indicate a lack of shared mental models and closed-loop communication. The failures observed in flight by Helmreich and colleagues indicate the need for more and better CRM training.

NOTECHS. NOTECHS (nontechnical skills) was initiated by the European Joint Aviation Authorities to develop a methodology to test the nontechnical skills (i.e., CRM skills) of pilots (Flin et al., 2003). Nontechnical skills are defined as "the cognitive and social skills of flight crew members in the cockpit, not directly related to aircraft control, system management, and standard operating procedures" (p. 96). Based on a review of existing systems, literature, and meetings with subject matter experts, the researchers developed a framework containing four categories that were divided into two social skills (cooperation, and leadership and management) and two cognitive skills (situation awareness and decision making). Each contains a subset of elements that correspond to behaviors that well-trained and high-performing pilots should exhibit (see Flin et al.).

We found two studies that assessed the validity, reliability, and usability of NOTECHS and that also examined the impact of cultural differences on NOTECHS (Flin et al., 2003; O'Connor et al., 2002). Results from these studies provided general support for the reliability, validity, and usability of the NOTECHS framework. Instructors reported high levels of satisfaction with the rating system and the consistency of the methodology. Instructor ratings met an acceptable level of agreement with previous ratings by trained experts. Cultural differences were not found to be more important than other variables, such as proficiency with the English language and experience with CRM evaluation.

Another study examined the validity of an adaptation of NOTECHS to health care, called Anaesthetists' Non-Technical Skills (ANTS; see Fletcher et al., 2003). The ANTS prototype has four categories: task management, team working, situation awareness, and decision making. For each element within these categories, a rating of 1 = *poor*, 2 = *marginal*, 3 = *acceptable*, or 4 = *good* is given by raters. Fifty consultant anaesthetists participated in the validation study. Results indicated that the ANTS prototype has acceptable validity, reliability, and usability. However, these results hold true only in an experimental setting; ANTS needs to be tested in a practical application.

Team Dimensional Training tool. Throughout the 1990s, the U.S. Navy conducted a comprehensive program examining tactical decision making under stress (known as TADMUS). Out of this work came the Team Dimensional Training (TDT) technique, in which team members convene after a training exercise or event to discuss and correct their teamwork processes and performance (Smith-Jentsch, Zeisig, Acton, & McPherson, 1998). The team discussion is structured so that it follows an expert model of teamwork and is facilitated by an instructor or leader. After the discussion, team members agree on a set of process-related goals on which they need to improve. Although TDT is a generic measurement technique, it can be embedded within the CRM training process.

TDT focuses on four teamwork dimensions: information exchange, communication, supporting behavior, and initiative/leadership. These dimensions are highlighted during the first step of the TDT process—Prebrief—and also clarify the purpose of the mission. The purpose of the next step, Observe Performance, is to observe trainees performing in the training or operational setting. Instructors are on hand to observe performance and document positive and negative events that occur. Step three, Diagnose Performance, occurs at the completion of the exercise when instructors come together to discuss what they observed, diagnose problems, and prepare for the next step, Debrief. During this final step, the core of the TDT process, the team is led by a facilitator and critiques its own performance. Key events are recapped, TDT dimensions are redefined, the team is guided through the self-correction process, feedback is provided, and performance improvement goals are set.

The benefit of the TDT technique over other team performance measurement techniques is that it enables team members to correct their own performance weaknesses, which will ultimately help them learn what went wrong and how to improve performance in the future.

Communication analysis. As noted previously, communication is one of the critical skills required for effective CRM. Team communication analyses have been employed to determine the types of communications used by high-reliability teams such as cockpit crews.

The importance of understanding team communications should not be underestimated. Priest, Burke, Guthrie, Salas, and Bowers (under revision) suggested that communication analyses can provide insight into several team processes, among them decision making, coordination, teamwork, leadership, backup behavior, and situation awareness. Previous research has focused on the amount and frequency of communication, but more recent work has focused on analysis of the content and sequence of communication through exploratory sequential data analysis (ESDA; see Bowers, Jentsch, Salas, & Braun, 1998; Priest et al., under revision).

ESDA techniques have been used to analyze communication between air traffic control teams sequencing the landing patterns of aircraft (Bellorini & Vanderhaegen, 1995). Bowers and colleagues (1998) analyzed the communication of cockpit crews to determine differences between high- and low-performing teams in terms of the types of communications used. Given that communication is critical to the success of CRM, team communication analysis techniques such as ESDA can provide vital information to researchers and training developers regarding types of communications for which teams should be trained.

Cognitive-Based Instruments

Concept mapping. Concept mapping is a knowledge elicitation tool that utilizes graphical representations of the knowledge structure and content of an individual (e.g., mental models; Swan, 1995). Although developed as an individual measurement tool, concept mapping has been adapted to measure shared team knowledge. Concept mapping creates a graphic representation of mental models for domain-specific concepts (Cooke, Salas, Cannon-Bowers, & Stout, 2000). The map contains different concepts and shows the relationships among two or more of them, provides information on shared thinking, and provides an understanding of what knowledge is necessary for successful individual and team performance. In the aviation community, concept mapping has been used to evaluate knowledge acquisition by pilots over the course of a training program (Evans, Hoeft, Kochan, & Jentsch, 2005) and is a viable supplement to written exams for pilots.

Pathfinder. Pathfinder helps to scale related concepts through participant ratings of paired comparisons (Mohammed et al., 2000). It has been suggested that a minimum of 15 to 30 concepts be paired in order to get the best results using Pathfinder (Goldsmith, Johnson, & Acton, 1991). Once paired comparisons are made, Pathfinder employs an algorithm to produce a representation of the structure of the concepts and the links between them.

Pathfinder has been used to illustrate changes in knowledge structure and performance after training (Kraiger, Salas, & Cannon-Bowers, 1995) and is beginning to be used in aviation. Pathfinder analysis has been employed to show relationships between novice and expert pilot ratings of information priority in flight operations (Schvaneveldt, Beringer, & Lamonica, 2001). This tool could be used to make comparisons between CRM-related concepts as well.

> *Lesson 7: The assessment of team dynamics remains a critical component during training and on the job.*
> *Lesson 8: CRM-related performance is composed of a set of interrelated knowledge, skills, and attitudes.*
> *Lesson 9: There are a number of valid, reliable, and diagnostic measurement tools available to researchers and practitioners.*

The tools available, though not perfect, offer some useful ways to capture, assess, and diagnose CRM-related processes and outcomes. These tools still need to be refined, adjusted, and improved as we learn more about team dynamics in high-consequence settings.

CRM Training Research

Since the 1980s, much has been learned regarding the design, delivery, and evaluation of CRM training. Although some of the research pertains to team training in general, it has direct relevance to CRM training. We next discuss what has been learned regarding the implementation of CRM training. This discussion comes from the literature as well as anecdotal observations and the "talk" from practitioners in a number of forums in the CRM community.

CRM training design and delivery. The design and delivery of team (CRM) training has been studied extensively (e.g., Salas & Cannon-Bowers, 2000b). A number of studies address the key knowledge and skills for CRM (see the team competencies section in this chapter) as well as guidelines to assist in the design and delivery of CRM training (e.g., Salas & Cannon-Bowers, 2000a; Salas, Wilson, Burke, Wightman, & Howse, 2006). Like all training programs, CRM training design and delivery should follow the basic steps of assessing training needs, identifying teamwork competencies, setting training objectives and learning outcomes, determining delivery methods, and developing opportunities for practice, assessment, and feedback.

Research suggests that practice and feedback are critical to the success of CRM training (Anderson, 1983). Providing trainees with the opportunity to demonstrate mastery of the learned knowledge, skills, and attitudes enables them to apply what they have learned, receive feedback, and correct their performance before reaching the operational environment.

One way this can be achieved is by taking an event-based approach to training (EBAT). The purpose of EBAT is to link training requirements (e.g., targeted competencies), practice opportunities (e.g., simulations, games), and performance measurement and feedback (Fowlkes, Dwyer, Oser, & Salas, 1998). Furthermore, EBAT techniques provide opportunities to practice by embedding events in training that elicit the desired behaviors. These behaviors can then be observed using any of the team process and performance measurement tools discussed previously. Thus, observers can diagnose the strengths and weakness in trainees and provide feedback. Feedback should be provided regarding the team process rather than performance outcomes (Kluger & DeNisi, 1996). For example, just because the flight crew landed the plane safely in the simulation does not necessarily mean that they used the right process to achieve that outcome.

The success of team training (and therefore CRM training) is contingent not only on which competencies are trained (e.g., communication, decision making, situation awareness, leadership) and how they are trained. Factors external to the training program are also important. These factors include trainee characteristics (e.g., pretraining knowledge, skills, and attitudes), organizational support for training (e.g., learning culture), supervisor and trainee participation in training development, and the physical environment (e.g., temperature, lighting). These factors must be managed appropriately prior to training to minimize degrading influences. For example, Smith-Jentsch and colleagues (Smith-Jentsch, Salas, & Baker, 1996) demonstrated that trainees' negative experiences with CRM prior to training hindered them from applying CRM knowledge, skills, and attitudes following training.

Lesson 10: The desired CRM-related learning outcomes must drive which competencies to capture, assess, diagnose, and remediate.
Lesson 11: Simulation-based training is a powerful strategy for acquiring the CRM-related processes and outcomes.
Lesson 12: CRM simulation-based training is effective when CRM-relevant instructional features are embedded in it.
Lesson 13: CRM training is only one team training strategy. Additional team-focused strategies are available that should be used in conjunction with CRM training.

So much has been learned about the design and delivery of team training, and the CRM community must leverage these lessons. Research can benefit the CRM community by guiding how to design the training systematically, based on the science, and exploiting available strategies and tools. However, more must be done to ensure that training is effective and that the learned knowledge, skills, and attitudes are transferred to the operational environment. The research conducted in this area is discussed next.

CRM training evaluation and transfer. A final area of research that is heavily studied in the CRM literature is the evaluation and transfer of CRM training. Training evaluation has been defined as "the systematic collection of descriptive and judgmental information necessary to make effective training decisions related to the selection, adoption, value, and modification of various instructional activities" (Goldstein, 1993, p. 147). Systematic training evaluation is not an easy task, but it is the only way to ensure that training programs are having the desired effect and are a worthwhile investment for the organization.

Cannon-Bowers et al. (1989) argued that training evaluation may serve a number of important functions. First and most obvious, program evaluation results can indicate whether the goals and objectives of a program are appropriate to achieve the desired outcome. Second, evaluation can indicate whether the content and methods used in training will result in achievement of the overall program goal. Third, evaluation data can be used to determine how to maximize transfer of training. Fourth, it can serve as feedback at both the individual and team level to suggest areas in need of improvement or revision. Goldstein (1993) offered similar arguments concerning the benefits of evaluation.

One of the most popular frameworks for guiding training evaluations is a training evaluation typology developed by Kirkpatrick (1976). Kirkpatrick argued for a multilevel approach to training evaluation consisting of four levels: reactions, learning, behavior (i.e., extent of performance change), and results (i.e., degree of impact on organizational effectiveness or mission success). This typology has been expanded by several researchers (e.g., see Kraiger, Ford, & Salas, 1993). Kraiger et al. expanded Kirkpatrick's typology by arguing that learning is multidimensional and results in cognitive, affective, and skill-based-learning outcomes. They suggest potential methods that can be used to evaluate each of these outcomes: cognitive (verbal knowledge, knowledge organization, cognitive strategies), affective (attitudinal, motivational), and skill based (compilation, automaticity).

Goldsmith and Kraiger (1997) built on this work by describing a method for the structural assessment of an individual learner's knowledge and skill; this has been used successfully in aviation research efforts (see Kraiger et al., 1995; Stout, Salas, & Kraiger,

1997). The utilization of Kirkpatrick's typology and corresponding revisions serves several important functions within the training evaluation process: first, to organize the type of information that should be collected in the assessment of training; and second, to argue for the added benefit/importance of collecting more than one level of evaluation information. Although both functions are important, the second has been more difficult to put into practice than the first. Specifically, Alliger and Janak (1989) reported that less than 10% of organizations assess training programs at all four levels of evaluation, as argued for by Kirkpatrick (1976).

For training to be effective, trainees must transfer what they have learned to the operational environment. The transfer of CRM training is one area that the CRM community has generally ignored. However, CRM industries can learn and benefit from much that has been learned about the transfer of training. Specifically, several characteristics of the work environment must be present for the transfer of training to occur: (a) supervisor support (e.g., supervisor acts as role model), (b) organizational transfer climate (e.g., rewards, appropriate resources provided), and (c) a continuous learning culture (e.g., trainees learn from mistakes). Nevertheless, supporting empirical evidence of the impact of the work environment is limited (e.g., Ford & Weissbein, 1997; Rouiller & Goldstein, 1993). By taking these issues into consideration, the transfer of CRM knowledge, skills, and attitudes to the job will be more likely to occur successfully.

Several reviews have evaluated the impact of CRM training in aviation and beyond (see O'Connor, Flin, Fletcher, & Hemsley, 2003; Salas, Burke, Bowers, & Wilson, 2001; Salas, Wilson, Burke, & Wightman, 2006b). These reviews suggest that although CRM training is effective at changing attitudes and behaviors relating to CRM, its impact on safety is not yet known. These findings are discussed in the next section.

Lesson 14: The science of team training can (must) guide the design, delivery, and evaluation of CRM training.
Lesson 15: CRM training must be evaluated and transfer encouraged.

Does CRM Training Work?

The talk within the CRM training community is that CRM training works. But do we know this for sure? What evidence is there? A number of published studies have evaluated CRM training in a range of domains, such as aviation, health care, offshore oil production, correctional facilities, and the maritime industry. Several reviews have examined the literature to determine if CRM training works (O'Connor et al., 2002; Salas, Burke, Bowers, & Wilson, 2001; Salas, Wilson, Burke, & Wightman, 2006b). These reviews used Kirkpatrick's (1976) four-level typology (reactions, learning, behavior, and organizational impact) to organize the analysis because it is considered one of the most popular frameworks for guiding training evaluations. We found approximately 100 articles that evaluated the effects of CRM training (see Table 2.1 for a brief overview of results). Although much of the data has been found within the aviation and military fields, the medical community has also made an impressive effort to evaluate the impact of CRM training on those in the field. Unfortunately, evaluation studies in other domains were less than fruitful. We encourage others implementing CRM training (e.g., railroad, nuclear and electrical power, maritime, offshore oil) to evaluate their training programs and publish their findings so we can cumulate our knowledge and learn from one another.

So, does CRM training work? Yes and no! The majority of the studies evaluating CRM training programs were found in the commercial and military aviation domains. This came as no surprise; CRM training has been implemented in those areas for nearly two decades. The studies were distributed in the following manner: 19% evaluated training solely at the reaction level, 13% at the learning level, 25% at the behavioral level, 3% at the organizational level, and 39% examined training at multiple levels (Salas, Burke, et al., 2001; Salas et al., 2006b).

Overall, it appears that CRM training has had a positive impact on attitudes in most domains, as indicated by self-report measures, such as the CMAQ (or a modified version). Participants reported liking the training and found it useful in their jobs (e.g., Grau & Valot, 1997; Schiewe, 1995). As to whether learning took place during training, those studies that examined trainees' attitude changes suggest that learning did occur; that is, trainees had more favorable attitudes toward CRM following training (e.g., Irwin, 1991).

However, studies that investigated actual learning of trained competencies indicated some inconsistencies. Whereas some trainees improved their knowledge base (e.g., Hayward & Alston, 1991), other studies indicated that no learning occurred (e.g., Brun et al., 2000). Likewise, behavioral evidence across communities suggests inconsistencies as well. Whereas some behaviors were shown to transfer to the job or to a simulated environment (e.g., communication), others were not (e.g., leadership; Jacobsen et al., 2001). Although one could argue that perhaps the behaviors that transferred were more applicable to one domain than to another, some of these mixed findings were within the same training program for a particular domain.

Finally, the studies that evaluated training at the organizational level did suggest that CRM training had a positive impact on safety in terms of reduced errors or incidents (e.g., Robertson & Taylor, 1995). However, because of the difficulty of evaluating training at this level (for example, criterion measures are difficult to identify, the influence of extraneous variables is hard to control), it cannot be said with certainty that CRM training led to these results.

The variability of these results leads us to conclude that there is a need to evaluate CRM training at multiple levels. Data from only one level provide a rather limited picture of the true impact of CRM training on trainees and the organization. Reaction data provide a picture of trainee acceptance and satisfaction but do not indicate whether the trainees learned the goal competencies. Furthermore, the fact that trainees learn the competencies or experience a positive change in attitudes does not signify a behavioral change on the job.

Finally, a change in trainees' behaviors will not necessarily lead to a change in the safety of the organization. Collecting evidence at multiple levels will establish a clearer picture of the success of CRM training and where it might have gone wrong so one can correct it in the future. We realize that collecting data at multiple levels is easier said than done. However, the importance of conducting such evaluations is imperative if one is to determine the true impact of CRM training on the bottom line—safety.

(text continues on page 60)

Table 2.1. Summary of CRM Training Evaluation Review

Source	Community	Reactions	Learning	Behavior	Results
Alkov, 1989	Military – U.S.				+
Alkov, 1991	Military – U.S.	+		+	+
Alkov & Gaynor, 1991	Military – U.S.		+		
Arnold & Jackson, 1985	Aviation – U.S.		+	+	
Baker et al., 1991	Military – U.S.	+			
Baker et al., 1993	Military – U.S.	+			
Barker et al., 1996	Military – U.S.			~	
Brannick et al., 1995	Military – U.S.			+	
Brun et al., 2000	Military – Europe	+	~		+
Byrdorf, 1998, as cited in O'Connor et al., 2003	Maritime – Europe		+	+	+
Byrnes & Black, 1993	Aviation – U.S.	+			
Chidester et al., 1991	Military – U.S.		+/–		
Chute & Wiener, 1995	Aviation – U.S.	+			
Clark et al., 1991	Aviation – U.S.			+	
Clothier, 1991	Aviation – U.S.			+	
Connolly & Blackwell, 1987	University students			+/–	
Davis et al., 2003	University students			+	+
Diehl, 1991	Summary paper				
Dwyer et al., 1997	Military – U.S.				
Elliott-Mabey, 1999, as cited in O'Connor et al., 2003	Military – Europe		+		
Ellis & Hughes, 1999	Health care	+		–	
Fisher, 2000	Air medical	+		+	
Flin et al., 2003	Health care	+/–			
Fonne & Fredriksen, 1995	Maritime – Europe		+		

Study	Domain	Finding
Gaba et al., 1998	Health care	+/−
Gaba et al., 2001	Health care	+/−
Geis, 1987	Military – U.S.	+
Goeters, 2002	Aviation – Europe	+
Grau & Valot, 1997	Military – Europe	+
Grav, 2003	Prison – Europe	~
Gregorich, 1993	Aviation – U.S.	+
Gregorich et al., 1990	Aviation – U.S.	+
Grubb & Morey, 2003; Grubb et al., 2001, 2002	Military – U.S.	+
Grubb et al., 1999	Military – U.S.	+
Halamek et al., 2000	Health care	+
Halliday et al., 1987	Military – U.S.	+
Hansberger et al., 1999	Aviation – U.S.	+
Harrington & Kello, 1992	Nuclear power – U.S.	+
Hayward & Alston, 1991	Aviation – U.S.	+
Helmreich, 1991	Summary article	~/−
Helmreich & Foushee, 1993; Taggart, 1994	Aviation – U.S.	+
Helmreich & Wilhelm, 1991	Aviation – U.S.	+
Helmreich et al., 1990	Aviation – U.S.	+/~
Helmreich et al., 1999	Aviation – U.S.	−
Holt et al., 1999	Aviation – U.S.	+
Holzman et al., 1995	Health care	+
Howard et al., 1992	Health care	+/−
i Gardi et al., 2001	Health care	+/−
Ikomi et al., 1999	Aviation – U.S.	+

Note. + = Positive findings; − = Negative findings; ~ = Neutral findings or no change.
Based on Salas, Burke, Bowers, & Wilson (2001) and Salas, Wilson, Burke, Wightman (2006b).

(Continued on page 58)

Table 2.1. Summary of CRM Training Evaluation Review (continued)

Source	Community	Reactions	Learning	Behavior	Results
Incalcaterra & Holt, 1999	Aviation – U.S.	+	+	+	
Irwin, 1991	Aviation – U.S.		+/–		
Jackson, 1983	Aviation – U.S.	+	+	+	
Jacobsen et al., 2001	Health care			+/–	
Jentsch et al., 1995	General aviation			+/~	
Karp et al., 1999, 2000	Military – U.S.	–			
Katz, 2003	Military – U.S.			+	
Kayten, 1993	Aviation – U.S.				+
Kurrek & Fish, 1996	Health care	+			
Lainos & Nikolidis, 2003	Aviation – Europe	+			
Lassiter et al., 1990	University students		~		
Leedom & Simon, 1995	Military – U.S.		+/~	+	
Margerison et al., 1987	Aviation – Australia	+	+/~	+	
Maschke et al., 1995; Hormann et al., 1995	Aviation – Europe	+			
McKinney & Barker, 1999	Military – U.S.		+	+	
Morey et al., 1997	Military – U.S.		+	+/~	
Morey et al., 2002	Health care	~		+/~	
Mudge, 1983	Aviation – U.S.	+		+	
Naef, 1995	Aviation – Europe			+/–	
Nullmeyer & Spiker, 2003; Spiker et al., 1998; Silverman et al., 1997	Military – U.S.			+	
Nullmeyer et al., 2003; Spiker et al., 2003	Military – U.S.			+	
O'Connor & Flin, 2002; Flin et al., 1999	Offshore oil production	+	~		

Study	Domain	Findings
O'Donnell et al., 1998	Health care	+
Orasanu et al., 1999	Aviation – U.S.	+/–
Povenmire et al., 1989	Military – U.S.	~	+
Predmore, 1991	Aviation – U.S.		+
Rollins, 1995	Aviation – U.S.		+
Salas et al., 1999	Military – U.S.	+	~/+
Schiewe, 1995	Aviation – U.S.	+/–	
Sexton, 2000	Health care and aviation – U.S.	+/–	~/+
Simpson & Wiggins, 1995	General aviation		+/~
Small, 1998	Health care	+	
Small et al., 1998	Health care	+	
Smith, 1994	General aviation	+/–	~
Spiker et al., 1999	Military – U.S.	+	+
Stout, Salas, & Fowlkes, 1996; 1997	Military – U.S.	+	+
Stout, Salas, & Kraiger, 1997; Fowlkes et al., 1992; 1994	Military – U.S.	+	~
Taggart & Butler, 1989	Aviation – U.S.	+	+
Taylor & Thomas, 2003	Aviation maintenance – U.S.	–	+	+
Taylor, 1998; 2000; Robertson & Taylor, 1995	Aviation maintenance – U.S.	+	+	+
Taylor et al., 1993	Aviation maintenance – U.S.	+	+/~	+/–
Thompson et al., 1999	Military – U.S.	+	+
Vandermark, 1991	Aviation – U.S.	+	
Wilhelm, 1991; Butler, 1991; 1993	Aviation – U.S.	+		+/–
Woldering & Isaac, 1999, as cited in O'Connor et al., 2003	Air traffic control – Europe	+	+
Yamamori & Mito, 1993	Aviation – Asia		+
Young, 1995	University students	+	

Note. + = Positive findings; – = Negative findings; ~ = Neutral findings or no change.
Based on Salas, Burke, Bowers, & Wilson (2001) and Salas, Wilson, Burke, Wightman (2006b).

Lesson 16: CRM training leads to positive reactions, positive changes in attitudes toward CRM concepts, and the transfer of learned competencies to a simulated or operational environment.

Lesson 17: The impact of CRM training on organizational results (e.g., safety, error reduction) is inconclusive.

Lesson 18: More multilevel training evaluations are needed.

What Do the Accident Data Tell Us?

One way to determine the impact of CRM training on safety is to evaluate the accident data in conjunction with the application of CRM training. An examination of commercial aviation accidents from 1991 to 2000 indicated that approximately 41% of Part 121 (commercial aviation) and 23% of Part 135 (commuter aviation) accidents involved a breakdown in CRM in the cockpit (Wiegmann & Shappell, 2001). CRM deficiencies such as lack of coordination among cockpit crews, captain's failure to assign tasks to other members, and lack of effective crew supervision have been cited as contributing factors in approximately half of reported fatal accidents (U.S. General Accounting Office, 1997). Other researchers have suggested similar causal factors within cited accident reports (see Chidester et al., 1991; Leedom & Simon, 1995). These numbers have remained relatively stable over time, with a slight decline. This indicates that CRM training may not be having the intended impact.

These numbers are not unique to commercial aviation. A review of accident data in the U.S. Navy and Marine Corps revealed similar trends. Researchers found that approximately 60% of the accidents between 1991 and 2000 involved a failure in CRM (Wiegmann & Shappell, 2000), and these numbers, though declining slightly, have remained fairly stable as well. Nullmeyer et al. (2005) examined the accident data within the U.S. Air Force C-130, H-53, F-16, and A-10 aircraft from FY95 through FY04. Their findings were similar in that 60%–85% of the accidents and mishaps involved human error. Situation awareness, task management, and decision making appear to be weak areas in all four of the aircraft types examined.

Lesson 19: CRM training may not be having as great an impact on accident rates as once believed; more data are needed.

OBSERVATIONS ABOUT THE RESEARCH AND PRACTICE OF CRM

After more than 20 years of experience with research related to and application of crew resource management, we make several observations based on our assumptions regarding CRM training; our knowledge of the CRM literature; and our reading of what CRM practitioners discuss in a variety of forums, need, and want (i.e., high reliability teams); and our own discussions with those in the industry.

First, as noted, we believe that CRM training is about promoting teamwork in the cockpit, operating room, or the command and control center. It is about ensuring closed-loop communication and supporting behavior, team leadership, and adaptability. More important, it is about effective decision-making processes, actions, and steps in context.

Second, we look at CRM training as *one* team training strategy—not *the only* team training strategy. As noted, many others (see Salas & Cannon-Bowers, 2001) are as useful, relevant, and valid as CRM training. These should be used, tested, and adapted.

Finally, we believe that CRM training should be about the diagnosis of critical weaknesses, remediation, and ultimately improvement of required competencies. Our observations are presented as food for thought as the community continues to evolve and mature.

1. CRM Training (Still) Does Not Leverage the Science of Training

We said this already, but it is worth emphasizing. It has become evident that the CRM training community (mainly practitioners) is lacking a science to guide it. We argue this for several reasons. First, CRM appears to be more of a philosophy or a way of thinking than a science. Because of this, the community is flooded with opinions about what to do and what not to do. However, few of these opinions are theoretically or empirically supported. Second, despite what we know about the science of training and team training, organizations applying CRM still seem to ignore the available relevant literature (see Salas et al., 1999) to guide and manage their CRM training efforts. That is, the explanation, application, and integration of what we have learned about the science of training need to make their way into the design, implementation, evaluation, and institutionalization of CRM training programs.

Finally, CRM practitioners and scientists do not offer constructive criticism to one another. Rather, the CRM community (we exaggerate to make a point) portrays the image of a happy family without faults or problems. This could not be further from the truth. We thus argue that in order to develop a science of CRM, one must look at both the good and the bad and challenge those ideas and principles that lack a strong foundation to support them.

2. CRM Training Lacks Standardization

Standardization for CRM training is lacking—for example, what to train and how to train it (see Salas, Burke, et al., 2001; Salas et al., 2006b; Wilson-Donnelly & Shappell, 2004). The various names associated with CRM training (e.g., Aircrew Coordination Training, Crisis Resource Management) indicate a need for consensus. There is no standard set of competencies (knowledge, skills, and attitudes) that should be trained (see Table 2.2).

The fact that we do not know what impact CRM training is having on safety caused us to take a closer look at the skills that are being trained. Our review of the studies reported here illustrates the variability in what is trained under the term *CRM*. Furthermore, many of the studies reviewed failed to define what skills were included in training. Without this information, it is difficult for others to understand how the training led to the observed results. Those studies that did define the CRM skills trained contain a multitude of skills that organizations are categorizing as CRM and using to train their

Table 2.2. Most Commonly Trained CRM Content by Community

Content	Military	Commercial	Medical	University Students/GA	Offshore oil	Maintenance	Maritime	Prison	ATC	Total
Communication	11	10	3	3	1		1	1	1	31
Situation awareness/ assessment	10	4	1	2	1			1	1	20
Decision making	5	5	1	2	1	1		1	1	17
Leadership/followership	1	3	4	1	1			1		11
Preflight brief/planning	4	5	1							10
Stress awareness/ management	3	2		2		1	1		1	10
Assertiveness	6					1	1			8
Conflict resolution/ Resolving conflict	1	4		2						7
Teamwork		2		1	1	1	1		1	7
Mission analysis/ evaluation/debrief	4			2						6
Crew coordination	4			1						5
Pilot judgment	2	3								5
Risk assessment/ management	4			1						5
Task management	4									4
Workload management	3		1							4
Human performance behaviors	1		2							3
Problem solving	2	1								3
Resource management/use/ identification of available resources	2					1				3
Advocacy		2								2
Critique		2								2
Feedback/criticism	1	1								2
Group dynamics/crew topics	2									2
Human error	1		1							2

Content item						Total
Inquiry			2			2
Interpersonal skills			1			2
Occupational stress	1		1		1	2
Tactics employment	2					2
Adaptability	1					1
Allocation						1
Attention allocation		1				1
Command authority	1	1				1
Confidence/doubt	1					1
Cooperation		1				1
Declaration		1				1
Fatigue	1					1
Hazardous thought patterns				1		1
Help competence		1				1
Help hands		1				1
Management			1			1
Personal resources					1	1
Pilot attitudes	1					1
Policy and regulations	1					1
Problem definition			1			1
Psychological factors				1		1
Recognition of event		1				1
Reevaluation		1				1
Routines and procedures					1	1
Self-analysis				1		1
Start of initial treatment		1				1
Team building				1		1
Team roles					1	1
Work with inmates				1		1

Note. Within each cell are the number of studies found that trained a particular competency. Content items are listed from most to least frequently trained. Where an equal number of studies trained a particular content item, the items were listed alphabetically.

teams. The most commonly trained competencies are communication, situation awareness/assessment, and decision making.

It appears that despite efforts put forth in many domains to improve CRM training, the lack of standardization may be impeding its progress. We found inconsistencies not only across domains but also within domains. For example, although the FAA provides some guidance as to how to implement CRM training in commercial aviation (FAA, 2001), there continues to be a lack of agreement on what competencies to train and how to implement CRM training (e.g., Driskell & Adams, 1992; Wilson-Donnelly & Shappell, 2004). Most agree that team skills and performance are a training focus (Driskell & Adams), but the exact skills are in debate. The challenge here is that organizations such as airlines have the flexibility to develop a CRM training program that is believed to best fit with their organization (e.g., AQP). Because of this, not only is it difficult to learn from each other, but it is also challenging for team members who may switch, for example, to different fleets or airlines. Organizations need to identify which core set of CRM skills are necessary (allowing for adaptation of unique skills in some domains) for CRM to be a success (we may already have an idea), and those should be the core skills that are trained.

Additional challenges arise as one tries to determine what works in CRM training and what doesn't because there is no single source that one can go to in order to determine what works. If we are all training different skills and defining CRM differently, we can't accumulate knowledge and learn from one another.

3. CRM Training Designers May Fall Prey to a Number of Myths

A number of myths or misconceptions persist regarding the design, implementation, and evaluation of CRM training programs in organizations (Salas, Wilson, Burke & Bowers, 2002). These contribute to the inconsistent findings regarding CRM training effectiveness. Unfortunately, CRM training is usually designed based on many unsupported assumptions about how to optimize skill acquisition.

4. CRM Training Has Become the Only Answer to Safety Concerns

As CRM training has expanded to domains beyond aviation, it has become the sole team training solution for improving safety and reducing errors. However, a number of proven team training strategies are available that, if conducted in conjunction with CRM training, would likely improve safety more. Team training strategies discussed in the literature include team leadership training, team self-correction, cross training, and team building (see Salas & Cannon-Bowers, 2000b).

5. CRM Training Research Is Still Needed—It Must Go On

Some have argued that the CRM community has exhausted all possible research areas. We believe this is not true. The CRM community currently relies heavily (albeit not solely) on only two or three pockets of information conducted throughout the 1990s—specifically, the work of Helmreich and colleagues at the University of Texas, of Orasanu at NASA Ames, and of Salas and colleagues at the Naval Air Warfare Center-Training Systems Division (NAWC-TSD).

The fact that the impact of CRM training is still not known provides a good argument for this. The advancement of new technologies requires that research be conducted in these areas; for example, dynamic assessment, distributed teams, and intelligent technology. The increased use of multinational coalition teams within the military warrants a closer look at CRM training and other team training strategies for improving performance in these environments.

6. The CRM Community's Biggest Challenge Is Implementation

One of the greatest challenges still faced by the CRM community is how to move from a community of interest to a community of practice. CRM needs to be more about "walking the walk" and not just "talking the talk." To do this, sponsors, instructors, developers, researchers, aviators, doctors, commanders, and all others involved need to agree on what approach to take to implement CRM training in organizations.

7. CRM Skills Training Is Separate from Technical Skills Training

Much of the literature discussing how CRM skills are trained indicates that they are trained in isolation from technical skills needed for the job. Partly to blame for this is the common assumption that technical skills should be trained prior to nontechnical (or CRM) skills. Thus, this indicates that these skills should be taught separately. Although this may be true initially, trainees need to integrate the technical and nontechnical skills required by the job, and it is important that they practice this integration before actually applying it.

8. More Robust and Multilevel CRM Training Evaluations Are Needed

As we mentioned earlier regarding CRM training evaluations, we argue for the use of a multilevel approach to evaluating training outcomes (i.e., reactions, learning, behavior, and organizational impact). Our review of the CRM training literature indicates that much of the research focuses on only one or two of these levels—commonly the lower levels—of the typology. Few of these studies evaluated training at the highest level (i.e., results or organizational impact). If one is to gain an accurate picture of why CRM training is or is not working, one must have information at each of these levels. Although we understand that acquiring this information can be costly, the return on investment will be worth it.

9. CRM Training Supporters Must Make a Business Case for CRM Training

CRM training supporters continuously must struggle to make a business case to the upper echelons of the organization. Unless CRM training is translated into positive dollar values, it is vulnerable to cost reduction efforts. CRM training requires an investment from the organization to fund the program. Without adequate funding, the CRM program will be less than fully successful. In turn, if CRM is not successful, the organization will not choose to support it.

WHERE DO WE GO FROM HERE?

It appears that after two decades of CRM training research and practice, the industry continues to struggle over how to evaluate and institutionalize it. Since the inception of CRM training, a great deal has been learned about how to design, develop, implement, and evaluate training programs in organizations (see Salas & Cannon-Bowers, 2000a; 2001; Tannenbaum & Yukl, 1992). Yet, it remains unclear whether CRM training is working.

We have presented evidence that the true impact of CRM training is not yet known, but we do not mean to argue that CRM training has been a waste of time or ineffective. Research indicates that it is effective at changing attitudes and improving performance, at least initially, and the aviation accident data indicate a slight decline. Observed deficiencies presented earlier may be partly to blame for impeding CRM's potential. But what can be done? How can we move ahead and show the value of CRM training?

This community needs a mandate, access to data, and resources to determine the efficacy of CRM training. First, we need a mandate to compel researchers to conduct sound, systematic, and robust evaluations. In other words, a directive from the organization, institution, or agencies that are designing and delivering CRM training is needed, which will ensure that the training is evaluated at multiple levels on a continuous basis and in a standardized way.

Once a mandate is established, access must be granted to collect the data. This requires a change in climate that calls on the aviation community to open its cockpits and the medical community to open its operating rooms to researchers.

Finally, the CRM community needs resources (e.g., funding, time) to conduct the necessary training and subsequent evaluations. We are aware of the difficulty in establishing a credible, direct cause and effect between CRM training and safety. The low occurrence of mishaps prevents it. But we submit that the CRM community must continue to seek better and robust evaluations of safety, clinical, and performance outcomes.

CONCLUSION

In this chapter we have reviewed the 20-plus years of CRM research and training literature in aviation, health care, fire services, correctional facilities, and beyond. We identified a number of theoretical foundations for CRM, including team effectiveness, team competencies, shared mental models, and social-psychological constructs. From this a number of research areas have emerged, among them personnel selection, the use of simulations, and culture.

Furthermore, we found that CRM training has been labeled with various names, has been defined in different ways, and has encompassed a wide variety of skills. Because of this, we sought to determine whether CRM training has been effective in reducing errors and improving safety. Some interesting results were found in this regard. Although some researchers have argued that there is no evidence that CRM is effective (e.g., Besco, 1995; Simmon, 1997), this review and others conclude that some evidence does exist. And this is important. The picture that has emerged after reviewing the existing literature suggests that CRM training is effective at some levels (i.e., attitudes, learning, behavior). But as previously stated, the picture is not as clear as it should be after 20 years.

Our observations suggest that more robust research, training, and evaluations are needed so that we can fully grasp the impact that CRM is having in the community. At this point, we believe the tools to determine this are there; what we need are a mandate, access to data, and the resources to make it happen.

ACKNOWLEDGMENTS

We thank Bob Nullmeyer for his thoughtful comments on an earlier draft of this chapter. This research was partly funded by the U.S. Army Research Institute, Ft. Rucker, Alabama; and partly supported by the U.S. Army Research Institute for the Behavioral and Social Sciences through contract DASW01-03-C-0010. The research and development reported here were also supported in part by the FAA Office of the Chief Scientific and Technical Advisor for Human Factors, AAR-100. Dr. Eleana Edens was the contracting officer's technical representative at the FAA. All opinions are those of the authors and do not necessarily represent the opinion or position of the University of Central Florida, the U.S. Army Research Institute, or the Federal Aviation Administration.

REFERENCES

Alkov, R. A. (1989). The U.S. Naval aircrew coordination training program. In R. S. Jensen (Ed.), *Proceedings of the 5th International Symposium on Aviation Psychology* (pp. 483–488). Columbus: Ohio State University.

Alliger, G. M., & Janak, E. A. (1989). Kirkpatrick's levels of training criteria: Thirty years later. *Personnel Psychology, 42*, 331–342.

Anderson, J. R. (1983). *The architecture of cognition*. Cambridge, MA: Harvard University Press.

Baker, D., Prince, C., Shrestha, L., Oser, R., & Salas, E. (1993). Aviation computer games for crew resource management training. *International Journal of Aviation Psychology, 3*(2), 143–156.

Baker, D. P., Gustafson, S., Beaubien, J., Salas, E., & Barach, P. (2005). *Medical teamwork and patient safety: The evidence-based relation* (AHRQ Publication No. 05-0053). Rockville, MD: Agency for Healthcare Research and Quality.

Bass, B. (1990). *Bass & Stogdill's handbook of leadership* (3rd ed.). New York: Free Press.

Beaubien, J. M., & Baker, D. P. (2002). *A review of selected aviation human factors taxonomies, accident/incident reporting systems, and data collection tools*. Retrieved August 8, 2006, from http://www.air.org/teams/publications/human_error/asap_lit.pdf#search='LLC4%20%20line/los'.

Bellorini, A., & Vanderhaegen, F. (1995). Communication and cooperation analysis in traffic control. *Proceedings of the 8th International Symposium on Aviation Psychology* (pp. 1265–1271). Columbus: Ohio State University.

Besco, R. O. (1995). The potential contributions and scientific responsibilities of aviation psychologists. In N. Johnston, R. Fuller, & N. McDonald (Eds.), *Proceedings of the 21st Conference of the European Association for Aviation Psychology: Vol. 2. Aviation psychology: Training and selection* (pp. 141–148). Aldershot, England: Avebury Aviation.

Bowers, C. A., & Jentsch, F. (2001). Use of commercial, off-the-shelf, simulations for team research. In E. Salas (Ed.), *Advances in human performance* (Vol. 1, pp. 293–317). Amsterdam: Elsevier Science.

Bowers, C. A., Jentsch, F., Baker, D., Prince, C., & Salas, E. (1997). Rapidly reconfigurable event-based line operational evaluation scenarios. *Proceedings of the Human Factors and Ergonomics Society 41st Annual Meeting* (pp. 912–915). Santa Monica, CA: Human Factors and Ergonomics Society.

Bowers, C. A., Jentsch, F., Salas, E., & Braun, C. C. (1998). Analyzing communication sequences for team training needs assessment. *Human Factors, 40*, 672–679.

Bowers, C. A., Salas, E., Prince, C., & Brannick, M. (1992). Games teams play: A method for investigating team coordination and performance. *Behavior Research Methods, Instruments, & Computers, 24*, 503–506.

Brannick, M. T., Prince, C., & Salas, E. (2005). Can PC-based systems enhance teamwork in the cockpit? *International Journal of Aviation Psychology 15*(2), 173–187.

Brun, W., Eid, J., Jihnsen, B. H., Ekornas, B., Laberg, J. C., & Kobbeltvedt, T. (2000). *Shared mental models and task performance: Studying the effects of a crew and bridge resource management training program* (Project report: 1 2001). Amsterdam, Norway: Militaer Psykologi og Ledelse.

Burke, C. S., Fiore, S., & Salas, E. (2003). The role of shared cognition in enabling shared leadership and team adaptability. In J. Conger & C. Pearce (Eds.), *Shared leadership: Reframing the how's and why's of leadership* (pp. 103–122). London: Sage.

Cannon-Bowers, J., Salas, E., & Converse, S. (1993). Shared mental models in expert team decision making. In N. J. Castellan, Jr. (Ed.), *Individual and group decision making: Current issues* (pp. 221–246). Mahwah, NJ: Erlbaum.

Cannon-Bowers, J. A., Prince, C., Salas, E., Owens, J. M., Morgan, B. B., Jr., & Gonos, G. H. (1989). Determining aircrew coordination training effectiveness. *Proceedings of the 11th Annual Meeting of the Interservice/Industry Training Systems Conference* (pp. 128–136). Arlington, VA: National Defense Industrial Association.

Cannon-Bowers, J. A., Tannenbaum, S. I., Salas, E., & Volpe, C. E. (1995). Defining team competencies and establishing team training requirements. In R. Guzzo, E. Salas, & Associates (Eds.), *Team effectiveness and decision making in organizations* (pp. 333–380). San Francisco, CA: Jossey-Bass.

Chidester, T. R., Helmreich, R. L., Gregorich, S. E., & Geis, C. E. (1991). Pilot personality and crew coordination: Implications for training and selection. *International Journal of Aviation Psychology, 1*(1), 25–44.

Cooke, N. J., Salas, E., Cannon-Bowers, J. A., & Stout, R. J. (2000). Measuring team knowledge. *Human Factors, 42*, 151–173.

Cooper, G. E., White, M. D., & Lauber, J. K. (Eds.). (1980). *Resource management on the flightdeck: Proceedings of a NASA/Industry workshop* (NASA CP-2120). Moffett Field, CA: NASA Ames Research Center.

Croft, J. (2001). Research perfects new ways to monitor pilot performance. *Aviation Week & Space Technology, 155*(3), 76–81.

Damitz, M., Manzey, D., Kleinmann, M., & Severin, K. (2003). Assessment center for pilot selection: Construct and criterion validity and the impact of assessor type. *Applied Psychology: An International Review, 52*(2), 193–212.

Davis, D. D., Bryant, J., Liu, Y., Tedrow, L., & Say, R. (2003). National culture, team behavior and error management in US and Chinese simulated aircrews. In R. S. Jensen (Ed.), *Proceedings of the 12th International Symposium on Aviation Psychology* (pp. 273–277). Columbus: Ohio State University.

Dennis, K., & Harris, D. (1998). Computer-based simulation as an adjunct to *ab initio* flight training. *International Journal of Aviation Psychology, 8*(3), 261–276.

Department of the Army. (1992). *Aircrew coordination exportable training package* (Vols. 1–3). Fort Rucker, AL: U.S. Army Aviation Center.

Driskell, J. E., & Adams, J. A. (1992). *Crew resource management: An introductory handbook* (Report number DOT/FAA/RD-92/26). Washington, DC: Research and Development Service.

Evans, A. W., Hoeft, R. M., Kochan, J. A., & Jentsch, F. G. (2005, July). *Tailoring training through the use of concept mapping: An investigation into improving pilot training*. Paper presented at the Human Computer Interaction International Conference, Las Vegas, NV.

Federal Aviation Administration (FAA). (2001). *Crew resource management* [Advisory Circular 120-51D]. Retrieved August 10, 2006, from http://www1.airweb.faa.gov/Regulatory_and_Guidance_Library/rgAdvisoryCircular.nsf/1ab39b4ed563b08985256a35006d56af/3aa93c66abe3115c86256c050067e4bf/$FILE/ac120-51d.pdf.

Fletcher, G., Flin, R., McGeorge, P., Glavin, R., Maran, N., & Patey, R. (2003). Anaesthetists' Non-Technical Skills (ANTS): Evaluation of a behavioural marker system. *British Journal of Anaesthesia, 90*(5), 580–588.

Flin, R., Martin, L., Goeters, K., Hoermann, J., Amalberti, R., Valot, C., et al. (2003). Development of the NOTECHS (Non-Technical Skills) system for assessing pilots' CRM skills. *Human Factors and Aerospace Safety, 3*, 95–117.

Ford, J. K., & Weissbein, D. A. (1997). Transfer of training: An updated review and analysis. *Performance Improvement Quarterly, 10*, 22–41.

Foushee, H. C. (1982). The role of communications, socio-psychological, and personality factors in the maintenance of crew coordination. *Aviation, Space, and Environmental Medicine, 53*, 1062–1066.

Foushee, H. C. (1984). Dyads and triads at 35,000 feet: Factors affecting group processes and aircrew performance. *American Psychologist, 39*, 885–893.

Foushee, H. C., Lauber, J. K., Baetge, M. M., & Acomb, D. B. (1986). *Crew factors in flight operations III: The operational significance of exposure to short-haul air transport operations* (NASA Technical Memorandum 88322). Moffett Field, CA: NASA Ames Research Center.

Fowlkes, J. E., Dwyer, D. J., Oser, R. L., & Salas, E. (1998). Event-based approach to training (EBAT). *International Journal of Aviation Psychology, 8*(3), 209–221.

Fowlkes, J. E., Lane, N. E., Salas, E., Franz, T., & Oser, R. (1994). Improving the measurement of team performance: The TARGETs methodology. *Military Psychology, 6*, 47–61.

Gersick, C. J. G. (1988). Time and transition in work teams: Towards a new model of group development. *Academy of Management Review, 31*, 9–41.

Goldsmith, T., & Kraiger, K. (1997). Structural knowledge assessment and training evaluation. In J. Ford, S. Kozlowski, K. Kraiger, E. Salas, & M. Teachout (Eds.), *Improving training effectiveness in work organizations* (pp. 19–46). Mahwah, NJ: Erlbaum.

Goldsmith, T. E., Johnson, P. J., & Acton, W. H. (1991). Assessing structural knowledge. *Journal of Educational Psychology, 83*, 88–96.

Goldstein, I. L. (1993). *Training in organizations: Needs assessment, development, and evaluation* (3rd ed.). Monterey, CA: Brooks/Cole.

Gopher, D., Weil, M., & Bareket, T. (1994). Transfer of skill from a computer game trainer to flight. *Human Factors, 36*, 387–405.

Grau, J. Y., & Valot, C. (1997). Evolvement of crew attitudes in military airlift operations after CRM course. In R. S. Jensen & L. A. Rakovan (Eds.), *Proceedings of the 9th International Symposium on Aviation Psychology* (pp. 556–561). Columbus: Ohio State University.

Hackman, J. R., & Morris, C. G. (1975). Group task, group interaction processes, and group performance effectiveness: A review and proposed integration. *Advances in Experimental Social Psychology, 8*, 45–99.

Hawkins, F. H. (1987). *Human factors in flight.* Aldershot, England: Avebury Aviation.

Hayward, B., & Alston, N. (1991). Team building following a pilot labour dispute: Extending the CRM envelope. In R. S. Jensen (Ed.), *Proceedings of the 6th International Symposium on Aviation Psychology* (pp. 377–383). Columbus: Ohio State University.

Hedge, J. W., Bruskiewicz, K. T., Borman, W. C., Hanson, M., Logan, K. K., & Siem, F. M. (2000). Selecting pilots with crew resource management skills. *International Journal of Aviation Psychology, 10*(4), 377–392.

Helmreich, R. L. (1984). Cockpit management attitudes. *Human Factors, 26*, 63–72.

Helmreich, R. L. (1994). Anatomy of a system accident: The crash of Avianca Flight 052. *International Journal of Aviation Psychology, 4*(3), 265–284.

Helmreich, R. L., & Foushee, H. C. (1993). Why crew resource management? Empirical and theoretical bases of human factors in aviation. In E. L. Wiener, B. G. Kanki, & R. L. Helmreich (Eds.), *Cockpit resource management* (pp. 3–45). San Diego, CA: Academic Press.

Helmreich, R. L., Klinect, J. R., & Wilhelm, J. A. (1999). Models of threat, error, and CRM in flight operations. In R. S. Jensen, B. Cox, J. D. Callister, & R. Lavis (Eds.), *Proceedings of the Tenth International Symposium on Aviation Psychology* (pp. 677–682). Columbus: Ohio State University.

Helmreich, R. L., Klinect, J. R., Wilhelm, J. A., & Sexton, J. B. (2001, March). The Line Operations Safety Audit (LOSA). *Proceedings of the First LOSA Week* (pp. 1–6). Cathay City, Hong Kong: ICAO.

Helmreich, R. L., & Merritt, A. C. (1996, April). *Cultural issues in crew resource management. Proceedings of the ICAO Global Human Factors Seminar* (pp. 141–148). Auckland, New Zealand: ICAO.

Helmreich, R. L., & Merritt, A. C. (1998). *Culture at work in aviation and medicine: National, organizational, and professional influences*. Aldershot, England: Ashgate.

Helmreich, R. L., Merritt, A. C., & Wilhelm, J. A. (1999). The evolution of crew resource management training in commercial aviation. *International Journal of Aviation Psychology, 9*(1), 19–32.

Helmreich, R. L., Merritt, A., Sherman, P., Gregorich, S., & Wiener, E. (1993). *The Flight Management Attitude Questionnaire* (NASA/UT/FAA Technical Report 93-4). Austin, TX: NASA—University of Texas at Austin.

Helmreich, R. L., Wilhelm, J. A., Kello, J., Taggart, W. R., & Butler, R. E. (1991). *Reinforcing and evaluating crew resource management: Evaluator/LOS instructor reference manual* (NASA/UT Technical Manual 90-2). Austin, TX: NASA—University of Texas at Austin.

Helmreich, R. L., Wilhelm, J. A., Klinect, J. R., & Merritt, A. C. (2001). Culture, error and crew resource management. In E. Salas, C. A. Bowers, & E. Edens (Eds.), *Improving teamwork in organizations: Applications of resource management training* (pp. 305–331). Mahwah, NJ: Erlbaum.

Hutchins, E., Holder, B. E., & Pérez, R. A. (2002). *Culture and flight deck operations*. Unpublished internal report prepared for the Boeing Company (Sponsored Research Agreement 22-5003), University of California, San Diego.

Ilgen, D. R., Hollenbeck, J. R., Johnson, M., & Jundt, D. (2005). Teams in organizations: From Input-Process-Output models to IMOI models. *Annual Review of Psychology, 56*, 1–19.

Irwin, C. M. (1991). The impact of initial and recurrent cockpit resource management training on attitudes. In R. S. Jensen (Ed.), *Proceedings of the 6th International Symposium on Aviation Psychology* (pp. 344–349). Columbus: Ohio State University.

Jacobsen, J., Lindekaer, A. L., Ostergaard, H. T., Nielsen, K., Ostergaard, D., Laub, M. et al. (2001). Management of anaphylactic shock evaluated using a full-scale anaesthesia simulator. *Acta Anaesthesiologica Scandinavica, 45*, 315–319.

Jentsch, F., & Bowers, C. A. (1998). Evidence for the validity of PC-based simulations in studying aircrew coordination. *International Journal of Aviation Psychology, 8*(3), 243–260.

Kanki, B. G., & Palmer, M. T. (1993). Communication and crew resource management. In E. L. Wiener, B. G. Kanki, & R. L. Helmreich (Eds.), *Cockpit resource management*. San Diego, CA: Academic Press.

Kirkpatrick, D. L. (1976). Evaluation of training. In R. L. Craig (Ed.), *Training and development handbook: A guide to human resources development* (pp. 18.1–18.27). New York: McGraw-Hill.

Klein, G. (2000). Cognitive task analysis of teams. In J. M. Schraagen & S. F. Chipman et al. (Eds.), *Cognitive task analysis* (pp. 417–429). Mahwah, NJ: Erlbaum.

Kluger, A. N., & DeNisi, A. (1996). The effects of feedback interventions on performance: A historical review, a meta-analysis, and a preliminary feedback intervention theory. *Psychological Bulletin, 119*, 254–284.

Koonce, J., & Bramble, W. (1998). Personal computer-based flight traveling devices. *International Journal of Aviation Psychology, 8*(3), 277–292.

Kozlowski, S. W. J., & Bell, B. S. (2003). Work groups and teams in organizations. In W. C. Borman, D. R. Ilgen, & R. J. Klimoski (Eds.), *Comprehensive handbook of psychology: Industrial and organizational psychology* (Vol. 12, pp. 333–375). New York: Wiley.

Kraiger, K., Ford, J. K., & Salas, E. (1993). Application of cognitive, skill-based, and affective theories of learning outcomes to new methods of training evaluation. *Journal of Applied Psychology, 78*(2), 311–328.

Kraiger, K., Salas, E., & Cannon-Bowers, J. A. (1995). Measuring knowledge organization as a method for assessing learning during training. *Human Performance, 37*, 804–816.

Kuang, J. C. Y., & Davis, D. D. (2001). *Culture and team performance in foreign flightcrews*. Paper presented at the 109th Annual Convention of the American Psychological Association, San Francisco, CA.

Lassiter, D. L., Vaughn, J. S., Smaltz, V. E., Morgan, B. B., Jr., & Salas, E. (1990). A comparison of two types of training interventions on team communication performance. *Proceedings of the Human Factors and Ergonomic Society 34th Annual Meeting* (pp. 1372–1376). Santa Monica, CA: Human Factors and Ergonomics Society.

Lauber, J. (1984). Resource management in the cockpit. *Air Line Pilot, 53*, 20–23.

Law, J. R., & Wilhelm, J. A. (1995). Ratings of CRM skill markers in domestic and international operations: A first look. *Proceedings of the Eighth International Symposium on Aviation Psychology* (pp. 669–674). Columbus: Ohio State University.

Leedom, D. K., & Simon, R. (1995). Improving team coordination: A case for behavioral-based training. *Military Psychology, 7*, 109–122.

MacLeod, N. (2005). *Building safe systems in aviation: A CRM developers' handbook*. Burlington, VT: Ashgate.

Marks, M. A., Mathieu, J. E., & Zaccaro, S. J. (2001). A temporally based framework and taxonomy of team processes. *Academy of Management Review, 26*, 356–376.

Maurino, D. E. (1999). Crew resource management: A time for reflection. In D. J. Garland, J. A. Wise, & V. D. Hopkin (Eds.), *Handbook of aviation human factors* (pp. 215–234). Mahwah, NJ: Erlbaum.

McIntyre, R. M., & Salas, E. (1995). Measuring and managing for team performance: Emerging principles from complex environments. In R. Guzzo & E. Salas (Eds.), *Team effectiveness and decision making in organizations* (pp. 149–203). San Francisco: Jossey-Bass.

Merritt, A. C., & Helmreich, R. L. (1995). Culture in the cockpit: A multi-airline study of pilot attitudes and values. In R. S. Jensen (Ed.), *Proceedings of the 8th International Symposium on Aviation Psychology* (pp. 676–681). Columbus: Ohio State University.

Mohammed, S., Klimoski, R., & Rentsch, J. R. (2000). The measurement of team mental models: We have no shared schema. *Organizational Research Methods, 3*, 123–165.

Morgan, B. B., Jr., Glickman, A. S., Woodard, E. A., Blaiwes, A. S., & Salas, E. (1986). *Measurement of team behaviors in a Navy environment* (Technical Report Number 86-014). Orlando, FL: Naval Training Systems Center.

Nullmeyer, R. T., Stella, D., Montijo, G. A., & Harden, S. W. (2005, December). *Human factors in Air Force flight mishaps: Implications for change*. Paper presented at the Interservice/Industry Training, Simulation, and Education Conference, Orlando, FL.

O'Connor, P., Flin, R., & Fletcher, G. (2002). Techniques used to evaluate crew resource management training: A literature review. *Human Factors and Aerospace Safety, 2*, 217–233.

O'Connor, P., Flin, R., Fletcher, G., & Hemsley, P. (2003). *Methods used to evaluate the effectiveness of flight-crew CRM training in the UK aviation industry* (CAA Report 2002/05). Retrieved August 10, 2006, from http://www.caa.co.uk/docs/33/CAPAP2002_05.PDF.

Orasanu, J. (1994). Shared problem models and flight crew performance. In N. Johnston, N. McDonald, & R. Fuller (Eds.), *Aviation psychology in practice* (pp. 255–285). Aldershot, England: Avebury.

Orasanu, J., Fischer, U., & Davison, J. (1997). Cross-cultural barriers to effective communication in aviation. In C. S. Granrose & S. Oskamp (Eds.), *Cross-cultural work groups* (pp. 134–160). Thousands Oaks, CA: Sage.

Orasanu, J. M. (1990, October). *Shared mental models and crew performance*. Paper presented at the Human Factors Society 34th Annual Meeting, Orlando, FL.

Priest, H. A., Burke, C. S., Guthrie. J. W., Salas, E., & Bowers, C. A. (under revision). Are all teams created equal? An examination of team communication across teams. Submitted to *Human Factors*.

Prince, C., & Salas, E. (1993). Training and research for teamwork in the military aircrew. In E. L. Wiener, B. G. Kanki, & R. L. Helmreich (Eds.), *Cockpit resource management* (pp. 337–366). Orlando, FL: Academic Press.

Prince, C., & Salas, E. (1999). Team processes and their training in aviation. In D. Garland, J. Wise, & D. Hopkins (Eds.), *Handbook of aviation human factors* (pp. 193–213). Mahwah, NJ: Erlbaum.

Rentsch, J., & Klimoski, R. (2001). Why do great minds think alike: Antecedents of team member schema agreement. *Journal of Organizational Behavior, 22*, 107–120.

Robertson, M. M., & Taylor, J. C. (1995). Team training in aviation maintenance settings: A systematic evaluation. In B. J. Hayward & A. R. Lowe (Eds.), *Applied aviation psychology: Achievement, change, and challenge. Proceedings of the Third Australian Aviation Psychology Symposium* (pp. 373–383). Aldershot, England: Avebury Aviation.

Rouiller, J. Z., & Goldstein, I. L. (1993). The relationship between organizational transfer climate and positive transfer of training. *Human Resource Development Quarterly, 4*, 377–390.

Rouse, W. B., & Morris, N. M. (1986). On looking into the black box: Prospects and limits in the search for mental models. *Psychological Bulletin, 100*(3), 349–363.

Salas, E., Bowers, C. A., & Edens, E. (Eds.). (2001). *Improving teamwork in organizations: Applications of resource management training*. Mahwah, NJ: Erlbaum.

Salas, E., Burke, C. S., Bowers, C. A., & Wilson, K. A. (2001) Team training in the skies: Does crew resource management (CRM) training work? *Human Factors, 43*, 641–674.

Salas, E., & Cannon-Bowers, J. A. (2000a). Designing training systems systematically. In E. A. Locke (Ed.), *The Blackwell handbook of principles of organizational behavior* (pp. 43–59). Malden, MA: Blackwell.

Salas, E., & Cannon-Bowers, J. A. (2000b). The anatomy of team training. In S. Tobias & J. D. Fletcher (Eds.), *Training & retraining: A handbook for business, industry, government, and the military* (pp. 312–335). New York: Macmillan Reference.

Salas, E., & Cannon-Bowers, J. A. (2001). The science of training: A decade of progress. *Annual Review of Psychology, 52,* 471–499.

Salas, E., Fowlkes, J. E., Stout, R. J., Milanovich, D. M., & Prince, C. (1999). Does CRM training improve teamwork skills in the cockpit?: Two evaluation studies. *Human Factors, 41,* 326–343.

Salas, E., Stagl, K. C., Burke, C. S., & Goodwin, G. F. (in press). Fostering team effectiveness in organizations: Toward an integrative theoretical framework of team performance. In J. W. Shuart, W. Spaulding, & J. Poland, (Eds.), *Nebraska Symposium on Motivation: Vol. 51. Modeling complex systems: Motivation, cognition and social processes.* Lincoln: University of Nebraska Press.

Salas, E., Wilson, K. A., Burke, C. S., & Bowers, C. A. (2002). Implementing crew resource management (CRM) training: Myths to avoid. *Ergonomics in Design, 10*(4), 20–24.

Salas, E., Wilson, K. A., Burke, C. S., Wightman, D. C., & Howse, W. R. (2006a). A checklist for crew resource management training. *Ergonomics in Design, 14*(2), 6–15.

Salas, E., Wilson, K. A., Burke, C. S., & Wightman, D. C. (2006b). Does CRM training work? An update, extension, and some critical needs. *Human Factors, 48,* 392–412

Schiewe, A. (1995). On the acceptance of CRM-methods by pilots: Results of a cluster analysis. In R. S. Jensen & L. A. Rakovan (Eds.), *Proceedings of the 8th International Symposium on Aviation Psychology* (pp. 540–545). Columbus: Ohio State University.

Schvaneveldt. R. W., Beringer, D. B., & Lamonica, J. (2001). Priority and organization of information accessed by pilots in various phases of flight. *International Journal of Aviation Psychology, 11*(3), 253–280.

Sherman, P. J., Helmreich, R. L., & Merritt, A. C. (1997). National culture and flight deck automation: Results of a multinational survey. *Human Factors, 7,* 311–329.

Simmon, D. A. (1997). How to fix CRM. In R. S. Jensen & L. A. Rakovan, *Proceedings of the 9th International Symposium on Aviation Psychology* (pp. 550–553). Columbus: Ohio State University.

Smith-Jentsch, K., Salas, E., & Baker, D. P. (1996). Training team performance-related assertiveness. *Personnel Psychology, 49,* 909–936.

Smith-Jentsch, K., Zeisig, R. L., Acton, B., & McPherson, J. A. (1998). Team dimensional training: A strategy for guided team self-correction. In J. A. Cannon-Bowers & E. Salas (Eds.) *Making decisions under stress: Implications for individual and team training* (pp. 271–297). Washington, DC: American Psychological Association.

Stout, R. J., Cannon-Bowers, J. A., Morgan, B. B., Jr., & Salas, E. (1990). Does crew coordination behavior impact performance? *Proceedings of the Human Factors Society 34th Annual Meeting* (pp. 1382–1386). Santa Monica, CA: Human Factors and Ergonomics Society.

Stout, R. J., Cannon-Bowers, J. A., Salas, E., & Milanovich, D. M. (1999). Planning, shared mental models, and coordinated performance: An empirical link is established. *Human Factors, 41,* 61–71.

Stout, R. J., Salas, E., & Kraiger, K. (1997). The role of trainee knowledge structures in aviation team environments. *International Journal of Aviation Psychology, 7*(3), 235–250.

Sundstrom, E., DeMeuse, K., & Futrell, D. (1990). Work teams: Applications and effectiveness. *American Psychologist, 45,* 120–133.

Swan, J. (1995). Exploring knowledge and cognitions in decisions about technological innovation: Mapping managerial cognitions. *Human Relations, 48,* 1241–1270.

Swezey, R. W., & Salas, E. (1992). (Eds.), *Teams: Their training and performance.* Norwood, NJ: Ablex.

Tannenbaum, S. I., & Yukl, G. (1992). Training and development in work organizations. *Annual Review of Psychology, 43,* 399–441.

Tenney, Y. J., & Pew, R. W. (2006). Situation awareness catches on: What? So what? Now what? In R. C. Williges (Ed.), *Reviews of human factors and ergonomics, Volume 2* (pp. 1–34). Santa Monica, CA: Human Factors and Ergonomics Society.

United States General Accounting Office. (US GAO). (1997). *Human Factors: FAA's guidance and oversight of pilot crew resource management training can be improved* (GAO/RCED-98-7). Washington, DC: Author.

Weick, K. (1990). The vulnerable system: Analysis of the Tenerife air disaster. *Journal of Management, 16,* 571–593.

Westrum, R., & Adamski, A. J. (1999). Organizational factors associated with safety and mission success in aviation environments. In D. J. Garland, J. A. Wise, & V. D. Hopkin (Eds.), *Handbook of aviation human factors.* Mahwah, NJ: Erlbaum.

Wiegmann, D. A., & Shappell, S. A. (2000). Human error and crew resource management failures in Naval aviation mishaps: A review of U.S. Naval Safety Center data, 1990–96. *Aviation, Space and Environmental Medicine, 70,* 1147–1151.

Wiegmann, D. A., & Shappell, S. A. (2001). Human error analysis of commercial aviation accidents: Application of the human factors analysis and classification system (HFACS). *Aviation, Space and Environmental Medicine, 72,* 1006–1016.

Wiener, E. L., Kanki, B. G., & Helmreich, R. L. (Eds.). (1993). *Cockpit resource management.* San Diego, CA: Academic Press.

Wilson-Donnelly, K. A., & Shappell, S. A. (2004). U.S. Navy/Marine Corps CRM training: Separating theory from reality. *Proceedings of the Human Factors and Ergonomics Society 48th Annual Meeting* (pp. 2070–2074). Santa Monica, CA: Human Factors and Ergonomics Society.

Zohar, D. (2000). A group-level model of safety climate: Testing the effect of group climate on microaccidents in manufacturing jobs. *Journal of Applied Psychology, 85*(4), 587–596.

CHAPTER 3

Representation Aiding to Support Performance on Problem-Solving Tasks

By Philip J. Smith, Kevin B. Bennett, & R. Brian Stone

> A substantial literature has been developed that deals with the use of visual displays to support human problem solving. In particular, the cognitive systems engineering literature has emphasized the use of visual displays to improve performance on complex real-world tasks such as process control with the labels *direct perception, ecological interface design, representational design,* and *semantic mapping,* focusing on the use of representations that take advantage of powerful perceptual processes to support problem solving. Although the theoretical orientations and details of each approach are slightly different, they all share a fundamental core belief: The effectiveness of a graphical decision aid depends on relationships between the representation, the domain and associated task(s), and the characteristics of the agent (person). This review begins with a discussion of strategies for representation aiding, assuming that the appropriate domain semantics have been determined. It then discusses a core assumption in this literature that, in the design of complex real-world systems, designers cannot anticipate all the possible scenarios that could arise and must therefore design displays that support effective problem solving even when novel or unanticipated scenarios are encountered. Finally, the results of empirical studies of designs based on representation aiding are reviewed.

In this review, we focus on strategies for representation aiding, as discussed under several labels in the human factors/ergonomics literature, including *direct perception* (Moray, Lee, Vicente, Jones, & Rasmussen 1994), *ecological interface design* (Rasmussen & Vicente, 1989), *representational design* (Woods, 1991) and *semantic mapping* (Bennett & Flach, 1992). Zachary (1986) defined representation aids as "interface techniques which provide ways of presenting the decision problem in the aiding-system interface that are tailored to the needs and capabilities of human cognitive processes" (p. 46). Thus, representation aiding could apply to any design that seeks to improve performance on a task through some set of displays (visual, auditory, tactile, verbal, etc.) by providing interface resources that enhance human performance (Norman, 1993). Our focus will be on the design of visual representation aids as applied to the design of interfaces for operating complex, real-world systems that are designed with the following goals in mind:

- The design should enhance or leverage human cognitive and perceptual processes in support of problem solving rather than replacing these processes.
- For those tasks and scenarios that have been anticipated during the design process, displays (representations) should be developed to support skill- and rule-based processing (Rasmussen, 1983; Vicente, 2002), at least for experienced users.

- The display design should also support knowledge-based behaviors (Rasmussen, 1983; Vicente, 2002) for novel or unanticipated scenarios and for less experienced users.

These features are discussed in more detail below.

LEVERAGING HUMAN COGNITIVE AND PERCEPTUAL PROCESSES

Representation aiding as a design strategy can be contrasted with three alternatives:

a. A fully automated solution (in which the software does not merely organize information for presentation to the user and support manipulation of this information but searches for and identifies what it considers to be the solution).
b. A protocol or operating procedure that is supposed to be rigidly followed by a system operator when a given situation is encountered (e.g., Bennett, 1992).
c. A task-neutral information display in which so-called primitive data about the current state of a system or problem are displayed; the user must then combine these data in some way to make the necessary inferences to detect, understand, and solve the problem at hand.

The rationale for preferring representation aiding over automation in certain cases may be based on cost (it may be more expensive to develop and field an adequate automated software solution) or on performance (it may be difficult or impossible with existing technologies for the designer to develop algorithms that can perform as well as human perceptual and cognitive processes). In addition, it can be argued that by keeping the person involved in the routine operation of the system (through the use of representation aiding instead of automation), he or she may be more likely to detect and deal with an anomalous situation that would not have been adequately handled by the automation as designed.

Thus, one argument for using representation aiding rather than automation is that a human operator needs to be effectively involved in the task in order to provide adequate handling of cases in which an automated solution would have shown brittleness (Skirka, Mosier, & Burdick, 1999; Smith, McCoy, & Layton, 1997; Smith, Geddes, et al., 2006). The importance of this consideration is that many serious accidents are caused by circumstances that have not been predicted or planned for by the designers of a system (the so-called Achilles heel of system design).

MAPPINGS: THE AGENT, THE DOMAIN, AND THE INTERFACE

Representation aiding is an approach to display design that goes beyond simply considering the characteristics of human cognitive and perceptual capabilities. As described in the introductory section, representation aiding further emphasizes the need to consider three primary system components (domain, agent, interface) and the quality of the mappings among them. Woods and Roth (1988) referred to these three components as the *cognitive triad*. We consider each component in greater detail.

Domain

As has been emphasized in the cognitive systems engineering literature (Rasmussen, Pejtersen, & Goodstein, 1994; Vicente, 1999), analysis and description of the domain are absolutely essential in developing effective representation aids. The goal is to achieve a detailed understanding of the "landscape" on which the work occurs, regardless of who is doing it or how it gets accomplished. The term *constraint* is used to refer to the sources of regularity in a domain and to the associated invariants that are directly relevant to certain conclusions or responses. Such constraints are often referred to as *behavior shaping* because they represent the fundamental elements of the system that must be considered for effective control. In cognitive systems engineering approaches, these constraints are often modeled using the analytical tools of the abstraction and aggregation hierarchies, including categories of information encompassing the physical, functional, and goal-related characteristics of the domain, as well as the relationships among these categories (Rasmussen et al., 1994).

Interface

A variety of representational forms can be used to present critical information in the domain. Spatial representations provide detailed information regarding the physical configuration of a system (e.g., maps and mimic displays) or the abstract properties of a system (e.g., geometrical forms that provide spatial analogies to higher-order domain constraints). Metaphorical displays use graphical representations that relate critical domain constraints to more familiar objects or concepts (e.g., the desktop metaphor). Finally, alphanumeric representations provide detailed numerical values and verbal descriptions or labels. The literature on representation aiding emphasizes the need to determine which display design will most effectively communicate the state of the constraints for the domain; the design will accomplish this by triggering automatic perceptual and cognitive processes.

In designing the interface, it is important to realize that the representations that are chosen can shape user behavior in very powerful ways. Each type of representational form produces cognitive and perceptual demands that vary in terms of the nature, focus, and amount of resources involved in its interpretation, as do the details of how that form is implemented. Even within the same class of solution, alternative realizations of a design can produce cognitive demands that are substantially different.

Agent

The cognitive agents who make the control decisions also introduce a set of constraints. Agents can be either human or computer (e.g., automation). To simplify this discussion, we will consider only those constraints associated with human agents.

Decades of research in psychology have revealed a number of constraints that characterize the human as a system component. One fairly obvious constraint is the limitation associated with working memory: Human agents can actively maintain only a small number of items in working memory. On the other hand, human agents are equipped with powerful perceptual systems and can develop new automatic processes through experience. In terms of display design, considerations of visual attention and visual form perception are particularly critical. Similarly, general modes of behavior exhibited by human agents (i.e., skill-, rule-, and knowledge-based behaviors) must be considered when making design decisions. These modes of behavior and the implications for design will be discussed in more detail later.

The net result is that the design of effective graphical representations is a complicated and intricate activity that is extremely context specific. There are no short cuts, easy solutions, or checklists. Principles of design can be described, but they must be creatively applied and adapted to the specific circumstances at hand. The domain, the interface, and the human agent—each contributes a set of mutually interacting constraints. The effectiveness of graphical decision support will ultimately depend on the quality of very specific sets of mappings among these constraints (e.g., Bennett & Flach, 1992; Bennett & Walters, 2001).

Mappings

One set of mappings involves the relationships among the constraints of the work domain and the informational content encoded into the graphical representations. The quality of these mappings is determined by the extent to which the relevant categories of domain information and the relationships among them are available in the interface. Essentially, this set of mappings determines whether or not the interface contains the information necessary to easily make effective responses. These mappings have been referred to as *correspondence* elsewhere (e.g., Bennett, Nagy, & Flach, 1997; Vicente, 1999), but we will use the term *content mapping*.

A second set of mappings involves the relationship between the visual properties of the graphical representations and the perceptual/cognitive capabilities and limitations of the observer. Essentially these mappings determine whether or not the domain constraints have been encoded or represented in the interface in a form that can be processed easily by the human agent. This has previously been referred to as *coherence mapping* (e.g., Bennett et al., 1997; Vicente, 1999); we will use the label *form mapping*.

The final set of mappings involves the relationship between the human agent and the domain. The domains of interest are usually complex and dynamic. It follows that the representations that are developed will also be complex and dynamic. The human agent must be sufficiently knowledgeable about the domain to understand and interpret the information that is presented in the representation.

REPRESENTATIONS TO SUPPORT ANTICIPATED SCENARIOS

Rasmussen (1983) described a set of general modes of behavior that are useful in considering the cognitive resources required of a human cognitive agent to complete some task: skill-, rule-, and knowledge-based behaviors. Each mode of behavior needs to be considered in the design of a representation aid, as all three need to be supported if the representation aid is to improve performance in both anticipated and unanticipated scenarios.

Skill- and Rule-Based Behavior

Skill-based behaviors are high-capacity, sensorimotor activities that can be executed automatically with little attentional demand. As an example, for a skilled driver, the information about the state of the natural environment (e.g., the road) can be specified by a set of sensory inputs that, in turn, automatically activate or trigger certain perceptions and associated responses. Thus, the environment (road) provides a reference *spatio-temporal signal* that is automatically perceived by the agent (driver). In response, the agent then automatically produces motor activity (e.g., a control input such as turning the wheel) that minimizes deviation from this signal (Jagacinski & Flach, 2003).

In rule-based behavior, the person has similarly developed associated actions that lead to effective solutions based on prior experience or knowledge. The person reacts to cues or signs that determine when a stored procedure should be executed. Continuing with the driving example, suppose the driver had obtained a set of directions for a destination. These directions consist of a set of signs to be monitored for (e.g., "when you see the billboard with the singing chicken") and rules to be followed (e.g., "turn right at the next street—Elm Street"). The driver then continues performance that is largely skill based (e.g., steering along the new course) until the next set of signs and actions (e.g., the next turn) is encountered. The key observation is that conscious deliberation is required to select from alternative courses of action at key decision points, thereby linking sets of prelearned activities. In this case, the prior experience or knowledge required for rule-based behavior exists outside the person who performs it (i.e., the supplier of the directions). In the case of experts, the required rules and knowledge will have been previously acquired.

The cognitive systems engineering literature on representation aiding emphasizes the value of designing to support skill- and rule-based behaviors for scenarios that have been anticipated by the developer, which makes it easier for someone to complete predetermined tasks efficiently and effectively. The strategy is to identify the relevant constraint (based on the task and associated domain semantics) and to display it in a manner that allows a rich set of signals and signs (i.e., affordances for action) to be directly perceived.

This strategy is illustrated in a simple example by Vicente and Rasmussen (1990), who contrasted the use of rote sensor displays that depict "all of the elemental data that are directly available from sensors" with so-called smart displays, which are "able to directly pick up goal-relevant properties that are relevant to survival" (p. 212). Rote sensor displays "measure fundamental dimensions (e.g., mass, length, and time)," and then the person is required to use these fundamental metrics "to painstakingly derive the higher order properties of interest" (p. 215). This contrasts with a smart display, in which the critical data relationships are shown more directly on a display, indicating the goal-relevant properties as highly salient visual properties.

The assumption is that if the information about fundamental metrics is displayed separately across a number of displays (such as a hard-wired set of separate displays on a control panel), the person must shift attention from one display to another and use slower, more attention-demanding controlled processes to "recover the goal-relevant domain properties from the elemental data represented in the interface" (Vicente & Rasmussen, 1990, p. 211). As a result, "it is not possible for operators to go out and directly explore the status of the system using the powerful perceptual systems that serve them so well in the natural environment" (p. 211).

The cognitive science literature on problem solving similarly emphasizes the value of representations to support automatic, associative processes to enhance performance. Studies address direct perception in routine situations in which task-relevant information in the environment is "directly picked up without the mediation of memory, inference, deliberation, or any other mental processes that involve internal representations" (Zhang, 1997, p. 181), so that "the end product of perception is the end product of the whole problem solving process" (p. 187).

Thus, both the cognitive science and cognitive systems engineering literatures place a strong emphasis on the use of representations or displays to trigger automatic perceptual and associative processes (Shiffrin & Schneider, 1977) to enhance problem solution. Providing powerful illustrations of this, Scaife and Rogers (1996) described "the extent to which differential external representations reduce the amount of cognitive effort required to solve informationally equivalent problems" (see also Kotovsky & Simon, 1990; Simon & Hayes, 1976; Zhang & Norman, 1994).

However, Scaife and Rogers (1996) noted this caution in discussing graphical representations: "The value of diagrams...is strongly related to the experience and expertise of the individual having 'operators' that match the display... Novice physicists, for example, will not make the same inferences as experts from the same diagram" (p. 195).

An Example of Representations to Support Anticipated Scenarios

A simple example of representation aiding that takes advantage of automatic perceptual processes is provided by Cole (1986), who developed a display for monitoring patient respiration (see Figure 3.1). The relevant fundamental dimensions for this monitoring task are the volume or depth of air exchange and the rate of air exchange. However, the task-relevant concerns are the total amount of oxygen respired by the patient and imposed by the respirator and whether the patient is achieving alveolar gas exchange.

Figure 3.1 shows the display's appearance during a sequence of 11 periods. For each period, it could potentially display two rectangles: one representing the ventilator's contribution to breathing and one representing the patient's spontaneous contribution to breathing. For Periods 1–3, only the rectangle for the ventilator is shown, as the ventilator is acting alone during these times. The height of each rectangle is an indictor of the volume of air exchanged (larger rectangles indicate that more air was exchanged). Note that any given volume of air exchange can be achieved by either many rapid, shallow breaths or by fewer, deeper breaths. This is the mean rate of respiration and is represented by the width of each rectangle. Thus, for Period 6, the rectangle for the patient's contribution (right rectangle) is wide but short, representing many shallow breaths (which do not contribute as effectively to gas exchange) and a low volume of air exchange.

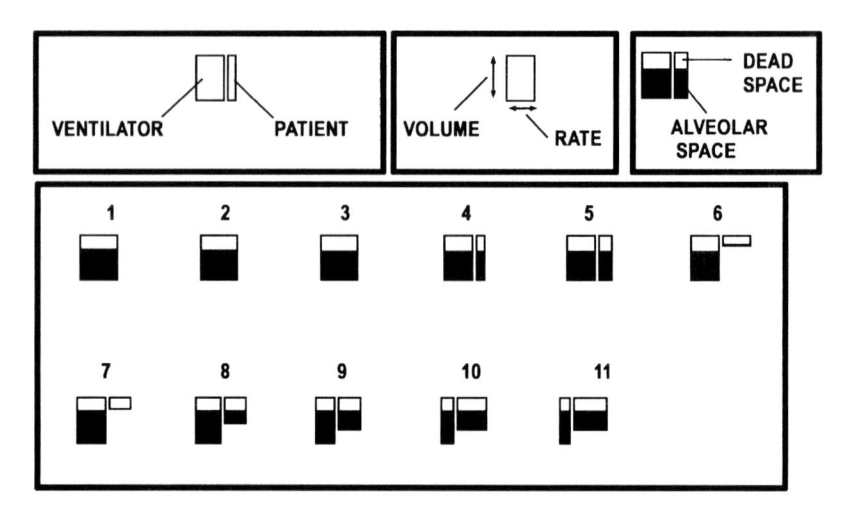

Figure 3.1. Representation aiding to monitor patient status on a respirator (after Cole, 1986). Reprinted with permission from Association for Computing Machinery, Proceedings of the 1986 ACM Conference on Computer-Human Interaction, *91–95.* Copyright 1986 by the Association for Computing Machinery. All rights reserved.

This display demonstrates some important principles for the design of representation aids and will be used to tie together several of the considerations that have been outlined thus far. Recall that form mapping refers to the extent to which the visual features that are present in the representations can be obtained by the human agent. A critical component of this mapping has been referred to as *emergent features* (Pomerantz, 1986). Emergent features are highly salient (e.g., notable, conspicuous, prominent) visual properties that arise from the interaction of the lower-level graphical elements of a representation. For example, Cole's display produces several different emergent features, including the width, height, area, and shape of the rectangles in Figure 3.1.

The presence of emergent features may be necessary for the design of effective representation aids, but it is not sufficient. In addition, "the emergent features must reflect the inherent data relationships that exist in the domain—that is, the highly salient emergent features must correspond to the information needed to complete domain tasks" (Bennett, Toms, & Woods, 1993, p. 73). In Cole's display, the form mapping is effective: The extent to which the air exchange is attributed to the patient or the ventilator is visually specified by the differences in the shape and size of the two rectangles. These displays are often referred to as *configural displays,* a term borrowed from the visual attention literature because of the overlap in theoretical considerations (Bennett & Flach, 1992).

Summary

Both the cognitive science and cognitive systems engineering literatures emphasize certain critical considerations in developing representations to aid problem solving for scenarios that have been anticipated by the designers. As we indicated earlier, representations need to be developed to support skill- and rule-based behavior under these conditions. In the case of Cole's display, these goals were achieved by mapping some of the key properties of this domain into geometric forms that provided spatio-temporal analogs.

The human agent interacts with this display largely at the level of skill- and rule-based control. The primary skill-based behavior is direct perception. The display has been designed so that it provides visual information that directly specifies the state of the system. For example, the shape and size of the two rectangles are spatio-temporal signals (i.e., emergent features) that specify whether the patient (e.g., Figure 3.1, Period 11) or the ventilator (e.g., Figure 3.1, Periods 1–3) is the major contributor to air exchange. Thus, the display shown in Figure 3.1 allows the human agent to directly perceive many of the domain constraints directly. This could be contrasted with an alternative representation aid in which each low-level datum was measured and presented in isolation, in the form of a digital value. Such a display would not support skill-based behavior and in fact would produce a much more difficult set of demands: The human agent would have to engage in knowledge-based behavior (e.g., computing relative rates of air exchange from the raw digital values).

The display in Figure 3.1 provides a rich set of visual cues that support rule-based behavior as well. The visual properties of the display serve as signs that suggest the appropriate control input. For example, a configuration such as that appearing in Figure 3.1, Periods 1–3 (i.e., air exchange is attributed exclusively to the ventilator) could serve as a sign that additional medical intervention is needed, especially if this appeared over a long period. Similarly, a configuration such as that represented in Figure 3.1, Period 11 could be an indication that it is time to take the patient off the ventilator.

REPRESENTATIONS TO SUPPORT PROBLEM SOLVING IN UNANTICIPATED SCENARIOS

Cognitive systems engineering researchers have posited that during unanticipated scenarios, even an expert in a complex dynamic domain may have to rely on actual problem-solving behavior (knowledge-based behaviors) when he or she is faced with a set of circumstances that have not been anticipated in the design of the operating system or in preparation through preplanned guidance, training, or operating experience. Rasmussen (1983) referred to this mode of performance as *knowledge-based behavior*. Under these circumstances, the individual must detect and identify the cause of the abnormality and consider a variety of information about the system being controlled (e.g., goals, constraints, functions, physical configuration) to compensate for the abnormality. Thus, in addition to designing to support skill- and rule-based behaviors to deal with these anticipated scenarios, the design also needs to support knowledge-based behavior (e.g., problem solving) for novel scenarios.

Vicente (2000) provided a simple but illustrative example that suggests how the process of designing for knowledge-based behavior should proceed. He contrasted verbal directions on how to travel from one point to another with the use of a map. In this analogy, the directions provide a very efficient but brittle description of how to accomplish the desired goal. The map requires more effort (reasoning) but is more flexible in the face of obstacles that might render the directions unusable. Building on this analogy, Vicente noted:

> This decrease in efficiency [of maps vs. directions] is compensated for by an increase in both flexibility and generality. Like maps, work domain representations are more flexible because they provide workers with the information they need to generate an appropriate response online in real time to events that have not been anticipated by designers. This is particularly useful when an unforeseen event occurs because task representations, by definition, cannot cope with the unanticipated. (2000, p. 115)

Before we consider how this might be achieved, we first discuss an example from the cognitive science literature that demonstrates the power of alternative representations on solving problems.

Supporting Knowledge-Based Processing by Representing the Constraints

The cognitive science literature provides some powerful examples that help in understanding the impact of external representations or displays on solving problems. An example from this literature is discussed, followed by consideration of an example from the human factors/ergonomics literature concerned with process control.

The digits game—Underconstraining the problem. The digits game (Perkins, 2000) demonstrates a design or presentation of a problem that fails to make salient the critical constraints, as a result making it difficult for a person to solve the problem (win the game).

Two players compete. Each player alternates in picking a number between 1 and 9. Once a player has selected a number, it cannot be picked again (sampling without replacement). To win, a player must be the first person to select a set that contains three numbers that add up to 15. If neither player accomplishes this, the game is a draw.

As an example, consider the following sequence in which Player 1 wins:

Player 1 – selects 5
Player 2 – selects 9
Player 1 – selects 2
Player 2 – selects 8
Player 1 – selects 6
Player 2 – selects 4
Player 1 – selects 7 (Player 1 wins!)

Even with a scratch pad, this is a challenging task as framed in these instructions because, given this framing or representation of the problem, each player has to compute and consider the implications of all the various possible outcomes that could result from selecting one of the still-available digits.

An alternative representation is to embed the digits in a Tic-Tac-Toe matrix (see Figure 3.2) and to tell the competitor that he or she is playing Tic-Tac-Toe, alternating selection of particular cells in the matrix with the opponent, and trying to get a line of three

Representation Aiding to Support Performance on Problem-Solving Tasks 83

2	7	6
9	5	1
4	3	8

Figure 3.2. Isomorphic representations: The digits game versus Tic-Tac-Toe. The three numbers along each vertical, horizontal, or diagonal line through the Tic-Tac-Toe matrix add up to 15.

Xs or three Os before the opponent does. Thus, in terms of picking the best sequence of three numbers, playing Tic-Tac-Toe (and implicitly identifying the selected digits along the way) is equivalent to playing the digits game (i.e., they are problem isomorphs).

However, as demonstrated in a study by Zhang (1997), the necessary cognitive processes are quite different. In particular, the Tic-Tac-Toe representation implicitly indicates all the sets of three digits between 1 and 9 that add up to 15 (the triad of numbers in any row, column, or diagonal), thus displaying the constraints relevant to this problem in the external world (Hutchins, 1995; Norman, 2002). In doing so, it constrains the set of possibilities that the competitor needs to consider in trying to pick three numbers to add up to 15. It also constrains the set of possibilities that the competitor needs to consider in order to block the opponent. In short, representation of the game as Tic-Tac-Toe helps to limit search by making "the domain constraints directly perceptible" (Flach & Bennett, 1992), focusing attention on those paths that could lead to success.

As a result, even for a novice player for whom this is a novel task, the representation of the digits game as Tic-Tac-Toe significantly reduces the complexity of the reasoning necessary to successfully play the game.

The digits game—Supporting different cognitive operations. This illustration further demonstrates how the external representation of a problem can change the mental processes necessary to solve that problem. Representing this game as Tic-Tac-Toe makes it unnecessary to do any arithmetic because the goal is stated in terms of spatial reasoning instead of numeric calculations. This supports "perceptual operations, such as searching for objects that have a common shape and inspecting whether three objects lie on a straight line" (Zhang, 1997, p. 185). For in Tic-Tac-Toe, "to identify a winning triplet is to search for three circles lying in a straight line," which "can be visually inspected," and the "winning invariant [the number of potential wins for a circle] . . . is represented externally by the number of straight lines connecting a circle: 4, 3 and 2 for the center, the corners, and the sides, respectively" (p. 190).

The Tic-Tac-Toe representation also illustrates aiding through the use of an external memory aid. This aspect of representation aiding has been studied extensively; for example, Sciafe and Rogers (1996) noted that the role of diagrams as external memory aids enables "a picture of the whole problem to be maintained simultaneously, whilst allowing the solver to work through the interconnected parts" (p. 193).

Likewise, Scaife and Rogers discussed the use of external memory aids to change long-term memory demands, citing O'Malley and Draper (1992), who differentiate between

> the knowledge users need to internalize when learning to use display-based word processors (e.g., MacWrite) with that which they can always depend upon being available in the external display. The tendency, therefore, appears to be for users to learn only what is necessary to enable them to find the information they require in the interface display. Information represented in such displays is viewed as an external memory aid. (Scaife & Rogers, 1996, p. 202)

Such use of representations to support external memory aids is also highlighted in the design literature in studies of performance on real-world tasks. Hutchins (1995), for example, cited numerous examples in which external representations (such as Post-It® notes) are used as memory aids to support individual cognition and as communication tools to support distributed cognition (Zhang & Norman, 1994).

A Process Control Example

In the illustration that follows of a simple domain analysis, we discuss alternative representation aids for operating a process control system analogous to an example presented in Vicente and Rasmussen (1990). The constraints in this domain have a high degree of regularity; the cognitive demands that are produced arise from the physical, functional, and goal-related properties of the domain itself. In essence, the behavior of the system is determined primarily by the laws of nature (e.g., conservation of mass) as opposed to the intentions of the agents who control it. Using the terminology applied by Rasmussen et al. (1994), this type of domain is *law driven*.

An effective representation aiding strategy for law-driven domains is to develop abstract geometrical forms that reflect the inherent goals as well as the functional and physical properties of the domain. The critical design approach involves the mapping of the domain constraints into graphic representations that provide spatial analogies to domain properties. The quality of these mappings in turn determines the effectiveness of the display. We next discuss two examples (Figure 3.3) that provide different sets of mappings and different degrees of effectiveness. (See Bennett et al., 1997, for a more comprehensive discussion.)

Consider the simplified process control system illustrated in Figure 3.3a. There is a reservoir for storing fluid, two input sources for adding fluid to the reservoir (shown on top of the reservoir), and an output source for removing fluid from the reservoir (shown underneath the reservoir). Each input has a pipe to carry the fluid, a valve to control the flow of fluid (V_1 and V_2), and a sensor to monitor the flow of fluid (I_1 and I_2) and the output (V_3 and O).

To identify critical constraints in this law-driven domain, Rasmussen et al. (1994) used an abstraction hierarchy. Such an abstraction hierarchy is an analytical tool that provides a framework for describing the critical categories of information in a domain and the relationships among these critical categories. In terms of representation aiding, they

Representation Aiding to Support Performance on Problem-Solving Tasks 85

Sample Process

Low Level Data	**High Level Properties**
(Process variables)	(Process constraints)

T = Time
V_1 = Setting for Valve 1
V_2 = Setting for Valve 2
V_3 = Setting for Valve 3
I_1 = Flow rate through Valve 1
I_2 = Flow rate through Valve 2
O = Flow rate through Valve 3
R = Volume of reservoir

$K_1 = I_1 - V_1$ Relation between commanded
$K_2 = I_2 - V_2$ flow (V) and actual flow (I or O)
$K_3 = O - V_3$

$K_4 = \Delta R = ((I_1 + I_2) - O)$
    Relation between reservoir
    volume (R), mass in ($I_1 + I_2$),
    and mass out (O).

G_1 = Volume goal
G_2 = Output goal (demand)

$K_5 = R - G_1$ Relation between actual states
$K_6 = O - G_2$ (R, O) and goal states (G_1, G_2)

3a. Mimic display.

3b. Configural display.

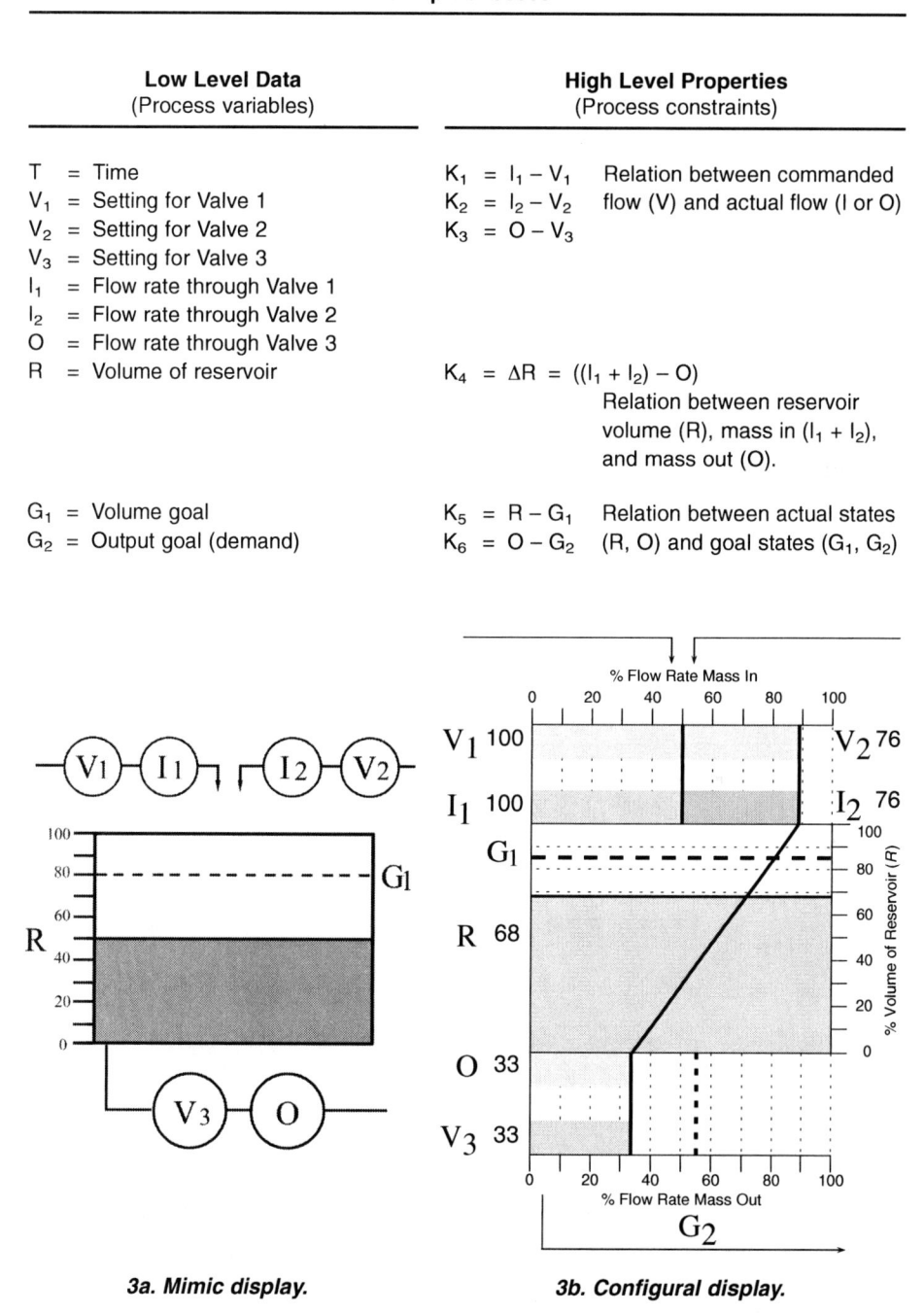

Figure 3.3. Mimic display and configural display for a process control system.

can be thought of as the different perspectives from which the domain will need to be considered by the agents who are controlling it.

For this process control domain, there are five levels of abstraction, listed in the left column of Table 3.1. The general properties that are represented in each level of abstraction are listed in the center column of the table. These labels and general properties are adapted from Rasmussen et al. Finally, the specific properties of this simple process are listed for each level of abstraction in the right column of Table 3.1.

There are only two goals for this control system: (a) maintain the reservoir at a particular level to ensure that a sufficient volume of fluid exists and (b) maintain a specified output flow rate to provide fluid to a downstream process. Nevertheless, even this simple system has some fairly complicated dynamics. A listing of critical information is provided at the top of Figure 3.3 in two columns. The column on the left lists the low-level data associated with the process. These are the process variables that can be measured by sensors and set by the human operator using valves in order to control the process.

The column on the right lists the high-level constraints. These constraints are the important relationships among the low-level variables that are determined by the process constraints; they are critical because they are relevant to certain tasks that may have to be performed by the operator of this system (such as deciding how much to increase the flow rate using V_1 in order to regain the desired output flow rate O when it is too low).

Table 3.1. Levels in the Abstraction Hierarchy for the Process Control Example

Means-Ends Relations	*General Properties Represented*	*Specific Properties for Process*
Purposes and constraints	Reasons for design Coupling to environment	Volume goal (G_1) and output goal (G_2) Constraints K_5 and K_6 (the difference between actual states and goal states)
Abstract functions and priority measures	Intended proper functioning Flow of information, resources, or other commodities through the system	Conservation of mass; Flow of mass through the system; K_1, K_2, K_3, and K_4 constraints
General functions	General functions to be coordinated Independent of physical implementation	Storage, sources, sink
Physical processes and activities	Physical properties necessary for control of system (adjusting, controlling, predicting)	Two feedwater streams, a reservoir for storage, a single output stream; moment to moment values and settings (V_1, V_2, V_3, I_1, I_2, O, R).
Physical form and configuration	Physical configuration, form, appearance, location, etc.	Physical locations and descriptions of components (e.g., causal connections pipe length, physical location of valves, size of reservoir)

Representation Aiding to Support Performance on Problem-Solving Tasks

For example, K_4 is determined by a law of nature: conservation of mass. This constraint essentially indicates that the changes in the volume of the reservoir (ΔR) will be determined by the relationship between the rate of mass flowing into and the rate of mass flowing out of the reservoir. For example, if the amount of mass flowing into the reservoir ($I_1 + I_2$) is greater than the amount of mass flowing out of the reservoir (O), then the volume of the reservoir (R) will be increasing (ΔR) at a rate that is proportional to the size of the positive net inflow (and vice versa). Alternatively, if the inflow through V_1 is 5 gallons per minute and through V_2 it is 10 gallons per minute, and the outflow through V_3 is 15 gallons per minute, then the volume stored in the reservoir will remain constant.

Representing physical form and configuration along with physical processes and activities. Suppose we start this process control task with the volume in the reservoir (R) at the desired level or goal (G_1). If the goal is to monitor the system to determine whether the current volume remains at G_1 over time, then the display shown in Figure 3.3a, using a simple analogical representation in which the geometric form of a rectangle is used to provide spatial analogies for the reservoir and its volume level, is effective. The operator can simply monitor to see whether the top of the gray rectangle (representing the filled portion of the tank) remains at the dashed line (the desired level).

Abstractly, then, the display in Figure 3.3a is a spatial representation that emphasizes the physical components and relations that characterize the system. The primary physical components are represented by the lines and labels at the top (i.e., the feedwater streams), the rectangular form in the middle (i.e., the reservoir), and the lines and labels at the bottom (i.e., the output stream). The physical relationships among the various components are emphasized (e.g., the physical placement of the valves relative to the sensors in the input and output streams).

This type of representation is often referred to as a *mimic display* because the representation mimics the physical structure of the system. Note that this type of representation provides a critical perspective for dealing with certain aspects of the system: It illustrates the physical components, the logical relationships among them, and the causal connections and structure.

Representing abstract functions and priority measures as well as purposes and constraints. Although the mimic display, representing physical form and configuration along with physical processes and activities, is effective for supporting performance for the monitoring task described earlier, it is not as useful in helping the operator to select an appropriate control input. The problem is that the information about higher-level properties or constraints is not provided in a direct fashion by the mimic display.

To make this point explicit, consider the constraint, K_4, which indicates that changes in the volume of the reservoir will be determined by the relationship between the rate of mass flowing into and out of the reservoir. For example, if the amount of mass flowing into the reservoir ($I_1 + I_2$) is greater than the amount of mass flowing out (O), then the volume of the reservoir (R) will be increasing at a rate (ΔR) that is proportional to the size of the positive net inflow. Knowledge of the state of this constraint is essential for the effective control of the system.

Consider the situation depicted in Figure 3.3a. The constraint K_5 (the difference between the volume goal and the current level) is represented graphically in the mimic display—that is, the difference between the dashed line, or G_1, and the top of the gray-filled rectangle, or R. Thus, there is direct visual evidence specifying that a system goal is not being met (i.e., G_1 is greater than R). On the other hand, although the operator could detect an imbalance in the input and output flows by watching to see if the top of the gray-filled rectangle (representing R) goes up or down, this would take time and is therefore an inefficient method for detecting a problem with the balance between the flow in and the flow out (K_4).

The low-level data that are necessary to determine the answer (i.e., I_1, I_2, and O) are present in the mimic display, assuming there are displays embedded in the mimic display indicating each of these flow rates. However, to use these data, the human agent is forced to collect the relevant information from each of these separate displays and complete the computations necessary to derive the answer. (Note that the situation is even worse for the second system goal [G_2] and its associated constraint [K_6]: Neither the output goal nor the difference between the current value and the output goal is represented in the mimic display.)

Figure 3.3b illustrates an alternative display using the funnel display concept described in Vicente (1991). In this configural display, the variables and relationships are represented using analog geometrical formats. When the underlying information changes, it is reflected in analogical changes to these geometrical formats. At the top of the display are two sets of horizontally stacked bar graphs (showing $V_1 + V_2$ and $I_1 + I_2$). The valve settings to control the rate of mass flowing into the system (V_1 and V_2) are represented by the top stacked bar graph. The combined horizontal extent of the two bars (i.e., the length of the bars in the horizontal axis) corresponds to the total "commanded" rate of mass flow into the system.

The sensed mass flow rates into the system (i.e., the rates measured by sensors) are represented in a similar fashion by the horizontally stacked bar graph directly below ($I_1 + I_2$). In this case, the commanded and sensed flow rates are exactly the same (88%). The commanded (V_3) and sensed (O) flow rates out of the system are represented similarly in the two bar graphs that appear at the bottom of the display. The reservoir volume (R) and the goal for reservoir volume (G_1) are represented in the same manner as previously described for the mimic display.

Such a configural display can be particularly useful in representing the states of the constraints that are determined by the relationships among the low-level parameters of a law-driven domain (Bennett & Flach, 1992). Each relational invariant is represented by a high-level visual property (i.e., an emergent feature, as discussed previously for the Cole, 1986, display).

The configural display in Figure 3.3b makes extensive use of such emergent features. First, consider the mapping between one of its emergent features and the fundamental domain constraint, K_4.

The horizontal length of the stacked bar graph for the rate of mass in ($I_1 + I_2 = 88\%$), relative to the horizontal length of the stacked bar graph for mass out (O = 33%) provides a visual indication of whether the flow in equals the flow out (i.e., mass balance). To make this relationship more visually salient, a bold line (more specifically, a contour) has been

Representation Aiding to Support Performance on Problem-Solving Tasks 89

added to explicitly connect these two stacked bar graphs (the value of $I_1 + I_2$ is connected by a line to the value of O). A perfectly vertical indicator (connecting line) specifies that mass flow is balanced. The direction of the indicator's deviation from the vertical directly specifies whether more mass is flowing into or out of the reservoir. A clockwise rotation of the indicator from vertical, as illustrated in Figure 3.3b, specifies a positive net inflow; a counterclockwise rotation from vertical specifies a negative net inflow. Furthermore, the degree to which the indicator deviates from vertical is a direct visual indication of the magnitude of the difference between the rate of mass flow into and the rate of mass flow out of the reservoir, as shown in Figure 3.4. Thus, the slant or orientation of this line is an emergent feature that provides direct visual evidence specifying both the direction and magnitude of the net flow of mass into or out of the reservoir. This visual evidence allows the operator to directly perceive mass balance rather than requiring him or her to calculate it.

This configural display contains similar emergent visual features to indicate the status of most of the other relevant domain constraints. Three of these domain constraints (K_1, K_2, and K_3) describe the relationships between commanded rates of flow (i.e., valve settings) and sensed rates of flow (i.e., sensor readings). The emergent features corresponding

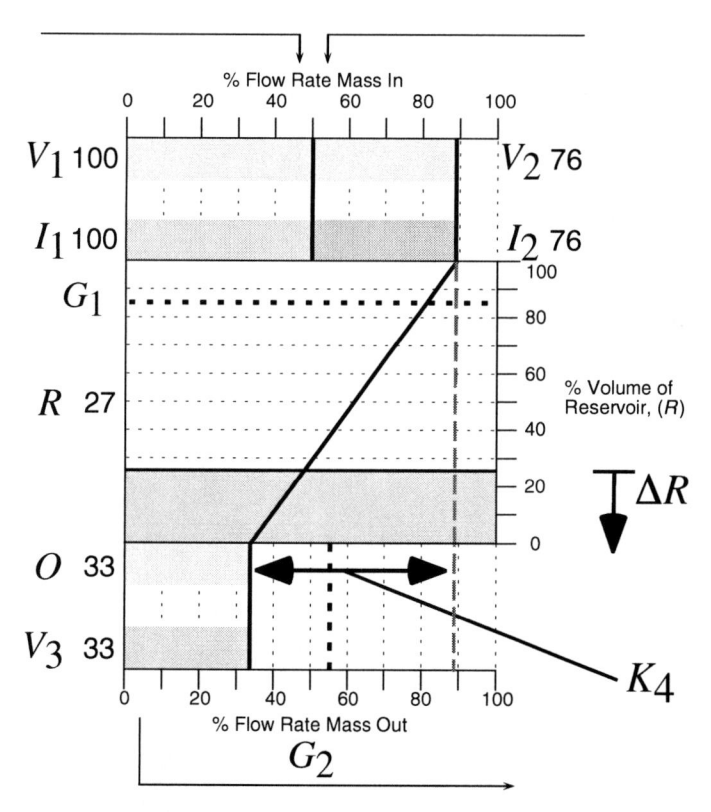

Figure 3.4. Example with the configural display for the process control system showing a positive net inflow (with more mass entering the reservoir than leaving the reservoir).

to these constraints are the bold lines connecting the ends of the associated bar graphs (i.e., V_1 to I_1, V_2 to I_2, and V_3 to O). Similar to the mass balance indicator, the orientation of each of these lines is an emergent feature that visually indicates the status of the corresponding constraint. When these lines are vertical (as in Figure 3.3b), it is a direct visual indication that the associated commanded mass flows are equal to the measured mass flows. See Figure 3.5 for an example in which K_2, the difference between the measured flow I_2 and the commanded flow based on the valve setting V_2, indicates a problem.

The final two constraints (K_5 and K_6) are goal related and describe the difference between a goal and the current value of the associated variable. For example, the constraint on mass inventory (K_5) is shown using the relative height of the filled area to represent volume within the reservoir and the bold, dashed horizontal line to represent the goal level G_1.

Supporting problem solving in anticipated scenarios. The configural display just described is clearly based in part on a desire to assist operators in dealing with certain predefined scenarios, such as detecting an imbalance between the total inflow and outflow.

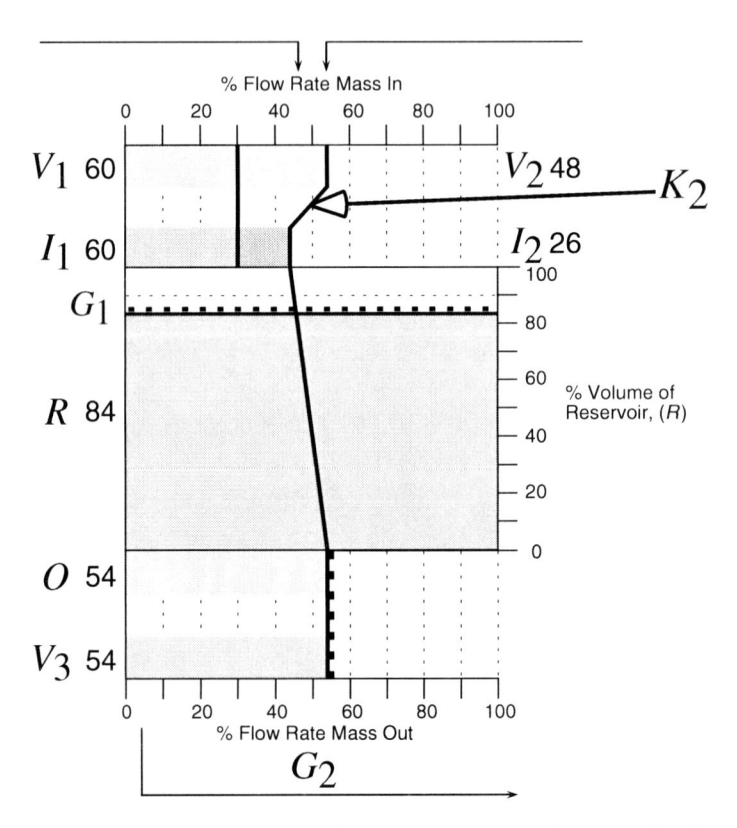

Figure 3.5. Example with the configural display for the process control system showing an imbalance in K_2 (the difference between the expected inflow V_1—based on the valve setting—and the sensed flow I_1).

It does so by providing salient perceptual cues (detecting vertical vs. slanted lines) that are direct indicators of certain imbalances. Thus, the user can directly perceive the constraints in the domain, making it unnecessary to perform the computations that would be required if the mimic display (with associated meters) in Figure 3.3a were used instead of the configural display in Figure 3.3b.

Supporting problem solving in unanticipated scenarios. A configural display like the one just described is also very helpful in solving unanticipated scenarios in this process control setting. As an illustration, contrast this configural display with a simple expert system that is brittle (Smith et al., 1997) because the designers developed it to correctly detect and diagnose the failure of a single sensor but not a simultaneous failure of two or more sensors. Although this might be unlikely for this simple process control system, it is not at all implausible that the designers of the expert system could overlook one of the myriad potential interactions that could occur within a complex, large-scale process control system. Similar to instructions on how to get to some unfamiliar location, the expert system could make the user's job easier for scenarios in which the software is fully competent. However, when a scenario arises that is outside the expert system's range of competence, the user could be faced with a very difficult problem-solving task, assuming he or she even recognizes that this has happened.

In contrast to the expert system, the configural display requires more active participation by the operator in terms of monitoring for and dealing with anticipated problem scenarios, because the operator has to visually scan the configural display for an indication of a problem instead of simply waiting to be alerted by the expert system. However, like the map used to get to an unfamiliar location, assuming that all the relevant constraints have been effectively represented, the configural display is likely to provide better support when dealing with an unanticipated scenario because the configural display has been explicitly designed to make salient a deviation of any of the critical constraints from its expected value (see Figure 3.5).

In terms of completeness, these critical constraints have been derived by representing the domain in a means-ends or abstraction hierarchy adapted from Rasmussen et al. (1994), with several levels of detail (see Table 3.1). These levels can be thought of as different perspectives on the domain that need to be considered by the operator who is controlling it. They provide a model, or a description, of the domain in terms of different types of information and the relationships among them. Ideally, displays should have complementary graphical representations of domain-related information at each level of the abstraction hierarchy if effective support is to be provided during knowledge-based behavior. These displays will then provide a graphical "explanation" of how the domain constraints are broken and the alternative resources that could be used to fix them.

Summary

The foregoing discussion introduced two additional points that offer concrete assistance to designers:

First is the use of a means-ends analysis to develop an abstraction hierarchy to identify the critical domain constraints and guide the development of a design that supports direct perception of violations of these constraints. This enables the use of skill- and rule-based processes to deal with many unanticipated scenarios even though they have not been explicitly considered in the design, because they have been implicitly considered through the identification of the domain constraints. Thus, for unanticipated scenarios, if the violation of critical constraints is made more salient via automatic perceptual processes, this should make it easier to apply knowledge-based processes to complete problem solving because the direct perception of the violated constraints helps to focus the attention of these controlled processes on critical aspects of the underlying abnormality.

Second is the use of geometric forms to produce configural displays that support automatic perceptual processing of information relevant to problem-solving goals; they reflect the constraints introduced by certain functional and physical properties of the domain.

A Caution

Powerful representation aids focus attention and influence the user's cognitive processes. As a result, if some skill- or rule-based process is triggered by a given design at an inappropriate time, the user may be induced to fixate on the wrong hypothesis (if it is a diagnosis task, for instance) or to perform the wrong action.

The magnitude of the biasing effects that visual representations can impose has been illustrated in a number of studies (Larson & Hayes, 2005; Smith, Geddes, et al., 2006). For instance, Smith et al. (1997) demonstrated that graphical presentation of a recommended reroute around weather by a flight-planning system to experienced airline dispatchers and pilots caused 35% of them to bias their situation assessment of the weather (a type of justification bias). This changed their mental model of the situation and thus influenced their evaluations of alternative routes around the weather.

Thus, if a representation emphasizes inappropriate aspects of the domain constraints for a particular scenario, it can actually hinder problem solving (the proverbial double-edged sword). For example, placing independent and noninteracting variables into a configural display (i.e., a single graphical object with contours driven by the individual variables) will produce emergent features that are highly salient, are difficult to ignore, and suggest higher-order properties that do not exist in the domain (e.g., Bennett & Fritz, 2005).

The intent of this caution is not to say that designs should not be developed that trigger automatic perceptual processes (that, in fact, is one of the goals of representation aiding). Rather, our point is that designers need to be aware of the influences their designs can have on how users solve problems, and they must think carefully about how to support performance for both anticipated and unanticipated scenarios.

INTERACTION DESIGN: CRAFTING THE DETAILS

The representation aiding literature reviewed in the earlier sections is concerned with certain high-level concepts, such as designing representations that support direct perception in dealing with anticipated scenarios and that also support effective controlled processing to deal with scenarios that were not specifically anticipated in the design. The representation aiding literature also provides guidance on the use of certain more specific design concepts, such as the use of configural displays (Bennett et al., 1993; Carswell & Wickens, 1987; Goettl, Wickens, & Kramer, 1991; Wickens & Andre, 1990).

Ultimately, however, this high-level design guidance must be translated into a detailed interaction design for a given application that specifies the underlying functionality of the product, as well as the look and feel of the interface with that functionality (Burns et al., 1997). The broader literature on display design provides a great deal of guidance when making these more detailed decisions to support skill- and rule-based behaviors. It also provides guidance on how to craft displays to support knowledge-based behaviors. This guidance comes from a number of different overlapping fields, which have produced literatures with the labels *human factors, human-computer interaction, industrial design,* and *visual communication*. Some of these findings are based on rigorous empirical studies and evaluations; others have developed into accepted principles through practice.

Based on analysis and accepted practice in much of the design world, Tufte (1983, 1990, 1997), for instance, suggested and illustrated a number of guiding concepts for visual display design:

- Often the most effective way to describe, explore, and summarize a set of numbers—even a very large set—is to look at pictures of those numbers. (Tufte, 1983, p. 9)
- Erase non-data-ink, within reason. (Tufte, 1983, p. 96)
- Reveal the data at several levels of detail, from a broad overview to the fine structure. (Tufte, 1983, p. 13)

Note that many of these types of design principles are based on judgments developed through practical experience rather than controlled empirical evaluations, and that such practice-based evaluation is how many fields develop such guidance. It should also be noted that such principles almost always require that the designer use judgment regarding whether and how to apply such principles to a specific application.

The human factors/ergonomics literature provides other empirically supported and generally complementary design concepts, summarized by Wickens, Lee, Liu, Becker, & Gordon (2004, pp. 187–191) with recommendations focusing on issues such as these:

- avoiding absolute judgment limits
- designing to support top-down processing
- designing consistent with pictorial realism (Roscoe, 1968)
- minimizing information access cost
- providing predictive aiding
- ensuring consistency.

In a review of the literature on visual displays, Bennett et al. (1997) further characterized display design based on four approaches concerned with aesthetic, psychophysical, attentional, and problem-solving perspectives. This review provides illustrations of how detailed design concepts developed from each of these perspectives can contribute to design, such as this: "To support the extraction of low-level data, the graphical elements of the [configural] display must be made more salient perceptually through a variety of techniques, including emphasis of scale, spatial separation, and color-coding" (Bennett et al., 1993, p. 71).

The cognitive systems engineering and human-computer interaction literatures further emphasize consideration of mental models in the design of complex systems. These literatures stress the importance of developing an interface or representation that helps the user to develop a correct mental model of how a product or tool functions and provides him or her with guidance about how to interact with it to complete various tasks. Researchers in these domains also discussed the importance of developing the design so that the user will maintain a correct mental model of the world that is viewed (in part) using the tool as an interface (Van Der Veer & Melguizo, 2003). This emphasis on mental models applies to the design of representation aids to support knowledge-based processing to deal with novel scenarios.

A substantial literature under the heading of *heuristic analysis* provides guidance on developing effective display designs (Badre, 2002; Krug, 2000; Nielsen, 1994, 1997, 2000; Silver, 2005; Spool, Scanlon, Snyder, DeAngelo, & Schroeder, 1998). For instance, Smith, Stone, and Spencer (2006) reviewed heuristics such as these:

a. How is the "focus of attention influenced by the display? Does it support completion of the alternative goals that users may have when looking at that display?"
b. "Are external memory aids provided to help the user remember or determine what actions to take or to remember what steps in some process have already been completed?"
c. "Are the relationships among associated controls and displays indicated through some form of functional grouping?"
d. "Does the product look and behave consistently?"
e. "Is the navigation robust enough to support the easy completion of alternative tasks, while still clear enough to help the user navigate along the correct paths without getting lost?"
f. "Are landmarks provided to help the user remember where he or she is within the system (relative to the overall navigational structure), and to understand where he or she can go next to complete different tasks?" (p. 24.13)

Note that such heuristics were developed to support a variety of display design contexts, not with a focus on the design of representation aids as discussed in this review. However, many of these heuristics are applicable to the design of the specific features of representation aids within a display, as well as the selection of strategies for embedding representation aids within some larger system design.

In short, as noted in the introduction, the goal of this review is not to review the entire display design literature. Rather, it emphasizes the unique aspects of the literature that have developed within the cognitive systems engineering literature under the label of *representation aiding*. However, the details of a specific design still must be crafted and are critical to successful representation aiding. It is not enough simply to ensure that information on the relevant domain constraints is available somewhere within an interface. This information must be presented in a fashion that helps the user to make use of the appropriate skill-, rule- and knowledge-based processes when alternative scenarios arise.

REPRESENTATION AIDING: EMPIRICAL EVALUATIONS

As discussed earlier, the success of a particular representation aid depends on two critical aspects of the design:

- identification of the semantics of the domain and the constraints that need to be represented in the display to support skill-based, rule-based, and knowledge-based behaviors;
- crafting of the interface through which the design strategies based on representation aiding are realized in a specific implementation.

It is therefore important to keep in mind that both of these aspects of the design could affect the results of an evaluation of a given design that incorporates representation aiding. In addition, it is important to understand the focus of the evaluation itself.

Regarding the nature and focus of the evaluation, Rasmussen et al. (1994) provided a framework that consists of five levels for categorizing alternative types of evaluation settings. These boundary levels are defined in terms of the *boundary conditions* or *constraint envelopes* that are present.

In describing boundary conditions, Rasmussen et al. (1994) stated that "the innermost boundary [Level 1] corresponds to the evaluation of actor-related issues in an environment that corresponds most closely to the traditions of experimental psychology. The remaining boundaries successively 'move' the context further from the actor to encompass more and more of the total work content" (p. 205). Next, we describe these five levels briefly, then follow with the results from a set of specific evaluations of certain representation aiding designs, which are framed in terms of these five levels to identify the extent to which they can be used for generalization.

Boundary 1: Controlled Mental Processes

Rasmussen et al. (1994) described the evaluations performed at this level as controlled laboratory investigations. The goal is to investigate the relationships between specific design features and the basic capabilities and limitations of participants. The experimental tasks to be performed are somewhat artificial, in the sense that there is little "direct concern for the eventual contexts of the end users" (p. 218). Because of the simplicity of these tasks, the strategies required for their completion are well defined and extremely limited. In Rasmussen et al.'s words,

> The formulation of the subject's instruction is at the procedural level and is very explicit. It serves to define the constraint boundary around the experimental situation and isolate it from (1) the general, personal knowledge background, and performance criteria of the subject, and (2) any eventual higher level considerations within the experimental domain itself. (p. 218)

In terms of display design, evaluations conducted at this level often examine the relationship between particular visual features that have been used to encode information into a display or graph and how well participants can extract or decode this information. Thus, the tasks to be performed are defined in terms of the physical characteristics of the display itself; performance is usually measured in terms of the accuracy and latency of responses. Prototypical examples of evaluations performed at this level can be found in the work of Cleveland and his colleagues (Cleveland, 1985; Cleveland & McGill, 1985); they systematically varied the visual features employed in alternative representations and ranked participants' abilities to make discriminations using these visual features.

Boundary 2: Controlled Cognitive Tasks

This level of evaluation is designed to assess performance at experimental tasks that are approximations of those found in real-world domains. The focus is on isolated "decision functions, such as diagnosis, goal evaluation, planning and/or the execution of planned acts" (Rasmussen et al., 1994, p. 219). These experimental tasks are typically more complex than those found at the previous level of evaluation.

To complete these tasks, a participant must consider more than the physical characteristics of the display alone—that is, he or she must consider the information presented and determine what this information means in the context of the task to be performed. At this level of evaluation, a participant's general knowledge of the task and the particular strategies that he or she develops and employs will become more important (relative to Boundary 1). Therefore, individual differences will have a more pronounced influence on the levels of performance that are obtained.

Performance assessment typically involves measures other than accuracy and latency. Prototypical examples of display evaluations performed at this level can be found in MacGregor and Slovic (1986) and Goldsmith and Schvaneveldt (1984). MacGregor and Slovic employed a multicue probability judgment task that required participants to consider multiple cues (age, total miles, fastest 10 km, time motivation) with varying degrees of diagnosticity to predict the amount of time runners would take to complete a marathon. The display formats used to present this information were varied, and performance measures (e.g., an achievement index) were calculated and analyzed using Brunswick's (1956) lens model.

Boundary 3: Controlled Task Situation

In contrast to the previous boundary levels, Boundary 3 evaluations incorporate a very direct link between the experimental tasks to be performed and the specific tasks that exist in a particular real-world domain. Analyses of real-world scenarios are used to develop causal or mathematical simulations that capture some portion of their inherent complexity.

Correspondingly, the experimental tasks to be completed at this level will be more complex than those encountered in the previous two boundary levels. The tasks may involve consideration of physical/functional characteristics of the domain, competing goals, limited resources to achieve these goals, and performance trade-offs. As a result, an individual's general knowledge of the domain, specific knowledge of the task(s) to be performed, and efficiency of the strategies employed will play a more important role in the findings that are obtained. The measures that are used to assess performance will be defined by the domain itself and will therefore be relatively domain specific.

Prototypical examples of display evaluations performed at this level can be found in the work of Moray, Lootsteen, and Pajak (1986) and Mitchell and Miller (1983, 1986).

Boundary 4: Complex Work Environments—Microworlds

Rasmussen et al. (1994, p. 224) described this evaluation boundary in the following fashion:

Representation Aiding to Support Performance on Problem-Solving Tasks

A more recent category of experiments has been focused on human problem-solving behavior in complex simulated work environments in which the entire decision process is activated, including value formation, goal evaluation, and emotional factors.

Simulations at this level could certainly be more complex than those at the previous level. However, additional complexity alone is not the primary distinguishing factor. Rather, it is critical that the boundary conditions of the evaluation be set up to determine the influence of the interface on the participant's goal formulation and performance criteria. Typically, the participant is presented with a relatively open-ended task and will be free to formulate the task and goals on his or her own. For example, operators might be instructed to "run a power plant" in a full-scale simulator without explicit specification of the task situation.

Note that the utilization of work domain experts is essential for evaluation at this boundary level.

Boundary 5: Experiments in Actual Work Environments

At this boundary, the display or interface evaluation is conducted using real practitioners in the actual domain (field studies). Rasmussen et al.'s (1994) evaluation of a library information retrieval system, the BookHouse, is an excellent example. This system assists librarians and patrons in the selection of fiction books from a library. This system was evaluated in a public library during a six-month period. Numerous measures of performance were obtained, including "(a) a questionnaire, (b) on-line logging of all dialogue events (mouse clicks, etc.), (c) observation, and (d) interviewing by the librarians who (e) also kept a logbook with reports of user responses, system behavior, and so on" (p. 319).

As these descriptions indicate, moving from one boundary to another involves changes in many dimensions, including complexity, fidelity, knowledge, strategies, and values. Inherent trade-offs are involved when conducting evaluations at the various levels. The primary concerns center on experimental control and the generalization of results. At lower levels in the framework, more experimental control can be exerted. The tasks are well defined, the knowledge to complete them is straightforward, and the strategies that can be executed are limited. Thus, the chances of achieving statistically meaningful results are increased.

The situation is reversed at higher levels of evaluation. On the other hand, the chances that the results that are obtained will actually transfer to applied settings are increased as experiments are conducted at higher levels of evaluation.

To summarize, these five levels of evaluation represent a continuum of settings that needs to be considered in the evaluation of displays and interfaces that are intended for the real world. Different issues in design are addressed at each level. No single level is, a priori, more or less important than any other level. Regardless of the boundary level, effective performance will depend on whether or not the constraints introduced by the three system components mentioned earlier (i.e., domain, agent, interface) are well matched. Furthermore, the generalization of results between boundary levels will depend on extremely complex relationships among these three components.

The bottom line is that care must be taken in choosing a particular level to evaluate a display. Even more care must be taken when one attempts to generalize beyond the confines of the specific situational factors that were involved in a particular evaluation.

Evaluations of Designs Based on Representation Aiding

To illustrate these points, we next describe the results of several evaluations in the context of the framework outlined earlier.

Example 1. A number of factors in display design were evaluated in three empirical evaluations contrasting a configural display based on representation aiding concepts with other display designs (Bennett et al., 1993, 2000; Bennett & Walters, 2001). The factors included the general representational format employed (configural displays, bar graph displays, and digital displays) and a variety of display design techniques (color coding, perceptual layering, visual separation, graphical extenders, and display grids). All three display evaluations used the same experimental task environment: a part-task simulation of a steam generator in a nuclear power plant.

The evaluations were conducted at two of the levels outlined earlier. The experimental tasks in the Boundary Level 1 evaluations were defined in terms of the physical characteristics of the display: Participants were required to provide quantitative estimates of either individual variables (e.g., steam flow) or relationships between variables (e.g., mass balance). Performance on these tasks was graded with respect to how well the reported estimates corresponded to the displayed information (accuracy) and how long the estimates took (latency).

The second level of evaluation was conducted at Boundary Level 3. The tasks to be performed, and therefore the measurements that were obtained, were defined in terms of a part-task simulation as opposed to the display itself. Thus, the quality of the displays was determined by whether or not they assisted the agent in controlling the system under normal operating conditions (e.g., time on task, root mean square error) or detecting the presence of faults (e.g., sensitivity and false alarms) and compensating for them.

The task constraints associated with the Boundary 1 evaluation were simple: Provide a quantitative estimate of a variable or the difference between two variables. The display design techniques that contributed a matching set of constraints were effective in improving performance. More specifically, performance was improved to the extent that a technique contributed visual structure, allowing perceptual processes to be substituted for controlled cognitive processes. The digital value technique (the annotation of the analog geometrical format with digital values) provided the best support for the Boundary 1 task: The need for any visual estimates or mental computations was eliminated. In contrast, the visual structure provided by color-coding and layering techniques (i.e., chromatic and luminance contrast) did not eliminate the mental estimates or mental computations and therefore did not improve performance.

The task constraints associated with the Boundary Level 3 evaluation were much more complicated. To complete the system control and fault detection/compensation tasks successfully, the participant had to consider individual variables, higher-level properties/relationships, competing goals/constraints, and the physical/functional characteristics of the system. Far fewer significant effects were found at this boundary level because of the factors outlined earlier. The primary findings were that the presence of the analog configural display improved performance relative to the digital-only display (9 of the 10 significant results).

The analog configural format supported performance at Boundary Level 3 because it represented the critical domain semantics directly. For example, the critical system properties of energy and mass balance were represented by highly salient emergent features (i.e., the height and width of a rectangle). Additional emergent features (shape, size, and location of the rectangle) made it possible to view critical properties in the context of system goals (e.g., how close is the indicated level to the goal or trip set points?). In essence, participants could utilize powerful pattern recognition capabilities to assess current system state and, to some degree, to determine the correct control input.

In contrast, the digital-only display (this display had only digital values) imposed a truly severe set of constraints. The route to underlying meaning was much less direct. Participants could not use pattern recognition capabilities to complete tasks because the domain semantics (relationships, properties, goals, and constraints) were not directly visible in the digital format. Instead, the participants were forced to derive the current system state mentally using the digital values in conjunction with their conceptual knowledge about the system. (See Bennett & Flach, 1992; and Bennett et al., 1997, for a more detailed discussion of similar considerations.) Thus, the constraints introduced by the digital-only display made it much more difficult to assess system state, to determine appropriate control input, and to gauge the appropriateness of the system dynamics. As a result, performance suffered.

The collective results of these evaluations provide a fairly clear message: The evidence supporting the generalization of results between boundaries was extremely limited. There was only one significant contrast indicating that a single display manipulation produced improved performance at both boundaries, and this result failed to be replicated in the second experiment. The overall lack of generalization across boundaries is particularly striking given the five experiments, dozens of display manipulations, and hundreds of statistical comparisons.

The implication is twofold: First, designers must ask which of the boundary levels considered in the study best match the generalizations they wish to make. If the goal is to gain insight into how alternative display designs are likely to affect performance by trained practitioners on tasks such as fault diagnosis and system control in a process control system, then it could be argued that the Boundary Level 3 results in this study are more predictive than Boundary Level 1 results. This conclusion is supportive of the concepts underlying representation aiding—such as the representation of critical system properties like energy and mass balance by highly salient emergent features (i.e., the height and width of a rectangle), and the use of additional emergent features (shape, size, and location of the rectangle)—such that the critical properties can be viewed in the context of system goals. It suggests that these design features become more dominant determinants of performance when the system operator must consider individual variables, higher-level properties and relationships, competing goals and constraints, and the physical and functional characteristics of the system.

Example 2. Reising and Sanderson (2002) evaluated the design of a pasteurization microworld based on representation aiding and used configural displays that presented information on all the constraints identified in an abstraction hierarchy for that domain. This supported inquiries "at any level of abstraction or decomposition" (see Figure 3.6). They contrasted this with a mimic display. The participants were students (domain nonexperts) who were given only limited training on the use of these alternative interfaces. Thus, this would be categorized as a Boundary Level 3 evaluation. They concluded that "fault diagnosis was better with the ecological [representation aiding] interface than with a conventional (or mimic-only) interface" (p. 242).

Results on a follow-up questionnaire suggested that the reason was that although "the mimic display within the ecological interface was the most highly valued" (Reising & Sanderson, p. 242), the indication of boundary regions in the energy flow and goals displays were very important. Participants further indicated that although they perceived the straight line emergent features of the mass display to be useful for telling whether or not the mass flow was normal, this information was less important overall for control. The researchers also noted, however, that "some participants found certain configural displays hard to understand at first but learned to make good use of them"; participants made comments such as "At first, the mass and energy displays were hard to understand, but later it became one of my most reliable interfaces" (p. 243).

This Boundary Level 3 evaluation thus further supports the use of representation aids to support higher-level tasks (such as fault diagnosis) in a process control setting. However, remember that this is still just a Level 3 evaluation—meaning that generalizations to knowledgeable operators in the field should be made with caution—and that the focus was on scenarios that had been anticipated during the design.

Example 3. Burns et al. (2003) evaluated a representation aiding design for a network management system. This included several displays meant to support performance at different levels in the abstraction hierarchy described in Table 3.1. Figure 3.7 illustrates a polar display meant to support diagnosis of faults at the generalized function level.

The evaluation compared this tool against an industry tool (which provided a variety of displays based on tabular, timeline, and network topology formats) for both detection tasks and diagnosis tasks using students with a background in computer science and computer networks. For this Level 3 evaluation, detection times were faster using the industry display, but diagnoses were faster and more accurate using the displays based on representation aiding design strategies.

Example 4. In a final example of the evaluation of representation aiding, Jamieson (2002a, 2002b) conducted a Level 4 evaluation using an industry simulator of a petrochemical system with professional operators. New interfaces were integrated into this simulator, which was part of the operational system for the testing and training of operational staff (Miller & Vicente, 1998). The researchers studied 30 professional operators in this domain, who operated within this industry simulator using three different interfaces: the current display used in actual operations, a representation aiding display developed using a domain analysis based on ecological interface design methods, and a combination of this ecological interface with displays designed to support performance on specific tasks. This latter display is shown in Figure 3.8.

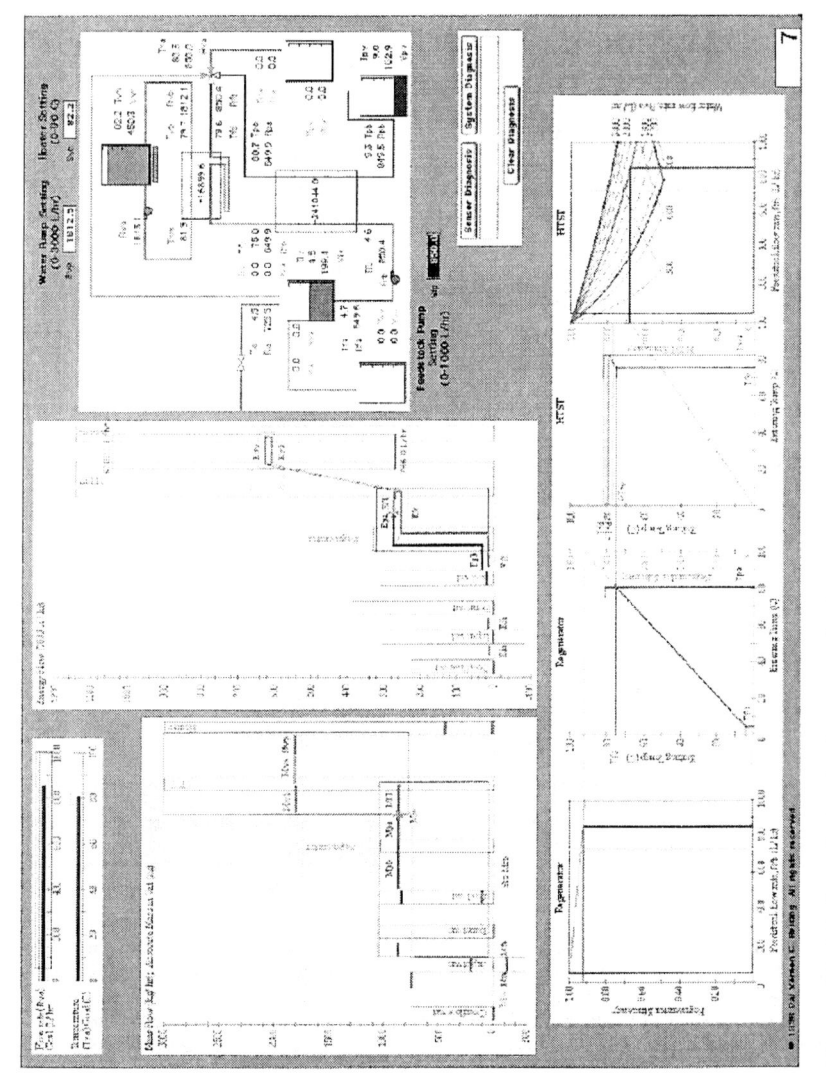

Figure 3.6. Interface design for the pasteurization system, reproduced at smaller size than viewed during actual operation to show the overall layout. From Reising and Sanderson (2002). Reprinted with permission from Human Factors, 44, 222–247. Copyright 2002 by the Human Factors and Ergonomics Society. All rights reserved.

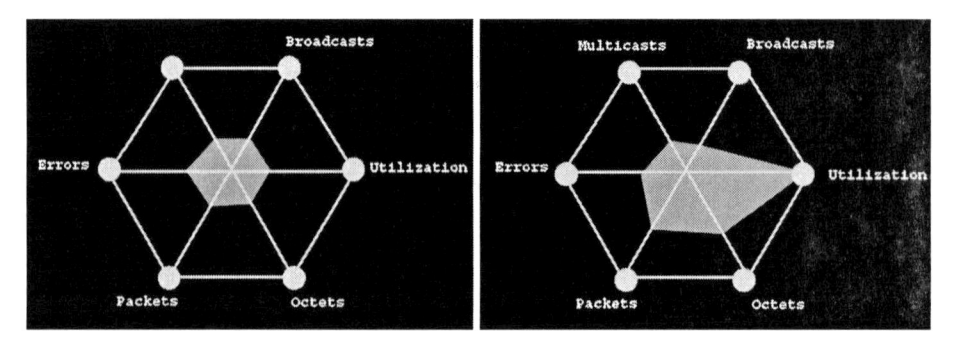

Figure 3.7. Polar display from a network management system to support diagnosis of faults at the generalized function level. The polar graphic shows the states of critical variables sensitive to network functioning along the different axes. Different shapes of the graphic are diagnostic of significantly different system states. Based on figure from Computer Networks, 43, Burns, C., Kuo, J., & Ng, S., Ecological interface design: A new approach for visualizing network management, 369–388, copyright 2003, with permission from Elsevier.

These participants were studied when operating the plant under normal and abnormal conditions. The results showed that the times for operators to complete their tasks were slowest using the current display and fastest for the combination of the ecological interface with displays designed to support performance on specific tasks, with average time for the ecological display in between. Only the difference between the current display and the combination of the ecological interface with displays designed to support performance on specific tasks was significant.

The results further demonstrated improved fault diagnosis with the design that was a combination of the ecological interface with the displays designed to support performance on specific tasks. There was no significant difference in the accuracy of fault diagnosis for the current design (the design used in actual operations) alone and the ecological display alone. These results regarding fault diagnosis applied both to scenarios considered during the design of the ecological interface and the task-based interface and to scenarios that were not anticipated during the design process.

Summary

Studies like those just reviewed, and others summarized in Burns and Hajdukiewicz (2004), provide support for the potential value of representation aiding as a design strategy to support the operation of complex systems for tasks such as fault diagnosis and process control. Although they do not provide data based on Boundary Level 5 evaluations, which would provide the greatest face validity for the results, there are findings that are supportive at Levels 3 and 4.

However, although all the aforementioned studies are supportive of the value of representation aiding as a design strategy, Jamieson's study also provides evidence that additional, complementary design strategies (such as the design of displays based on cognitive task analyses) may further enhance performance when combined with designs based on the principles discussed in this review on representation aiding.

Representation Aiding to Support Performance on Problem-Solving Tasks 103

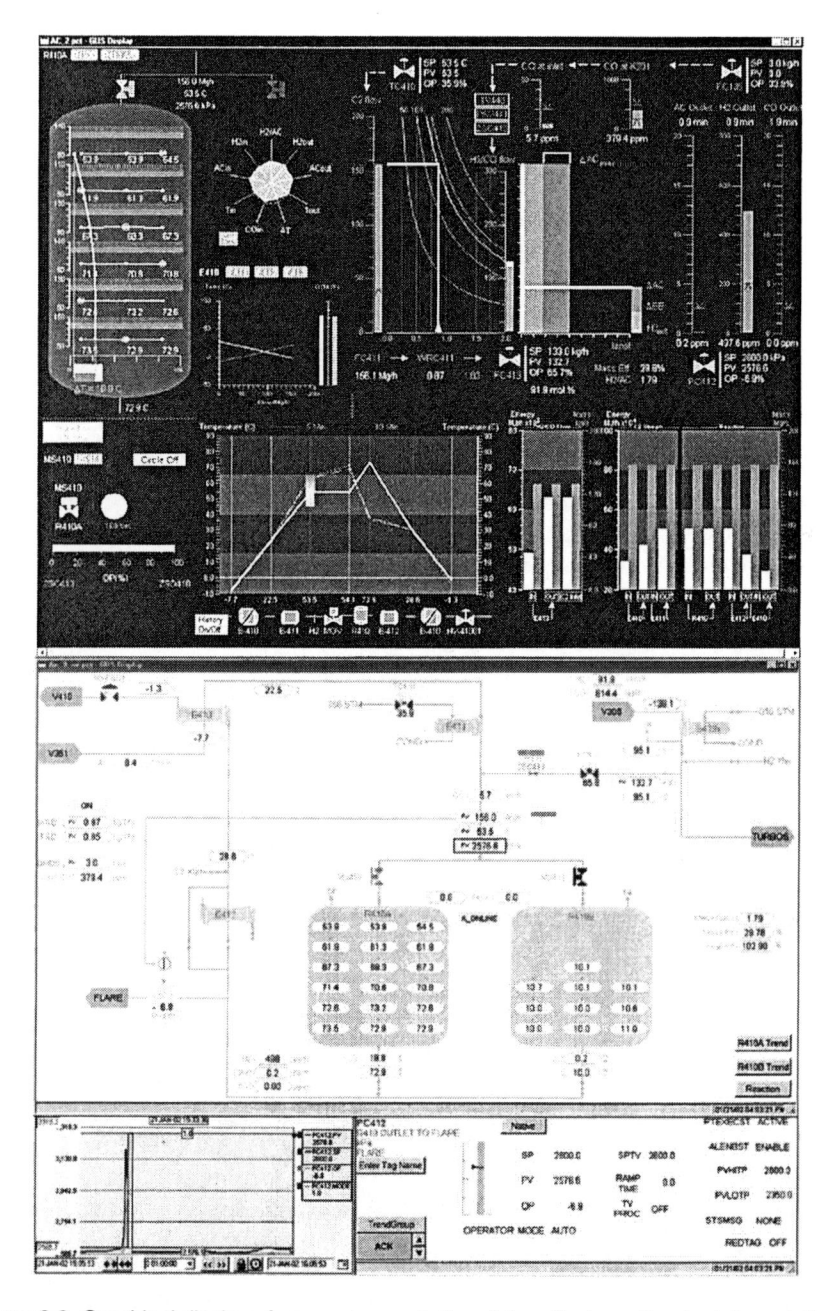

Figure 3.8. Graphical displays from a representation aiding display with a task-based display for process control of an acetylene hydrogenation reactor. The display presents information about the physical state of the system and its functioning, as well as information based on procedures to support specific tasks. From Jamieson, 2002a. Reprinted with permission from Proceedings of the Human Factors and Ergonomics Society 46th Annual Meeting, *2002.* Copyright 2002 by the Human Factors and Ergonomics Society. All rights reserved.

CONCLUSIONS AND FUTURE RESEARCH

The literature on representation aiding reviewed in this chapter is grounded on a model that identifies direct perception as a key process underlying expert performance. This model asserts that in many real-world tasks, experts become attuned to patterns in the environment that afford direct perception (Vicente & Rasmussen, 1990) of the diagnosis or solution to a problem, without the need for slower, controlled reasoning processes that must be applied to some internal representation of the problem.

Designs Based on Representation Aiding for Anticipated Scenarios

Based on this model, this literature provides the following guidance to designers: For known scenarios, develop representation aids that enable direct perception, allowing the user of the system to employ skill- and rule-based processes to conduct routine operations and to detect, diagnose, and deal with abnormalities. Such representation aids support performance when the system is operating correctly or when some malfunction arises, because they support direct perception of the diagnoses or responses necessary to maintain proper functioning and to respond to abnormalities.

This literature goes a step further and provides illustrations of how to develop such representations. One of the most significant examples is the use of configural displays (Bennett et al., 1993), in which the critical features that support direct perception emerge from carefully designed data displays that indicate the state of relevant domain properties or constraints.

There is strong empirical evidence that attention to this guidance on the use of representation aiding to support direct perception in known scenarios can lead to the design of more effective displays. These findings have repeatedly shown that in a variety of simulated worlds for different domains, for those scenarios that have been predicted by the designer, displays incorporating representation aiding lead to better fault diagnosis than do systems in use in these domains and better than other alternatives, such as mimic displays (Burns & Hajdukiewicz, 2004; Vicente, 2002).

Evidence regarding the use of representation aiding to support performance in anticipated scenarios would be further strengthened by future research if Boundary Level 5 evaluations (studying experienced practitioners at work using representation aiding designs for their everyday operations) were conducted and provided similar results.

Designs Based on Representation Aiding for Unanticipated Scenarios

The ecological interface design literature on representation aiding goes beyond helping designers develop displays to deal with anticipated scenarios, however. It starts with the premise that we need to develop decision support systems (Smith et al., 2006) that keep the human expert in the loop because of the potential brittleness of the technologies used to operate complex systems (Larson & Hayes, 2005; Smith et al., 1997). It then provides a method for doing so, first by identifying the critical constraints for the domain of interest (Vicente, 2002) and then by supporting direct perception to enable the user to directly perceive a constraint that has been violated.

In some cases, knowledge that a particular constraint has been violated may be sufficient to directly diagnose or solve the problem, even in an unanticipated scenario. In other cases, however, direct perception of which constraint or constraints have been violated helps focus the user's attention as controlled, knowledge-based processes are employed to reason about the underlying problem.

Thus, representation aiding makes the violation of critical constraints associated with the domain more salient via automatic perceptual processes. This, in turn, should make it easier to apply controlled processes to complete problem solving, with critical aspects of the problem solving that once required knowledge-based behavior translated into skill- and rule-based behaviors and thereby made more efficient. This is consistent with findings in the cognitive science literature regarding the value of diagrams for problem solving.

Scaife and Rogers (1996), for instance, discussed the value of diagrams for improving performance "through directing attention to key components that are useful or essential for different stages of a problem-solving or a learning task" (p. 207). Similarly, this is analogous to the representation of the digits game as Tic-Tac-Toe, as discussed earlier, in which the interface indicates key constraints and further serves as an "externalized mental model for problem solving" (Vicente, 2002, p. 64).

However, with regard to the use of representation aiding to make violations of domain constraints salient, note that we said they "should make it easier to apply controlled processes to complete problem solving." We include this caveat because there is a need for further research along three dimensions in order to further support this hypothesis.

First, more research is needed at Boundary Levels 3–5 using test cases in which scenarios that were not anticipated in the design are developed.

Second, more detailed models are needed to characterize the nature of the knowledge-based processes that are engaged when an unanticipated problem arises in a system that supports direct perception to detect violations of domain constraints, and that provides an externalized mental model to further support these controlled reasoning processes. Much of the literature on representation aiding was developed in contrast to traditional symbolic reasoning models based on models of direct perception to support performance instead of on models involving the applications of cognitive operators to internal models (Hegarty, 2004; Keodiger & Nathan, 2004; Koffka, 1935; Newell & Simon, 1972; Ormerod, 2002; Sternberg & Davidson, 1995; Zhang & Wang, 2005). Vicente (2002) noted, however, that "knowledge-based behavior involves serial, analytical problem solving based on a symbolic mental model" (p. 64). This raises an interesting question: To what extent does the cognitive science literature on problem solving as symbolic reasoning provide insights into knowledge-based processing when using representation aids to deal with an unanticipated scenario?

Third, at present, discussions of representation aiding provide examples of specific forms that can be used to support direct perception. There is, however, a need for stronger guidance and supporting empirical data on how to craft the interface for a specific application, especially with respect to supporting both skill- and rule-based behaviors for anticipated scenarios and knowledge-based processes for novel or unanticipated scenarios.

Additional Research Issues

Most work to date on representation aiding has focused on the operation of a system by a single person. In order to deal with cognitive complexity, however, many complex systems involve teams of operators. Thus, a wide-open research question concerns how and when to apply representation aiding concepts to support distributed work.

In addition, most research on representation aiding has compared either the design of an operational system or a system based on some so-called pure design strategy (such as the use of a mimic display) with a pure design based on representation aiding. An exception was the study by Jamieson (2002a, 2002b), described earlier, in which the best performance on fault diagnosis tasks was found to be provided by a design that combined displays based on representation aiding and task-based displays developed using a cognitive task analysis. This exception raises the interesting question of whether and when hybrid solutions are likely to be most effective and indicates the need to develop a model that helps to understand and guide such design decisions.

REFERENCES

Badre, A. N. (2002). *Shaping web usability, interaction design in context*. Boston: Pearson Education.
Bennett, K. (1992). Representation aiding: Complementary decision support for a complex, dynamic control task. *IEEE Control Systems*, August, 19–24.
Bennett, K., & Flach, J. (1992). Graphical displays: Implications for divided attention, focused attention, and problem solving. *Human Factors, 34*, 513–533.
Bennett, K., Nagy, A., & Flach, J. (1997). Visual displays. In G. Salvendy (Ed.), *Handbook of human factors and ergonomics* (pp. 659–696). New York: Wiley.
Bennett, K., Toms, M., & Woods, D. (1993). Emergent features and graphical elements: Designing more effective configural displays. *Human Factors, 35*, 71–97.
Bennett, K. B., & Fritz, H. I. (2005). Objects and mappings: Incompatible principles of display design—A critique of Marino and Mahan. *Human Factors, 47*, 131–137.
Bennett, K. B., Payne, M., Calcaterra, J., & Nittoli, B. (2000). An empirical comparison of alternative methodologies for the evaluation of configural displays. *Human Factors, 42*, 287–298.
Bennett, K. B., & Walters, B. (2001). Configural display design techniques considered at multiple levels of evaluation. *Human Factors, 43*, 415–434.
Brunswick, E. (1956). *Perception and the representative design of psychological experiments*. Berkeley: University of California Press.
Burns, C., & Hajdukiewicz, J. (2004). *Ecological interface design*. Boca Raton, FL: CRC Press.
Burns, C., Kuo, J., & Ng, S. (2003). Ecological interface design: A new approach for visualizing network management. *Computer Networks, 43*, 369–388.
Burns, C., Vicente, K., Christofferson, K., & Pawlak, W. (1997). Towards viable, useful and usable human factors design guidance. *Applied Ergonomics, 28*, 311–322.
Carswell, C. M., & Wickens, C. D. (1987). Information integration and the object display. *Ergonomics, 30*, 511–527.
Cleveland, W. (1985). *The elements of graphing data*. Monterey, CA: Wadsworth.
Cleveland, W. S., & McGill, R. (1985). Graphical perception and graphical methods for analyzing scientific data. *Science, 229*, 828–833.
Cole, W. (1986). Medical cognitive graphics. *Proceedings of the 1986 ACM Conference on Computer-Human Interaction* (pp. 91–95). New York: Association for Computing Machinery.
Flach, J. M., & Bennett, K. B. (1992). Graphical interfaces to complex systems: Separating the wheat from the chaff. *Proceedings of the Human Factors Society 36th Annual Meeting* (pp. 470–474). Santa Monica, CA: Human Factors and Ergonomics Society.

Goettl, B., Wickens, C., & Kramer, A. (1991). Integrated displays and the perception of graphical data. *Ergonomics, 34,* 1047–1063.

Goldsmith, T. E., & Schvaneveldt, R. W. (1984). Facilitating multiple-cue judgments with integral information displays. In J. C. Thomas & M. L. Schneider (Eds.), *Human factors in computer systems* (pp. 243–270). Mahwah, NJ: Erlbaum.

Hegarty, M. (2004). Diagrams in the mind and in the world: Relations between internal and external visualizations. *Lecture Notes in Artificial Intelligence, 2980,* 1–13.

Hutchins, E. (1995). *Cognition in the wild.* Cambridge: MIT Press.

Jagacinski, R. J., & Flach, J. M. (2003). *Control theory for humans: Quantitative approaches to modeling performance.* Mahwah, NJ: Erlbaum.

Jamieson, G. (2002a). Empirical evaluation of an industrial application of ecological interface design. *Proceedings of the Human Factors and Ergonomics Society 46th Annual Meeting* (pp. 536–540). Santa Monica, CA: Human Factors and Ergonomics Society.

Jamieson, G. (2002b). *Ecological interface design for petrochemical process control: Integrating task and system-based approaches* (CEL-02-01). Cognitive Engineering Laboratory, University of Toronto.

Keodiger, K., & Nathan, M. (2004). The real story behind story problems: Effects of representations on quantitative reasoning. *Journal of the Learning Sciences, 13,* 129–164.

Koffka, K. (1935). *Principles of Gestalt psychology.* New York: Harcourt, Brace and World.

Kotovsky, K., & Simon, H. (1990). Why some problems are really hard: Explorations in the problem space of difficulty. *Cognitive Psychology, 22,* 143–183.

Krug, S. (2000). *Don't make me think—Common sense approach to web usability.* Indianapolis: New Riders.

Larson, A., & Hayes, C. (2005). An assessment of WEASEL: A decision support system to assist in military planning. *Proceedings of the Human Factors and Ergonomics Society 48th Annual Meeting* (pp. 287–291). Santa Monica, CA: Human Factors and Ergonomics Society.

MacGregor, D., & Slovic, P. (1986). Graphic representation of judgmental information. *Human-Computer Interaction, 1,* 179–200.

Miller, C., & Vicente, K. (1998). *Abstraction decomposition space analysis for NOVA's E1 acetylene hydrogenation reactor* (CEL-98-09). Cognitive Engineering Laboratory, University of Toronto.

Mitchell, C. M., & Miller, R. A. (1983). Design strategies for computer-based information displays in real-time control systems. *Human Factors, 25,* 353–369.

Mitchell, C. M., & Miller, R. A. (1986). A discrete control model of operator function: A methodology for information display design. *IEEE Transactions on Systems, Man and Cybernetics, 16,* 343–357.

Moray, N., Lee, J., Vicente, K. J., Jones, B. G., & Rasmussen, J. (1994). A direct perception interface for nuclear power plants. *Proceedings of the Human Factors and Ergonomics Society 38th Annual Meeting* (pp. 481–485). Santa Monica, CA: Human Factors and Ergonomics Society.

Moray, N., Lootsteen, P., & Pajak, J. (1986). Acquisition of process control skills. *IEEE Transactions on Systems, Man and Cybernetics, 16,* 497–504.

Newell, A., & Simon, H. A. (1972). *Human problem solving.* Englewood Cliffs, NJ: Prentice Hall.

Nielsen, J. (1994). *Usability engineering.* San Francisco: Morgan Kaufmann.

Nielsen, J. (1997). Usability testing. In G. Salvendy (Ed.), *Handbook of human factors and ergonomics* (pp. 1543–1567). New York: Wiley.

Nielsen, J. (2000). *Designing web usability: The practice of simplicity.* Indianapolis: New Riders.

Norman, D. A. (1993). *Things that make us smart: Defending human attributes in the age of the machine.* New York: Pegasus.

Norman, D. A. (2002). *The design of everyday things.* New York: Doubleday.

O'Malley, C., & Draper, S. (1992). Representation and interaction: Are mental models all in the mind? In Y. Rogers, A. Rutherford, & P. Bibby (Eds.), *Models in the mind* (pp. 73–92). London: Academic.

Ormerod, T. (2002). Dynamic constraints in insight problem solving. *Journal of Experimental Psychology—Learning, Memory and Cognition, 28,* 791–799.

Perkins, D. (2000). *Archimedes bathtub: The art and logic of breakthrough thinking.* New York: W. W. Norton.

Pomerantz, J. (1986). Visual form perception: An overview. In H. C. Nusbaum & E. C. Schwab (Eds.), *Pattern recognition by humans and machines Vol. 2: Visual perception* (pp. 1–30). Orlando, FL: Academic.

Rasmussen, J. (1983). Skills, rules, knowledge, signals, signs and symbols, and other distinctions in human performance models. *IEEE Transactions on Systems, Man and Cybernetics, 13,* 257–266.

Rasmussen, J., Pejtersen, A. M., & Goodstein, L. P. (1994). *Cognitive systems engineering.* New York: Wiley.

Rasmussen, J., & Vicente, K. J. (1989). Coping with human errors through system design—Implications for ecological interface design. *International Journal of Man-Machine Studies, 31*, 517–534.

Reising, D., & Sanderson, P. (2002). Ecological interface design for Pasteurizer II: A process description of semantic mapping. *Human Factors, 44*, 222–247.

Roscoe, S. (1968). Airborne displays for flight and navigation. *Human Factors, 10*, 321–332.

Scaife, M., & Rogers, Y. (1996). External cognition: How do graphical representations work? *International Journal of Human-Computer Studies, 45*, 185–213.

Shiffrin, R., & Schneider, W. (1977). Controlled and automatic information processing: II. Perceptual learning, automatic attending, and a general theory. *Psychological Review, 84*, 127–190.

Silver, M. (2005). *Exploring interface design.* Clifton Park, NY: Delmar Learning.

Simon, H., & Hayes, J. (1976). The understanding process: Problem isomorphs. *Cognitive Psychology, 8*, 165–190.

Skirka, L., Mosier, K., & Burdick, M. (1999). Does automation bias decision making? *International Journal of Human-Computer Systems, 51*, 991–1006.

Smith, P. J., Geddes, N., & Beatty, R. (2006). Human-centered design of decision support systems. In A. Sears & J. Jacko (Eds.), *Handbook of human-computer interaction* (2nd ed., pp. 656–675). Mahwah, NJ: Erlbaum.

Smith, P. J., McCoy, E., & Layton, C. (1997). Brittleness in the design of cooperative problem-solving systems: The effects on user performance. *IEEE Transactions on Systems, Man and Cybernetics, 27*, 360–370.

Smith, P. J., Stone, R. B., & Spencer, A. (2006). Design as a prediction task: Applying cognitive psychology to system development. In W. S. Marras & W. Karwowski (Eds.), *Handbook of industrial ergonomics* (2nd ed., pp. 24.1–24.18). New York: Marcel Dekker.

Spool, J., Scanlon, T., Snyder, C., DeAngelo, T., & Schroeder, W. (1998). *Web site usability: A designer's guide.* San Francisco: Morgan Kaufmann.

Sternberg, R. J., & Davidson, J. (Eds.). (1995). *The nature of insight.* Cambridge: Bradford Books/MIT Press.

Tufte, E. (1983). *The visual display of quantitative information.* Cheshire, CT: Graphics Press.

Tufte, E. (1990). *Envisioning information.* Cheshire, CT: Graphics Press.

Tufte, E. R. (1997). *Visual explanations.* Cheshire, CT: Graphics Press.

Van Der Veer, G., & Melguizo, M. (2003). Mental models. In A. Sears & J. Jacko (Eds.), *Handbook of human-computer interaction* (pp. 52–80). Mahwah, NJ: Erlbaum.

Vicente, K. (1991). *Supporting knowledge-based behavior through ecological interface design* (Tech. Report EPRL-91-1). Urbana-Champaign, IL: Engineering Psychology Research Laboratory and Aviation Research Laboratory, University of Illinois.

Vicente, K. (1999). *Cognitive work analysis: Toward safe, productive, and healthy computer-based work.* Mahwah, NJ: Erlbaum.

Vicente, K., & Rasmussen, J. (1990). The ecology of human-machine systems II: Mediating "direct perception" in complex work domains. *Ecological Psychology, 2*, 207–249.

Vicente, K., & Rasmussen, J. (1992). Ecological interface design: Theoretical foundations. *IEEE Transactions on Systems, Man and Cybernetics, 22*, 589–606.

Vicente, K. J. (2000). Work domain analysis and task analysis: A difference that matters. In J. Schraagen, S. Chipman, & V. Shalin (Eds.), *Cognitive task analysis* (pp. 101–118). Mahwah, NJ: Erlbaum.

Vicente, K. J. (2002). Ecological interface design: Progress and challenges. *Human Factors, 44*, 62–78.

Wickens, C., & Andre, A. (1990). Proximity compatibility and information display: Effects of color, space, and objectness on information integration. *Proceedings of the Human Factors and Ergonomics Society 32nd Annual Meeting* (pp. 1335–1339). Santa Monica, CA: Human Factors and Ergonomics Society.

Wickens, C. D., Lee, J., Liu, Y., & Gordon-Becker, S. (2004). *An introduction to human factors engineering* (2nd ed.). Upper Saddle River, NJ: Pearson/Prentice Hall.

Woods, D. D. (1991). The cognitive engineering of problem representations. In G. R. S. Weir & J. L. Alty (Eds.), *Human-computer interaction and complex systems* (pp. 169–188). London: Academic Press.

Woods, D. D., & Roth, E. M. (1988). Cognitive systems engineering. In M. Helander (Ed.), *Handbook of human-computer interaction* (pp. 1–41). Amsterdam: Elsevier.

Zachary, W. (1986). A cognitively based functional taxonomy of decision support techniques. *Human-Computer Interaction, 2*, 25–63.

Zhang, J. (1997). The nature of external representations in problem solving. *Cognitive Science, 21*, 179–217.

Zhang, J., & Norman, D. (1994). Representations in distributed cognitive tasks. *Cognitive Science, 18*, 87–122.

Zhang, J., & Wang, H. (2005). The effect of external representations on numeric tasks. *Quarterly Journal of Experimental Psychology Section A—Human Experimental Psychology, 58*, 817–838.

CHAPTER 4

Usability Assessment Methods

By Joseph S. Dumas & Marilyn C. Salzman

Usability assessment methods evolved from traditional human factors/ergonomics techniques beginning in the early 1980s. Following a brief historical introduction, we describe the four major categories of these methods: usability testing; usability inspections; surveys, interviews, and focus groups; and field methods. We describe the basic techniques of these methods, their strengths and weaknesses, their measurement characteristics, their validity and reliability, and how they are being applied to product development and assessment. We discuss the new challenges to usability assessment: (a) using the methods thoughtfully, (b) expanding the scope of assessment beyond usability and productivity to affect end user experience, (c) and emphasizing design solutions to problems in addition to simply finding the problems themselves.

One can find precursors to usability and its evaluation in the human factors/ergonomics (HF/E) literature before 1980 (e.g., Chapanis, 1959; Kirk & Ridgeway, 1971), but the emergence of usability assessment as a discipline really began in the early 1980s. It was in 1982 that the first articles on usability appeared at the annual Human Factors Annual Meeting (Goodwin, 1982). That year also marked the first conference that was wholly dedicated to the design and evaluation of human-computer interaction, "Human Factors in Computer Systems" held in Gaithersburg, Maryland. There, Lewis and Mack (1982) described the use of think-aloud protocols for diagnosing usability problems; Sheppard, Kruesi, and Bailey (1982) looked at both performance and preference in evaluating documentation; and Roberts and Moran (1982) introduced an "easy-to-use" method to evaluate text editors. Many believe that it was at this conference that the usability profession was born.

During the 1980s, most of the assessment methods we discuss here—usability testing, interviews, surveys, inspections—were formulated and introduced into the design process. By the early 1990s, the first didactic texts describing usability assessment methods appeared (Dumas & Redish, 1993; Hix & Hartson, 1993; Nielsen, 1993; Rubin, 1994; Shneiderman, 1987). Throughout the 1990s, usability assessment methods were widely disseminated and refined to make them more cost-effective, easier to use, and productive (e.g., Bias & Mayhew, 1994; Brooke, 1996; Nielsen, 1994; Olson & Moran, 1995; Virzi, 1992). The usability assessment toolkit was also expanded to include a variety of field methods, which afforded more naturalistic study of designs in use (e.g., Beyer & Holtzblatt, 1998; Hughes, King, Rodden, & Andersen, 1995; Wixon & Ramey, 1996). With the turn of the 21st century came increasing attention on leveraging and refining methods to better support the full design cycle, accelerate design creativity, and deliver value to users and businesses (Blythe & Wright, 2003; Jordan, 2000).

Today, the concept of usability is pervasive. Usability can be found in most dictionaries, is sometimes referenced in casual conversations, and is used to market products. There is even an international standard that defines *usability* as "[the] extent to which a product can be used by specified users to achieve specified goals with effectiveness, efficiency and satisfaction in a specified context of use" (ISO DIS 9241–11, 1998, p. 2). *Effectiveness* is defined as "accuracy and completeness with which users achieve specified goals"; *efficiency* as "resources expended in relation to the accuracy and completeness with which users achieve goals"; and *satisfaction* as "freedom from discomfort, and positive attitudes towards the use of the product" (p. 2).

Since the year 2000, however, there has been a growing sense that this so-called traditional definition of usability is limiting, especially as usability professionals have sought to improve the efficiency and effectiveness of their methods, to integrate usability assessment into earlier stages of design, and to cost-justify their practices. These efforts have led usability professionals to realize that task effectiveness, efficiency, and satisfaction are only part of the story; affective aspects of the user's interaction are also critically important to ensuring the success of a product. The result has been an expansion of what usability professionals focus on and, unfortunately, a clouding of what is meant by *usability* and *usability assessment*.

Some have advocated expanding the definition of *usability*. For example, on her Web site and in several articles and books, Quesenberry (e.g., 2004, 2005) broadens the ISO definition by adding *engaging*: "how pleasant, satisfying, or interesting an interface is to use" (Quesenberry, 2004, p. 5). Other experts have also advocated looking beyond traditional views of usability to consider "user experience," which is shaped not only by usability but by aesthetic, emotional, social, and business factors (Hancock, Pepe, & Murphy, 2005; Jordan, 2002; Karat, 2003; Teague & Whitney, 2002). For example, both Jordan and Hancock et al. draw on Maslow's hierarchy of needs to create a hierarchy for *consumer needs*. In this case, functionality is essential, usability is important (and now expected), and pleasurable interactions are what users really want. Hancock et al. have called this concept *hedonomics*.

We share the view that usability is the attribute of a product that makes it easy to understand and use, and that other product attributes—aesthetics, desirability, ability to motivate—combined with individual, contextual, and social factors help to create a user's experience. This broadened view of what it takes for a product to be successful has had two important implications for usability practice. First, traditional usability measures are being adapted to assess the broader notions of usability and the user experience. Second, new methods are being used to supplement the more traditional ones (e.g., Karat, 2003; Murphy, Stanney, & Hancock, 2003; Pagulayan, Keeker, Wixon, Romero, & Fuller, 2003). Usability practitioners are supplementing traditional measures with value-based metrics and methods drawn from the marketing, anthropology, and psychology disciplines. Questions such as "Is it fun?" "Is it motivating?" "Does it provide enough variety (as opposed to consistency)?" are a few examples of what usability practitioners are asking today.

In this chapter, we discuss traditional assessment methods: usability testing, expert reviews, surveys, interviews, focus groups, and field methods. We conclude by highlighting how these methods are being extended to examine the broader user experience.

We also note a few promising new methods that warrant further consideration beyond usability.

USABILITY TESTING

Usability testing is an empirical method for uncovering the strengths and weaknesses in the usability of a product or system and, less commonly, for measuring or comparing its usability. There are three book-length descriptions of how to apply the method (Barnum, 2002; Dumas & Redish, 1999; Rubin, 1994). Usability testing has sometimes been referred to as *think-aloud testing* because of the importance of thinking aloud to the objectives of testing.

Valid usability tests have the following six characteristics:

1. The focus is on the usability of the product. Usability tests are not useful for answering marketing questions, such as how many customers will buy a product, nor are they research experiments designed to add to scientific knowledge.
2. The participants are end users or potential end users of the product. This characteristic separates testing from inspection methods in which end users are not included.
3. The participants perform tasks with the product. The tasks are selected because they are typical or are critical to product use, or because they probe an expected strength or weakness of product use. The use of tasks separates testing from other user-based methods such as surveys.
4. The participants are usually asked to think aloud as they perform tasks or immediately afterward.
5. The data are recorded and analyzed. Data typically include qualitative measures such as user satisfaction ratings and quantitative measures such as task success and error rates.
6. The results and often recommendations for improvements are communicated to appropriate audiences, such as user interface designers, product managers, and programmers.

How Is Usability Testing Used?

As with all usability assessment methods, early application in the development cycle is now considered the ideal. In its early days, testing was compared with quality assurance testing, which often is used to find bugs when a product design nears completion. This view of usability testing changed as professionals realized that major structural problems uncovered late in development may not be fixed because of time constraints.

Testing conducted throughout the product's development in order to guide design is called *formative*. Those tests focus on the usability strengths and weaknesses and how they can be improved. Testing conducted at or near the end of product development is called *summative*. Those tests focus on measuring the product's efficacy and whether it can be used as designed. The development of software prototyping tools has made it possible to conduct formative tests earlier in the development process. Studies examining how prototype fidelity affects product assessment (Cantani & Biers, 1998; Virzi, Sokolov, & Karis, 1996; Walker, Takayama, & Landay, 2002) offer important guidance for how best to leverage prototypes. In general, participants are not biased against a product because it does not have a polished look or complete interactivity, and, some practitioners believe, users may be more open to providing constructive feedback if the prototypes are of lower fidelity (Snyder, 2003).

Early concept testing. Particularly with new products, there is a short period during which alternative user interface design concepts are explored. In the past, this exploration was done "in the back room" by one or a few designers with no user input. With the development of easy-to-use sketching tools and paper prototypes, it has become possible to explore alternative designs with users before the designs become final, and it is difficult to discard them (Snyder, 2003). Concept tests are formative and may have only a few tasks, and often the test administrator sits with participants and actively probes the concepts with them. The decision about which concept or part of a concept to keep developing is based more on the administrator's sense of the participants' confusion or lack of it rather than on measures such as task time and success.

Diagnostic testing. As the product design develops, the testing process becomes more structured and quantitative. Tasks are sampled to cover the core functions and to probe problem areas. Although session length can vary based on objectives and task complexity, as a rule, typically there are 12–18 tasks with a session length of 90–120 minutes. Participants are selected with a recruiting screener, and there are six to eight users in a market segment.

There may be several iterations of testing as the design develops. More quantitative measures, such as average task completion or the number of assists from the administrator, may be used to document usability problems. There is also some form of subjective measure, often a set of Likert scale rating forms. The final product of the test is a list of strengths and weaknesses and recommendations for improvements. A meeting or briefing is commonly used to communicate results. Sometimes a written report is created, but increasingly it may be used only to archive the results for future reference.

Benchmark and comparison testing. Not all usability tests have a diagnostic purpose. Some have a measurement focus, either for benchmarking a product's usability or for comparing the usability of different products or versions. These performance-based tests tend to be summative and more like research experiments than typical diagnostic tests.

At present, the usability specialist's interpretation of summative usability test data plays a large role in evaluating the product's usability. Experienced usability professionals believe they can make a relatively accurate and reliable assessment of a product's usability when the product is stable, the number of participants is sufficiently large (larger than for most diagnostic tests), participants are discouraged from making lengthy comments or evaluative statements in their think-aloud protocol, and the test administrator makes minimal interruptions to the flow of tasks.

A noteworthy supplement to the usability professional's judgment for measurement-based tests is McGee's (2003, 2004) Usability Magnitude Estimation (UME) technique. Based on a traditional psychological measurement technique—magnitude estimation—the UME helps usability professionals assess and compare product attributes, tasks, interactions, and other usability attributes on a single continuum. This method produces a measure of product usability that is user based—a ratio scale—and does not require large samples of respondents.

An important variation on the benchmark usability test is one whose focus is primarily on *comparing* usability. Here the intention is to measure how usable a product is relative to some other product or to an earlier version of itself.

There are two types of comparison tests: (a) a diagnostic comparison test focused on finding as much as possible about a product's usability relative to a comparison product, and (b) a summative comparison test intended to produce results that measure comparative usability and/or to find the winner over the others. In both tests, there are two important considerations: The test design must provide a valid comparison between the products, and the selection of test participants, the tasks, and the way the test administrator interacts with participants must not favor any of the products.

As soon as the purpose of the test moves from diagnosis to comparison measurement, the test design moves toward becoming more like a research design, focusing on two questions: Will each participant use all of the products, some of the products, or only one product? How many participants are enough to detect a statistically significant difference?

In the research methods literature, a design in which participants use all the products is called a *within-subjects* design, and in a *between-subjects* design, each participant uses only one product. If one uses a between-subjects design, one avoids having any contamination from product to product, but one needs to make sure that the groups who use each product are equivalent in important ways, and the sample size must increase.

Because it is difficult to match groups on all the relevant variables, between-subjects designs need to have enough participants in each group to wash out any minor differences. An important concern to be aware of in the between-subjects design is the situation in which one of the participants in a group is especially good or bad at performing tasks; Gray and Salzman (1998) called this the *wildcard effect*. If the group sizes are small, one superstar or dud could dramatically affect the comparison. With larger numbers of participants in a group, the wildcard has a smaller impact on the overall results. This phenomenon is one of the reasons that summative tests have larger sample sizes than do diagnostic tests. The exact number of participants depends on the design and the variability in the data. Sample sizes in summative tests are closer to 20 in a group than the 5–8 common in diagnostic tests.

If one uses a within-subjects design in which each participant uses all the products, it eliminates the effect of groups not being equivalent and can have a smaller sample. However, one then has to worry about other problems, the most important being order and sequence effects. (See Dumas [1998] for rules on counterbalancing.) One also has to be concerned about the test session becoming so long that participants get tired.

Additional Considerations for Usability Testing

In addition to testing basics, some special issues have surfaced in the literature: the variations and importance of thinking aloud, the controversy over number of participants, the new world of remote testing, the reliability of usability testing, and the focus on fixing problems.

The variations and importance of thinking aloud. Although concurrent thinking out loud is normally done as part of a diagnostic usability test, it is really a method of its own. It has been used in conjunction with many other methods. Thinking aloud has been used in psychological research since the turn of the 20th century, but it is best known as a cognitive psychology method for studying short-term memory (Ericsson & Simon, 1993). Thinking aloud provides usability testing with most of its drama. Without thinking aloud, it is unlikely that usability testing would have become the most influential usability engineering method (Rosenbaum, Rohn, & Humburg, 2000).

When usability testing was being developed, thinking aloud was borrowed from cognitive psychology without much thought. It was not until shortly after 2000 that usability specialists began to look at it more closely. Independently, Boren and Ramey (2000) and Dumas (2001) went back to examine what Ericsson and Simon described as the method used in research and whether usability practitioners were really following that method. The two reviews showed that the descriptions of how to use the think-aloud method provided to usability testing practitioners by Dumas and Redish (1999) and Rubin (1994) were in direct contradiction to the instructions used in cognitive psychology research. In research studies, participants are *discouraged* from reporting on feelings or expectations or making any verbal diversions over and above the content of their actions. In usability testing, participants are *encouraged* to report on their feelings and expectations and additional relevant issues. In addition, Boren and Ramey observed usability test sessions at two labs and found a great deal of variability in how thinking aloud was implemented.

A few research studies have been done on the think-aloud method in a usability testing context. Krahmer and Ummelen (2004) compared typical usability testing think-aloud instructions with the instructions proposed by Ericsson and Simon (1993) for research studies and found that the research instructions do not work well in a testing context. Ebling and John (2000) traced usability problems that were identified in a usability test back to their source in the test measures. They found that more than half the problems identified in their test came from the think-aloud protocol *alone*. Their study supplements an earlier one by Virzi, Sorce, and Herbert (1993), who showed that fewer problems are identified when the participants do not think aloud.

For the practitioner, this research confirms the importance of thinking aloud as a method to catch the eye and ear of visitors as well as to uncover usability problems.

Are five enough? The controversy over number of participants. Part of the popularity of usability testing comes from its ability to find usability problems with only a few participants. Anyone who watches multiple test sessions with the same set of tasks notices that the same issues begin to repeat and that, with the same user population, somewhere in the 5–8 test participant range it begins to be unproductive to test more participants. So it was with great joy that testers greeted the research studies by Virzi (1990, 1992) showing that 80% of the total number of usability problems that will be uncovered by as many as 20 participants will be found by as few as 5 and that those 5 will uncover most of the problems judged as severe. This finding has been confirmed several times (Faulkner, 2003; Law & Vanderheiden, 2000). Practitioners who are confident that it is normally a waste

Usability Assessment Methods   115

of resources to run additional participants continue to conduct diagnostic tests with small numbers of participants.

There have been some challenges to the generality of this finding, most notably by Lewis (1994, 2001). The challenges make the reasonable case that not all tests are the same. A moderately complicated product being tested for the first time might indeed yield most of its problems to 5–8 participants. But what about a product that is being retested after most of its problems have been fixed? Or a small piece of a large product? Or what about a very large and complicated product? In each of these cases, one might expect that it would take more participants to find most of the problems. Lewis (2001) detailed a formula for determining how many participants are needed for a variety of situations.

The new world of remote testing. A relatively recent innovation in usability testing is the ability to test when the administrator and the participant are not colocated. There are several advantages to remote testing: Testers can reach a larger population of participants because they are not limited to the local area; it is easier to get participants to volunteer because they do not have to travel; participants work at their desk in their own work environments; the test does not require a usability lab; and often participants do not have to be compensated (Perkins, 2001).

In the past, technology made it difficult to conduct remote sessions (Dumas, 2003), but that is no longer true. The emergence of faster processors, increased storage capacity, the Internet, collaborative sharing tools, broadband connections, and recording software make remote testing feasible.

Remote testing takes two forms: (a) *synchronous*, in which the administrator and the participant work together, usually communicating over the phone, sharing their desktops via collaborative software; and (b) *asynchronous*, in which participants work without the direct guidance of an administrator. Each of these variations has its strengths and weaknesses.

Synchronous testing has the same time limitations as laboratory testing; thus, it typically has the same number of participants as used in lab testing. Unless a special Web camera is used, the administrator cannot see the participant. We do not yet know what the impact of not seeing the participant is, but one laboratory study indicates that administrators judge usability problems as less severe when they cannot see the participant's face (Lesaigle & Biers, 2000).

One type of asynchronous testing involves a participant with two browser windows, one with the product in it and the second with the instructions for the participant to follow. The instructions window includes the tasks to be attempted, buttons to click at the beginning and end of a task, a free-form comment area, and questions or ratings to answer during or after each task. One advantage of asynchronous testing is a larger sample size. Tullis, Flieshman, McNulty, Cianchette, and Bergel (2002) were able to test 88 participants over a short period. The disadvantage is that one cannot see or interact directly with participants. In the Tullis et al. study, participants provided a good deal of feedback in the free-form comment field, which provided insight into the usability problems with the product. Tullis et al. reported no substantial difference between asynchronous testing and laboratory testing in terms of performance measures and the number and types of problems identified.

An additional form of asynchronous testing can be done by instrumenting a Web site. The instrumentation captures key data and can prompt users for answers and comments via pop-up forms (Hong & Landay, 2001).

Although there is very little research on the strengths and weaknesses of remote testing relative to laboratory testing, its advantages make it likely to increase in popularity.

The reliability of usability testing. In the early 1990s, several studies were conducted to understand the reliability of inspection methods relative to usability testing. (We discuss these in the section on inspections method.) However, these studies revealed little about the reliability of usability testing itself. Jacobsen, Hertzum, and John (1998) were the first to conduct a research study on reliability in usability testing—that is, how evaluators differ when analyzing the same usability test sessions. Four evaluators—all usability specialists—independently analyzed the same set of videotapes from four usability test sessions. Each session involved a user thinking out loud while solving tasks in a multimedia authoring system. As many as 46% of the problems were uniquely reported by single evaluators and 20% by only two evaluators. Consequently, the results suggested there is a substantial evaluator effect in think-aloud studies.

A series of studies conceived by Rolf Molich have pursued this issue of reliability of testing further (Molich et al., 1998; Molich, Meghan, Ede, & Karyukin, 2004; Molich & Dumas, in press). Known as the Comparative User Evaluation (CUE) studies, these papers report studies in which multiple independent usability tests of the same product are done by professional usability teams. The CUE-1 and CUE-2 studies (Molich et al., 1998, 2004) documented differences in reported problems, methods, tasks, and usability reports. In the CUE-4 study, nine teams conducted usability tests of the same hotel reservations site (Molich & Dumas, in press). These teams were chosen because they included well-known and well-published usability professionals. The teams found a total of 340 usability issues, but the overlap between the usability problems was disappointing. Not one of the serious usability problems was found by all nine of the usability test teams. The average percentage overlap between pairs of teams that conducted usability testing was only 11.5%. Two of the testing teams started with the same task set, though one of them did modify some tasks. Those two teams did find more problems in common than the average pair (24%) but not as many as would be expected if dissimilar tasks were the main cause of the unreliability.

Hertzum and Jacobsen (2001) did an analysis of comparative studies of testing (and inspections) to date and termed their lack of reliability the *evaluator effect.* They recommended multiple independent evaluations regardless of which assessment method is used.

Although there are now half a dozen studies showing similar results, there is a good deal of scepticism about the results of these studies. Many usability specialists believe that their own tests find most of the important usability problems. Because the testing teams in the CUE studies had the freedom to choose their own tasks and because the teams could *not* interact directly with developers during or after testing, practitioners who can work with developers believe they are finding most of the important problems. At this time, not much is known about the conditions that would lead to more reliable tests. From the CUE studies one can draw two conclusions: (a) usability specialists should not conduct

Usability Assessment Methods

assessments when they cannot interact with developers during the assessment, and (b) there are hundreds of usability problems in moderately complex products.

The focus on fixing problems. The focus of usability evaluation methods in the early to mid-1990s was on finding as many problems as possible. A group of usability specialists at Microsoft created a new method that may look similar to traditional usability testing but is, in fact, different (Medlock, Wixon, McGee, & Welsh, 2005; Medlock, Wixon, Terrano, Romero, & Fulton, 2002). Known as the Rapid Iterative Test and Evaluation Method (RITE), this method focuses on fixing designs rather than just finding problems. In outline, the method consists of the following:

a. Key decision makers for the product participate in the study with the usability specialists.
b. The team selects the tasks to be run and attends all sessions. As with traditional usability testing, users who are part of the target market for the product are recruited, and the think-aloud method is used in sessions.
c. After each session, the usability specialist identifies problems and their severity. Team members then decide whether they have enough data to verify each problem and how to refine the design to address the problem.
d. The design team refines the design and tests it with the next participants.
e. Problems are identified again, including whether the refinements have mitigated previous problems. If not, new refinements are created.
f. The team decides which problems they can fix and which need to be examined in more detail or require resources that are not currently available.
g. Additional participants are run until the major problems have been fixed or there are no more resources to continue.

In the tests that have been run to date with this method, the developers have become intensely involved in the process, which results in strong team bonds.

The Strengths and Weaknesses of Usability Testing

Usability professionals believe usability testing has been the most influential evaluation method (Rosenbaum et al., 2000). They have used it to convince managers and developers that usability is a worthwhile investment. These strengths contribute to usability testing's success: Testing directly involves end users in a face-to-face encounter; testing often takes place in a usability lab, adding to its credibility; participants' think-aloud protocol often provides drama; useful diagnostic tests can be run with a small sample of users; and testing is particularly good at uncovering initial ease-of-learning barriers to productivity.

But testing also has some limitations: It takes more time and resources than other evaluation methods; it is typically done with only one or two market segments of the user population; the scope of the test is limited by the sample of tasks chosen; longer-term ease-of-use problems are harder to identify; a single test may uncover only a small fraction of total problems with a product; and the laboratory setting often misses influences that the operational environment places on the usability of products.

INSPECTION METHODS

Human factors/ergonomics professionals have been using guidelines and checklists to evaluate the usability of products and systems since the mid-1970s. To this day, experience and professional judgment are still valued tools for evaluating product designs. The methods that emerged in the 1990s have moved in two directions: *expert reviews,* in which individual specialists inspect a user interface; and *walkthroughs,* in which small teams of developers led by a usability specialist use a group process to explore how tasks are performed.

In Rosenbaum et al.'s (2000) survey of methods used by practitioners, expert reviews were the most frequently used assessment method, having been employed by 70% of the respondents. In contrast, walkthroughs were used infrequently.

In this section, we examine these two inspection methods with their variations and then conclude with some comments on their strengths and weaknesses.

Expert Reviews

How are expert reviews used? Around the year 1990, there was a reaction against the proliferation of guidelines because they were becoming difficult to apply. Since then, the focus of expert reviews has been on their so-called discount characteristics: They can be done quickly and do not require resources outside the development team. One can conduct an expert review with only one or a few usability specialists. Reviews are touted over usability tests because they are less resource-intensive and do not require research design skills to plan or conduct. According to Nielsen and Molich (1990):

> Unfortunately, in most practical situations, people actually do not conduct empirical evaluations because they lack the time, expertise, inclination, or simply the tradition to do so… In real life, most user interface evaluations are heuristic evaluations but almost nothing is known about this kind of evaluation since it has been seen as inferior *by most researchers*. (p. 249, emphasis added)

All expert reviews have the following in common:

a. A user profile is created or used that defines the characteristics, skills, and knowledge of users or potential users of the product. The usability specialist is expected to examine the product from the point of view of this profile.
b. At least one usability specialist reviews the product from the point of view of the user profile. Note, however, that there has been no quantification of who qualifies as a usability specialist. Most published reports list only the years of experience of the specialist.
c. An objective determines the scope of the review. For example, a review might cover every screen or select screens or might focus on violations of organizational guidelines.
d. A report is produced that includes a list of usability issues with some indication of their severity. As with all usability evaluations, there is no standardized method for determining the severity of violations; each published study has used its own method. There also are no published best practices for how to translate issues into meaningful design solutions.

Below we discuss some more common variations of the expert review method.

Heuristic evaluation. There has been some confusion in the use of the term *heuristic evaluation*. Some have used it as a synonym for *expert review*, in which an experienced usability specialist examines a user interface and identifies problems without using explicit rules (Jeffries, Miller, Wharton, & Uyeda, 1991). In this chapter, heuristic evaluation means an expert review in which a small set of heuristics—typically about 10—is used by a usability specialist to identify and classify usability issues. The heuristics are described in a sentence or two, such as "Visibility of system status. The system should always keep users informed about what is going on, through appropriate feedback within a reasonable time" (Nielsen, 1994, p. 30).

Early studies were tried with nonusability specialists performing heuristic evaluations. In those studies, programmers, developers, and computer science students conducted the evaluations (Nielsen, 1992; Nielsen & Mack, 1994; Nielsen & Molich, 1990). If usability experts could teach others how to inspect a product effectively using a short set of rules, usability evaluation could be done at a "discount" compared with a method such as usability testing. The other professionals, however, did not find as many usability problems as did trained usability professionals. Consequently, since the early studies, heuristic evaluations have almost always been performed by people with training and experience in human factors/ergonomics or usability evaluation.

Team reviews. In team reviews, multiple independent expert reviews are performed. Early in the development of this method, it was recognized that any one specialist finds only a proportion of the total set of problems. Early studies indicated that a typical specialist working alone finds less than 40% of problems that a larger group finds. Based on these finding, Nielsen (1994) recommended that 3–5 specialists independently inspect the product and then pool their issues into a common list. This practice is commonly reported as the preferred approach.

Some practitioners do not have inspectors work independently but require them to work as a group. They believe that usability specialists find more problems working together, and they also invite developers to attend the reviews (Sawyer, Flanders, & Wixon, 1996).

Task-based reviews. Some reviews require the inspectors to attempt a set of user tasks as part of the inspection. The assumption behind this variation is that it gives the independent reviewers a common basis for their reviews.

User-based reviews. Although it is not commonly done, some reviews include a user on the review team (Muller, Matheson, Page, & Gallup, 1998).

Reviews using impact ratio. Contrary to the common focus of reviews on finding problems, Sawyer et al. (1996) proposed that the quality of a review be judged by an *impact ratio*, which is a measure of the proportion of the problems identified that a development team commits to fixing. This measure was an early example of trying to change the focus of evaluation from finding problems to finding solutions.

Strengths and weaknesses of reviews. Expert reviews conducted by experienced usability specialists are considered a discount method—quick to complete and less expensive than usability tests or surveys with large samples. This strength has made expert reviews the most popular method for product evaluation. The reviews produce lists of issues, sometimes grouped by heuristics. If instructed to do so, the specialists can also uncover design inconsistencies and violations of industry or company style guides. In a recent study, reviews were found to be just as effective at finding problems as were usability tests (Molich & Dumas, in press).

But the credibility of reviews has been lower than that of usability tests. Developers and managers are less likely to make changes based on a reviewer's list. The fact that the review process is hidden does not give recipients any insight into how the problems were uncovered or the reasoning behind problem statements.

Walkthroughs

How are walkthroughs used? We devote less time to this method in this chapter because it is less common, though it does have strong advantages. Here, we will use the term *usability walkthrough* to refer to the general walkthrough process. Usability walkthroughs have the following in common:

- In preparation for a walkthrough session, a usability specialist working with the development team (a) profiles the characteristics of the users of the product; (b) selects the tasks and steps within the tasks that will be analyzed; (c) determines the correct path through each task; (d) defines the questions that will be asked at each step in each task; and (e) establishes the rules for the walkthrough, especially whether error paths will be examined and whether discussions of solutions to problems will be allowed.
- The product being analyzed is prepared. If the product is a working system or prototype, there may be little preparation. But if the product is a paper prototype or even a specification for a product, the team may need to put it into a form that can be analyzed.
- The sessions are conducted in a group setting. A usability specialist leads the session, keeping the discussion focused on the steps and tasks. Without strong leadership, the walkthrough discussion can become an unproductive argument (Bias, 1994). A scribe records the results of the analysis of each step, key comments from the team, and any suggestions for fixes. Additional team members such as other usability specialists, developers, and project managers participate in the group discussions.
- As the session proceeds, the leader asks questions at each step of each task, and the answers are determined and recorded. As they appear, usability issues are listed. The session may or may not examine error paths or solutions to problems.
- Following the session, the scribe creates a report listing the usability issues and, when they are included, solutions to problems.

Below we discuss some variations on the basic method.

Cognitive walkthrough. In a cognitive walkthrough, the focus is on ease of learning in walk-up-and-use products (Wharton, Bradford, Jeffries, & Franzke, 1992). The cognitive walkthrough appeared about the same time that heuristic evaluations were emerging. It was based on a theory—cognitive engineering—of how people learn to use relatively simple products, such as a shopping center kiosk (Polson & Lewis, 1990). The theory proposes that success is a trial-and-error process that begins with a goal, then a step-by-step process in which a user tries a likely action and then evaluates the results of the action in terms of whether there is progress toward the goal. In the initial studies, no users were involved in a cognitive walkthrough, though developers were. The cognitive engineer prepared a set of four questions about each step in the process toward the goal (Polson & Lewis, 1990). A team of inspectors that included the cognitive engineer and product developers then reviewed a task by asking questions about each step in the path toward success. Although these inspections did uncover problems, they required long sessions even though the session participants avoided lengthy discussions about error paths or solutions to problems.

Since the inception of cognitive walkthroughs, there have been continuing attempts to simplify the process, such as reducing the number of questions for each step to two, and reducing the burden of recording (Rowley & Rhodes, 1992; Spencer, 2000).

Walkthroughs with users. A little-used variation includes users as part of the team conducting the walkthrough (Bias, 1994). Users get to go first and say what they would do at each step. Then the rest of the team discusses the action. Bias reported that this variation only works with a team that is willing to let the user speak without intimidation or condescension. It requires aggressive users who are open about their reactions to the product in this face-to-face environment.

Strengths and weaknesses of walkthroughs. Walkthroughs have not achieved the popularity of expert reviews, perhaps because they take more preparation and much more time to conduct and document. Consequently, the scope of each walkthrough needs to be limited. They are best suited for the early stages of product development, when alternative user interface concepts or even product specifications are being evaluated.

There are three important advantages of walkthroughs: (a) they do not require resources outside the product team, (b) they focus on performing steps and tasks with the product, and (c) they have the potential for team building. A team of developers works together under the leadership of a usability specialist. Developers see how usability specialists think about users and usability. Usability specialists can explain how local usability problems may be symptoms of a more general problem and how important users' goals are to how they approach tasks.

Additional Considerations for Inspection Methods

The acceptance of inspection methods. Although the Rosenbaum et al. (2000) survey showed that expert reviews are very popular, they were rated as having a much lower impact on product improvement than was usability testing. (Walkthroughs appear not to be popular.) We believe that there are at least two reasons for the reduced impact of expert reviews. First, typically, developers are not integrated into the review process, which reduces the opportunity of mutual trust and for educating developers about UI issues. Second, evaluation methods that do not include end users are not as persuasive as methods such as testing or user surveys. Because the recipients of the review do not see the process of inspection happening, they may conclude that the findings and recommendations are just opinion because they are not based on user input.

These problems can be mitigated by having an open inspection process in which development team members watch the review as it takes place and by having the usability specialists think out loud as they conduct their inspections.

The reliability of inspections. Beginning in 1998, a series of research and analysis papers questioned the reliability of inspection methods. Almost from the beginning, researchers had realized that having only one usability specialist inspect a product for usability was not enough (Nielsen, 1994; Nielsen & Molich, 1990). But with the publication of the Virzi (1990, 1992) studies, usability testing was assumed to find more problems than did inspections.

Desurvire, Kondziela, and Atwood (1992) conducted the first of several studies of inspection methods (heuristic evaluation and cognitive walkthrough), treating usability testing as the gold standard. They were interested in how well an inspection predicted the results from a usability test. Their assumption was that usability testing finds the so-called true problems. The three experts who conducted inspections in their study found fewer than half the problems that were uncovered by Desurvire et al. in a usability test of the same product. Furthermore, the authors described inspection problems as less severe than the usability test problems, though the validity of some of these findings has been called into question (Gray & Salzman, 1998).

Bailey, Allan, and Raiello (1992) also examined inspections (heuristic evaluations specifically) from the perspective of usability testing. They conducted multiple iterative usability tests of products that had been evaluated with heuristic evaluations. Whereas the heuristic evaluations had found many problems (43), usability test results suggested that only a couple (2) significantly affected performance. These researchers characterized the two problems as the only "real" problems and the remaining problems as "false alarms." Although there are some potential flaws in their approach (Gray & Salzman, 1998), the authors do call into question the reliability of inspections.

The use of usability test data as a benchmark against which to compare other evaluation methods continues to the present day (Andre, Hartson, & Williges, 2003; Sears, 1997). In these studies, any usability issue found by an inspection method is considered a false alarm if it is not also found through usability testing.

In the CUE-4 study, which did not treat usability testing as the gold standard, eight inspection teams (in addition to the nine usability testing teams discussed earlier) evaluated the same hotel reservations site (Molich & Dumas, in press). Not one of the serious usability problems reported was found by all eight inspection teams. Additionally, 27.9% of the issues were unique—that is, they were reported by only one of the inspection teams.

Even with extensive study, it is not understood how to account for the evaluator effect, which results in unreliable and incomplete problem discovery. It is known that usability experts work quickly, finding problems in a few hours, especially the problems they consider most severe (Dumas, Sorce, & Virzi, 1995; Molich & Dumas, in press). However, it seems unlikely that spending more time on the review would uncover substantially more problems. In the Molich and Dumas (in press) CUE-4 analysis, on average, teams that spent more time on their reviews did not find more problems.

There seems to be some limit to the number of problems that any one usability specialist or a team working together will find. The cause of this limit is not known, but what is known is that, like inspections, a typical usability test also finds only a small proportion of the total problems. Perhaps these evaluator effects appear both in inspections and in usability testing because of inherent limitations of usability specialists themselves. Woolrych and Cockton (2002) theorized that this is the case and have begun to examine analyst decision-making processes as a means for understanding the strengths and weaknesses of inspection methods. Their research might also help to better understand how to mitigate evaluator effects for testing.

The validity of inspections. As we discussed earlier, from the earliest days of using discount inspections, researchers have attempted to measure their validity against the results of usability tests. Unfortunately, these analyses fail to acknowledge the fact that most usability tests also find only a fraction of the total problems. In addition, they ignore the fact that the problems found by inspections (but not usability tests) may in fact be real problems. Molich and Dumas (in press) found few, if any, false alarms from inspections when the content of each problem statement was examined.

Understanding what an expert does to uncover problems in practice is key to assessing the validity of those problems. Unfortunately, not much is known about this process. To our knowledge, no think-aloud research study has ever been conducted to understand the thought processes experts use when reviewing a product. We know that some experts find more problems, but we do not know why. We do not know whether two experts using the same set of heuristics are applying them in the same way. The qualifications of usability specialists who have been used in published studies have not been measured. Most studies have reported no data on qualifications beyond average years in the profession. Woolrych and Cockton (2002) proposed that analysis of the inspectors is the next important step in helping to explain why inspections succeed or fail in different contexts.

The value of heuristics. One of the interesting aspects of the literature on reviews is that there has never been a study that focused on the value of explicit heuristic rules. No one knows what the result would be if a group of usability specialists reviewing a product with a set of heuristics were compared with an equivalent group of usability specialists reviewing the product without the heuristics. When eight teams of experienced usability professionals were asked to review a product, however, only one team used explicit heuristics (Molich & Dumas, in press). Apparently, explicit heuristics are not as highly valued as the attention paid to them in the literature would lead us to believe.

Strengths and Weaknesses of Inspections

Usability professionals use inspections (expert reviews and walkthroughs) in applying their knowledge and experience to product evaluations. Inspections can be done quickly and require neither resources outside the development team nor the research skills often required for usability testing. But typically they do not directly involve users and often do not involve developers. Also, they are designed to find potential problems, not solutions, and focus on potential dissatisfiers rather than probable satisfiers (Jordan, 2002).

Consequently, in their present form, inspections are not ideal for fostering innovation or creative design solutions. Additionally, inspections are less valuable than usability tests or surveys in persuading developers to make product improvements or in educating them about users and usability.

SURVEYS, INTERVIEWS, AND FOCUS GROUPS

Once human factors/ergonomics professionals began to involve users directly in product development, there was a need to tailor traditional methods for usability assessment. As we described earlier, this process began in the late 1980s and was fully established by 1995. There have been a few additions to these methods since then, including survey tools for evaluating Web sites and instrumenting surveys and pop-up questions as users do their work.

Traditionally, surveys have been used to reach large samples of users, interviews have been used to probe a small number of users more deeply, and focus groups have been used to stimulate users to express unique responses (Courage & Baxter, 2005). In this section, we focus on the adaptation of these traditional methods to assess the usability of products.

How Are Surveys Used?

Most usability groups have created and used their own surveys for product development. Although these surveys have unknown validity and reliability, they do not require costly licenses. Our impression is that most usability specialists consider a few Likert rating scales good enough to measure the subjective reaction to products. This strategy fits well with the current notions that discount methods are preferred but ignores the standardized surveys that are readily available.

Two types of standardized surveys exist to assess usability: (a) short questionnaires that can be used to obtain a quick measure of users' subjective reactions, usually to a product that they have just used; and (b) longer questionnaires that can be used alone as an evaluation method and that may be broken into several subscales.

Short questionnaires. A number of short questionnaires have been published, almost all of which use Likert-type rating scales. One of the first was a three-item questionnaire developed by Lewis (1991), which measured users' judgments of how easily and quickly tasks were completed. The Software Usability Scale (SUS) has 10 questions (Brooke, 1996); it can be used as a stand-alone evaluation or as part of a user test. SUS can be applied to any product, not just software, and yields a single 100-point scale. If SUS were a valid and reliable measure of subjective usability, the 100-point scale would make usability measurement or comparisons easy to interpret. A final advantage of SUS is that its use does not require a license.

Longer surveys. These surveys were initially created to be their own evaluation tool, though they only take a few minutes to complete. In several cases, they have subscales that can be employed separately for specialized usage. The Computer User Satisfaction Inventory (CUSI) was developed to measure attitudes toward software applications (Kirakowski & Corbett, 1988). It has 22 questions comprising two subscales: affect, the degree to which respondents like the software; and competence, the degree to which respondents feel they can complete tasks.

A licensed survey called the Questionnaire for User Interaction Satisfaction (QUIS) consists of a set of general questions that provide an overall assessment of a product and a set of detailed questions about interface components. Later versions included a set of demographic questions, a measure of overall system satisfaction, and hierarchically organized measures of 11 specific interface factors (Shneiderman, 1997). Each question uses a nine-point rating scale, with the end points labeled with adjectives, such as "confusing" to "clear" for online help text. Because the factors in QUIS are not always relevant to every product, practitioners often select a subset of the questions or use only the general questions. There is a long form of QUIS (71 questions) and a short form (26 questions; *http://www.lap.umd.edu/QUIS/index.html*).

The Computer System Usability Questionnaire (CSUQ) was developed at IBM (Lewis, 1995). It is composed of 19 statements with a rating on a seven-point scale of "Strongly Disagree" to "Strongly Agree" for each statement. The questionnaire has been subjected to numerous validity and reliability studies. Factor analysis of the questions revealed three primary factors: system usefulness, information quality, and interface quality.

The Software Usability Measurement Inventory (SUMI) was developed to evaluate software *only* (Kirakowski, 1996). It has six subscales: global, efficiency, affect, helpfulness, control, and learnability. The global subscale is similar to QUIS's general questions. The SUMI questionnaire consists of 50 statements to which the user has to reply either "Agree," "Undecided," or "Disagree" (e.g., "The way that system information is presented is clear and understandable."). Its developers have created norms from several thousand respondents so that software products can be compared against other, similar products. Despite its length, SUMI can be completed in about five minutes. It does assume that the respondents have had several sessions working with the software, so it is best for summative evaluations (Kirakowski, 2005). SUMI has been applied not only to new software under development but also to compare software products and to establish a usability baseline.

All the aforementioned surveys were developed as paper-and-pencil tools, though SUMI now is online. The Web Site Analysis and Measurement Inventory (WAMMI)—developed by the same people who created SUMI—is online and was developed to assess the usability of Web sites. It consists of 20 questions. Users access it from a link on the site they are using, and a report is generated with scores in six subscales. A full description of WAMMI is available at *http://www.wammi.com/*.

Additional Survey Considerations

The validity and reliability of usability surveys. It is ironic that most usability groups create their own survey questions for use in usability testing because the standardized surveys are the only usability assessment measures with established validity and reliability (Kirakowski, 2005). The SUS, CSUB, SUMI, and WAMMI surveys all have high reliabilities, and their concurrent validities have been established with significant correlations with other usability measures.

Sample sizes required. A recent study compared many of these questionnaires, including SUS, CSUQ, and QUIS (Tullis & Stetsen, 2004). Tullis and Stetsen were interested to know how stable the results would become as sample size increased. They asked 123 participants to use two Web sites—one poor in usability and one much better—and then fill out one of five questionnaires. The researchers then compared the results of samples of each questionnaire against the total set of data.

The important point is that none of the questionnaires could distinguish between the poor and much more usable site more than 80% of the time until the sample size reached at least 10 participants. The only questionnaire that could distinguish between the two sites with 100% accuracy was the SUS, but not until the sample size had reached 12. CSUQ was the second most reliable at that sample size. This study confirms the fact that questionnaires are not reliable with the small sample sizes typically used in usability testing. If practitioners want to use a questionnaire with larger sample sizes, the simple 10-question SUS was found to be the most reliable.

How Are Interviews Used?

Interviews are best for situations in which detailed or voluminous information is required from a relatively small number of respondents and a skilled interviewer is needed. Courage and Baxter (2005) listed eight considerations for when to use an interview instead of a survey:

1. The task to be accomplished requires two-way interaction with respondents.
2. The respondents would be unlikely to answer survey questions.
3. There is no mailing or contact list of qualified respondents.
4. The interaction must occur in a specific location.
5. The complexity of the questions/responses requires a skilled interviewer.
6. A relatively small sample size is acceptable.
7. The respondents are not widely dispersed geographically.
8. The content of the material is likely to create the need for anonymity.

An effective interview requires a trained, experienced interviewer and careful preparation. Although usability professionals have little formal training in interviewing skills, they are often asked by product developers to conduct them anyway. We therefore advise usability specialists to seek appropriate interview training.

Interviews are also conducted at the end of usability tests. Those interviews consist of a review of key events that occurred while participants were conducting tasks and a set of questions asking participants to sum up their experience and to give their final impression of the usability of the product. Interviewing skills become essential so that the test moderator does not ask biased questions, give body language cues, or force choices onto participants (see Dumas & Redish, 1999, Table 19-1).

How Are Focus Groups Used?

In Rosenbaum et al.'s (2000) survey of usability practice, focus groups were used almost as frequently as surveys and were rated about as useful in terms of their impact on product usability. Although focus groups are recognized as useful for understanding product definition, there is disagreement among usability professionals about their value for usability assessment. In a panel discussion on this topic (Rosenbaum, Cockton, Coyne, Muller, & Rauch, 2002), some experienced usability professionals found focus groups useful for product evaluation, whereas others did not.

If there is any consensus among professionals, it is that the group discussion must be supplemented by task performance, either during or just prior to the discussion (Sullivan, 1991). For example, Pagulayan et al. (2003) reported the value of conducting a focus group of participants who had just completed a game-playing session. Their group discussions provided additional information that had not been obvious from watching the session or even from the ratings provided by the participants.

The Value of Surveys, Interviews, and Focus Groups

These traditional human factor/ergonomics methods have been adapted (with varying degrees of success) for assessing product usability. Several surveys are tailored for usability measurement that have established validity. The reliability of all the surveys is low with samples sizes at 10 or below, so their use in typical usability tests is questionable. Interviews have been valuable for probing the details of usability problems and design alternatives more fully, though their length restricts their use to a small sample. The value of focus groups for usability assessment is still being debated but does depend on the respondents using the product under discussion.

FIELD METHODS

Field methods are a collection of techniques for studying users, their activities, and their interaction with products in real-world contexts. These methods are an important supplement to usability testing, heuristic evaluation, and surveys for the following reasons:

- Real-world contexts can be complex, changing how users interact with products.
- Field studies enable researchers to observe what people actually do, as opposed to what they say they do, which can sometimes differ. This offers more ecological validity than other methods.
- They enable researchers to study users in the environments that help to shape the usability and desirability of the products.
- Field methods adapt nicely throughout a product's development cycle, from conception through refinement and launch.
- They can help design teams see gaps between what existing products offer and what users need and value in their work or play.

There are a number of excellent resources on commonly used field methods, including Beyer and Holtzblatt (1998), Courage and Baxter (2005), and Wixon and Ramey (1996).

How Are Field Methods Used?

Field methods can be useful for requirements gathering, as well as for formative and summative evaluation. At their most fundamental level, they are designed to understand user needs, behaviors, and product interactions in a real-world context. Consequently, the actual deployment of those field methods is variable. They are heavily influenced by the study's objectives, by when the study occurs in the product development life cycle, and by the amount of time and money a company is willing to invest in the process. Some characteristics of field studies are described next.

Field studies can be flexed to accommodate a broad range of needs. They can be limited to a single visit or can be longitudinal in nature, examining how user needs, tasks, and product interactions evolve over time.

Field studies can be exploratory or evaluative (both formative and summative). Exploratory studies are often very open-ended and have the goal of understanding users, tasks, and contexts in the here and now. Exploratory studies are good for inspiring and defining new product concepts.

Formative field studies can be used during the iterative design process to address product and feature-specific questions. Summative studies such as beta testing can be used to determine a product's readiness for launch, or even to validate a design after it has launched. See Kirah, Fuson, Grudin, and Feldman (2003) and Hughes et al. (1995) for examples of how field studies can be adapted to address questions throughout a product's development.

Field studies can take a top-down (goal-driven) or bottom-up (data-driven) approach. Woods (1996) described how he used initial interviews to enable him to be more goal driven in his contextual inquiry and observations. Kirah et al. (2003) contrasted their user-driven ethnographic field method approach—used early in the development process—with the more product/feature-specific field studies conducted later in the development process.

Direct behavioral observation and interviews are the most common tools employed during field research. However, researchers can use a number of data collection tools to supplement these tools. Artifact collections, collages or photo journals, diaries, and call or usage logs help triangulate data. Several experienced field researchers have also found it useful to partner with users in design by incorporating participatory design methods into their fieldwork.

Last but not least, field studies can be more or less naturalistic in their approach. In a naturalistic field study, the goal is to minimally influence the subject of observation. By this definition, a purely naturalistic study is based on observation only and often extends over time. As one begins to introduce structure or interviews to the studies in order to help focus and control, the studies become less naturalistic and researchers must take precautions to minimize the biases they introduce. Today, most researchers employ moderately naturalistic field studies so they can be sufficiently focused and efficient without unduly influencing attitudes, activities, or context.

Field Method Tools

Next, we discuss several tools that field researchers employ in their work.

Behavioral observation. Data derived through direct observation is perhaps the most fundamental aspect of field study. The goal of behavioral observation is to systematically record and describe behaviors as they occur. There are a number of techniques for doing so. A common one is a semistructured note-taking guide that helps the observer characterize the end-to-end workflow, by which he or she can capture a sequence of activities, the people and tools involved, and any errors or critical incidents. Hackos and Redish (1998) provided a good example of how to prepare such a guide.

In addition to this semistructured note-taking approach, other naturalistic tools can help characterize behaviors. For example, narrative descriptions can be used to systematically record and describe the behaviors as they occur. These descriptions should be sequential and objective and rely on direct quotations (not paraphrases) and observable behaviors (not inferences). These descriptions help researchers capture rich descriptive and diagnostic information about the person-product-environment interaction, but they can be very time consuming.

Event and time sampling offer shorthand ways to describe behavioral patterns. With time sampling, the researcher captures what behaviors are happening at given time instants or intervals. Though it is easy to apply and analyze, it can lose important sequential information. Event sampling captures and characterizes behaviors or events as they happen (e.g., duration, sequences, situational context). It is a powerful tool and is similar to the data-logging approaches used in many usability tests.

The main limitation of event sampling is that it is not good for the study of relatively undefined behaviors or workflows. For a discussion of these techniques and some options for analyzing them, see Bakeman and Gottman (1986).

Interviews. As we discussed in the section on surveys, interviews, and focus groups, interviews can take many forms. When used in field studies, interviews can range from unstructured to structured, but the most common is semistructured. A semistructured interview is guided by both the goals of the study and behavioral observations, and it provides a framework or a set of topics for discussion rather than a detailed script. Done well, interviews can provide insights into thought processes, motivations, and needs in a way that strict behavioral observation cannot. Interviews can also help explain behaviors.

However, interviews are highly susceptible to order effects and biases (interviewer, social, self-serving, and conformity), and they can influence behaviors and thought patterns. Therefore, planning the interviews and training the field researcher in good interview practices are critical to preserving the ecological validity of the interview studies.

Artifacts. Artifacts are physical objects from the user's environment. Artifacts can supplement interview and observation data with tangible representations of the environment. Because it is impractical to remove objects from many environments, photos or video recordings are often used instead.

Collages or photo journals. Like artifacts, collages can describe the nature of the user's activities and environment. They are often created by the user rather than the researcher to tell a story about the nature and context of their activities, as well as to highlight the products, tools, or artifacts that have significance.

Diaries. Because directly observing people, activities, and product usage longitudinally is time consuming and logistically difficult, diaries are often used to characterize needs and activities when the field researcher is not present. Diaries track how users are using the product, report critical incidents (or "eureka" events), capture insights or ideas in the moment, and track changes in behaviors and attitudes over time.

Diaries take many forms. They can be structured or unstructured; event-, time- or even "beeper"-driven; or paper-, e-mail-, or voice mail–based. For some helpful discussions on diary methods, see Kirakowski and Corbett (1990), Kuniavsky (2003), and Palen and Salzman (2002b).

Customer support or product usage logs. Customer support and product usage logs are not field study tools per se, but they are useful for indirectly and objectively capturing needs and behaviors in the real world. Although it can be logistically challenging, monitoring customer support or product usage logs for individuals in a field study is a rich way to supplement the data collected by other means (e.g., Grinter & Eldridge, 2003; Jung, Persson, & Blom, 2005). Sampling and analyzing logs as the primary mechanism for studying real-world behaviors can also be powerful.

Common Field Method Variations

We next present several common variations of field methods. These are not mutually exclusive or exhaustive; rather, they represent a number of ways in which field methods have been adapted to successfully accommodate different research goals.

Contextual inquiry. Contextual inquiry is a popular field study technique combining semistructured interviewing with observation. Originally developed by Whiteside, Bennett, and Holtzblatt (1988) and popularized by Beyer and Holtzblatt (1998), contextual inquiry is defined as "a field data-gathering technique that studies a few carefully selected individuals in depth to arrive at a fuller understanding of work practice across all customers. Through inquiry and interpretation, it reveals commonalities across a system's customer base" (p. 37).

This approach has been described as using a master/apprentice model in which users guide the usability specialist through their work activities while the usability specialist listens and observes, probing for clarification when needed. Instead of a formal script or structured questionnaire, a list of topics, concerns, or activities is typically created in advance to help define the scope of the session and conversation.

Ethnographic interviews. An approach advocated by Wood (1996), the ethnographic interview is a variant of contextual inquiry. The ethnographic interview begins with semistructured interviews of key informants. These initial interviews focus on understanding the users, environments, terminology, and challenges. Information gathered in these interviews helps define the scope and objectives for subsequent observations. Wood claimed that these initial interviews allow for a more top-down approach than does contextual inquiry and that it makes for keener observations and questioning.

Ethnographic field studies. A form of fieldwork that emerged out of anthropology is the ethnographic field study. It is qualitative and naturalistic in nature, focusing on understanding core values and behavioral patterns of a small but carefully selected group of users. Ethnographic field studies tend to be longitudinal in nature (Spinuzzi, 2000). Behavioral observation and interviews (often semistructured) form the core of the methodology (Kirah et al., 2003). These interviews are often supplemented with one or more other data collection techniques, including artifact collection, video or photo documentation, diaries, and collages.

As Spinuzzi and Kirah et al. pointed out, there is confusion in the literature about what constitutes an ethnographic field study and other field methods, but the difference is important. The key distinction is the way in which the data are derived. In an ethnographic study, data collection tends to be holistic and driven by the end user; in other types of field studies, it is often centered on product-related questions. As such, ethnographic studies are best used for exploratory purposes, to help define requirements, and to inspire design ideas. Other types of field studies are more appropriate for usability assessment.

Evaluative field studies. Evaluative fieldwork tends to draw on approaches used in contextual inquiry and ethnographic field studies. What makes evaluative field studies unique is that they tend to be driven by product-specific research questions, and a product or prototype is used during the study. For some examples of evaluative field studies, see Kirah et al. (2003), Jung et al. (2005), and Rose, Shneiderman, and Plaisant (1995).

Beta field studies. Beta field studies emerged out of the quality assurance process and are often used for that purpose (Kuniavsky, 2003). Nevertheless, the beta field study is a form of evaluative field method and can help to ensure that products are launch-ready in a holistic way. By leveraging observation, interviews, and critical incident reports, as well as bug reporting, field researchers can use beta tests to uncover issues that emerge with the introduction of the product into the real world (Kirah et al., 2003). The beta field study can also help to identify and fix any kinks in support and business processes for the product prior to its launch.

Longitudinal panel studies. Longitudinal panel studies are a form of fieldwork that has been strongly influenced by human factors/ergonomics, ethnography, and marketing. Longitudinal panel studies can bring together the full range of ethnographic and field study techniques (interviews, observation) along with more traditional usability testing and even focus groups. Smaller panel studies allow for periodic observation sessions; larger panel studies tend to rely more heavily on interviews, surveys, and indirect behavioral measures such as diary reporting or customer call and product usage logs. For a few varied examples of longitudinal field trials, see Berg, Taylor, and Harper (2003); Malhotra, Gosain, and Hars (1997); and Palen and Salzman (2002a).

Whether large or small, panel studies offer the advantage of giving field researchers a long-range view and focus on triangulating data. Longitudinal panel studies are ideal for putting the usability of the product in a much broader context. Data from panel studies can enable practitioners to see how the product's interface, supporting materials, business processes, and product-related communications shape users' evolving needs and satisfaction with the product (Palen & Salzman, 2002a). Panel studies can ultimately help practitioners assess the underlying causes of a products success or failure and identify opportunities for product innovations.

Discount field studies. Because field studies are more resource intensive and time consuming than other usability methods, field researchers have sought ways to adapt the technique so that it is better suited for the rapid pace of iterative design. Millen (2000) described in detail how careful attention to scope, the use of multiple interactive observation techniques, and collaborative data analysis can be used for so-called rapid ethnography. Hughes et al. (1995) also describe how "quick and dirty ethnography" can be used at different stages of the development cycle to sensitize developers to a product's usability and desirability.

Finally, Wixon and Ramey (2002) highlighted how some practitioners are adapting field methods for quicker results.

Strengths and Weaknesses of Field Methods

As noted in the foregoing discussion on discount field methods, the key barriers to field method research are logistical in nature. Field methods—even discount field methods—tend to be more time consuming and resource intensive than other usability assessment methods. They take more effort to plan, recruit, and execute than other methods. Because field studies often involve more in-depth observation and longer study periods, they also require significant goodwill on the part of the participants. Last but not least, when studies are longitudinal in nature, attrition can become a problem.

There are also some statistical considerations when deciding whether to use a field study instead of other usability assessment methods. Because they happen in a real-world context, field studies tend to be more ecologically valid than other assessment methods. That ecological validity, though, comes at the cost of experimental control. Causal relationships (e.g., Feature X is causing Problem Y) are harder to establish. Finally, there are environmental and social confounds, though these are the very things that can add to the richness of the data.

Despite their challenges, field methods offer several unique advantages that make them very attractive and powerful tool for usability assessment and iterative design. They enable one to study users in their natural environments and to see what they actually do rather than what they report they do. Field methods can also inspire design by immersing design teams in the user's environment, work, or play. As such, they yield insights on how a product is used and how well it aligns with, supports, or enhances needs. Finally, field methods are the best way to evaluate how a variety of factors (product marketing, documentation, support, social and environmental interactions) come together to influence the user experience.

DISCUSSION

As can be seen from our descriptions of methods, there are many to choose from, and each has specific strengths and weaknesses. Examples of the use of each method at all stages of product development can be found. It is difficult to create general rules that do not have exceptions for when and how to use these methods; but there are a few rules of thumb that we believe are useful:

1. Most of the methods are labor intensive and, when they involve users, have used small samples. The exceptions are surveys, Web site instrumentation, and perhaps focus groups.
2. Use field methods early in development or after products are in use. Often their impact on design is indirect but profound.
3. Expert reviews and usability tests both create lists of usability issues, but when there is a choice, it is best to conduct an expert review before a usability test. The review uncovers issues that can be eliminated before users spend valuable time finding them.
4. There is no need to create home-grown questionnaires and rating forms for use in usability testing. There are several off-the-shelf versions with established reliabilities and validities.
5. Empirical methods that involve users—such as usability testing, surveys, and some field methods—have greater impact on product managers than do methods such as inspections, which depend only on the judgment of human factors/ergonomics professionals.
6. Although all methods benefit from involving developers in their planning and execution, methods such as RITE and walkthroughs, which by their nature require that developers be part of the process, increase opportunities for team collaboration with related disciplines.

Finally, using these methods for usability alone will surely be insufficient moving forward. So usability professionals—practitioners and researchers alike—face three important challenges: using the methods thoughtfully, focusing on fixing rather than only on finding problems, and expanding one's scope to look beyond usability to user experience.

Using Methods Thoughtfully

With the fast-paced, low-cost demands of today's corporate culture, it is sometimes difficult to go beyond the discount methods (rating scales, expert reviews, and small informal usability tests). Based on separate surveys of the profession, both Rosenbaum et al. (2000) and Mao, Vredenburg, Smith, and Carey (2005) reported that the expert review (and informal review in the case of Mao et al.) is the most commonly used method, although its perceived impact on design is lower than that of more formal methods. In contrast, field methods are perceived as highly effective, but they are not used very often. Similarly, Hudson (2000) found that informal and unstructured methods were more widely used than were more formal and structured ones.

Usability professionals have a breadth of flexible methods available to them and should not limit themselves to methods that are easy and convenient but have lower impact. Rather, we encourage readers to leverage the breadth and flexibility of the full toolkit of assessment methods and to be deliberate in their applications.

Finding Fixes Rather Than Fault

From the very early days of the human factors/ergonomics discipline, the issue of the relationship between HF/E professionals and people in other disciplines has been widely discussed, often in the context of product evaluation (Meister, 1999). When the discipline was being formed, during and just after World War II, HF/E professionals found themselves on project teams with engineers and were supervised by professionals from other disciplines. Meister took the opportunity to survey 49 human factors/ergonomics professionals who were important contributors in the early days of the profession and who went on to create and become leaders of the Human Factors and Ergonomics Society.

Most of the early HF/E professionals had a psychology background, and their training had focused on behavioral science research. Furthermore, of the skills needed for system development—namely, analysis, design, and evaluation—early HF/E professionals were most comfortable with evaluation (Meister, 1999). Engineers sometimes criticized them for being negative—that is, emphasizing the faults with designs rather than contributing to solutions. Consequently, these early HF/E professionals described their relationship to engineers and managers as difficult. They often felt misunderstood and ignored.

This theme of feeling isolated and undervalued has continued to the present day. Rosenbaum et al. (2000) reported that 25% of the usability professionals they surveyed saw resistance to user-centered design/usability from engineers and managers as an obstacle to the acceptance of a user-centered process.

More recently, Sklar and Gilmore (2004) noted that the perception of designers is that usability professionals are primarily critics and rarely point to the positive qualities of a design—the same evaluation of the human factors/ergonomics profession that engineers made 50 years ago. In a review of the usability issues listed by teams evaluating a product, Dumas, Molich, and Jeffries (2004) noted that less than 15% of the issues were positive and that the wording of usability problems was sometimes harsh and insulting to designers.

We believe that these recent calls for usability specialists to be more positive in their evaluations and to integrate themselves into the business processes (Siegel & Dray, 2003) of their organizations are a sign of healthy self-examination.

Going Beyond Usability to User Experience

Recently, there has been a change in the scope and emphasis of usability assessment to examine concepts such as hedonomics, "funology," and user experience in addition to usability. There is more emphasis on surveys, interviews, and field methods, as well as on measures of the affective reactions to product and even brand. Although they are used sparingly today, field methods offer usability practitioners a variety of ways to take a holistic view of the user experience. We encourage practitioners to make field methods a more integral part of usability assessment practice.

Surveys and rating scales also have begun to play an important role in assessing the user's holistic experience with products. For example, Hassenzahl, Beu, and Burmester (2001) reported using semantic differential scales that measure the affect a design evokes by using scales such as "Exciting—Dull" and "Interesting—Boring."

In addition, a number of other survey instruments can be useful when examining the user experience. Jordan (2002) created a scale of semantically polar adjectives that describe 17 dimensions of a product's "personality." Jordan (2000) also described a questionnaire that he used to assess a product's "pleasurability." A related, but nonverbal, measure is the use of "emocards," which depict drawn facial expressions of emotions (Desmet, 2002). Participants pick or rate the cards that best describe the feeling they get from a design.

Some of the most interesting concepts about dimensions beyond usability have come from the development of computer-based games. According to Pagulayan et al. (2003), games differ from productivity applications in several important ways, such as providing pleasure, being challenging and innovative, and avoiding simplicity.

Pagulayan et al. (2003) described the methods they used to evaluate games. For the most part, they are the methods we have described in this chapter: traditional usability tests and RITE tests, expert reviews, instrumenting games to collect data automatically, and surveys. Game designers also use panels or cohorts of players who play a game repeatedly over time to assess issues that go beyond early use.

Finally, Jordan (2000) described a number of useful variations on traditional methods and introduced a few that are notably unique. We briefly describe them here and refer readers to Jordan's book for more in-depth discussion of these methods, along with useful variations on the methods we have described here.

a. *Private camera conversations.* Users enter a private room with a camera, where they freely interact with and talk about the product. Although this method has some similarities to unstructured usability testing, such as the "listening labs" advocated by Hurst (2003), it is entirely unguided and unscripted, allowing the user to control the agenda entirely. Free from experimenter bias (and control as well), this technique has been found to be useful for getting users to go beyond rational reasons for liking a product to more emotional and hedonic features.

b. *Laddering.* A form of interview, the laddering technique allows the interviewer to extract a hierarchy of needs and priorities. It starts with what users like or dislike in a product and systematically moves to an in-depth discussion of experiential and emotional causes for those reactions. This method has been useful in mapping out the relationships among features, needs, and experiences.

c. *Kansei engineering and SEQUAM.* These rely primarily on structured interviewing and semantic scales to quantitatively link design features to dimensions of the user experience. Although labor intensive, both have been used successfully to deconstruct the hedonic aspects of design.

What does this all mean for the usability professional? This broadening of scope brings with it a change in what one assesses, more so than a change in underlying methods. Going beyond usability to user experience means that one needs to consider more than effectiveness, efficiency, and satisfaction and look beyond the use of a product in isolation. Aesthetic, emotional, social, and business factors must become a central part of the assessment. In usability tests, one can leverage free exploration and follow-up interviews to provide insights into the emotional impact and desirability of the product. One can expand one's surveys and interviews to address these hedonomic constructs. One can turn to field methods for a holistic view of how these factors come together to influence the user's experience. Finally, usability professionals need to recognize the limitations of inspections for understanding the user's experience and look to derive heuristics that can broaden the scope of their inspections.

We believe that this broadened view of usability assessment is critical to success within organizations. Products are now more usable than ever before, thanks in no small part to the last 25 years of usability assessment. With that success, though, come more stringent requirements for a product's success in the marketplace: Users now expect more from products, and the competitive landscape demands that usability professionals look beyond task performance to factors influencing brand loyalty. This challenge means tighter relationships with marketing and other disciplines. It also means that finding ways to make these relationships work—by finding fixes rather than fault—can help the usability profession thrive and grow.

REFERENCES

Andre, T., Hartson, H., & Williges, R. (2003). Determining the effectiveness of the usability problem inspector: A theory-based model and tool for finding usability problems. *Human Factors, 45*, 455–482.

Bailey, R. W., Allan, R. W., & Raiello, P. (1992). Usability testing vs. heuristic evaluation: A head-to-head comparison. *Proceedings of the Human Factors Society 36th Annual Meeting* (pp. 409–413). Santa Monica, CA: Human Factors and Ergonomics Society.

Bakeman, R., & Gottman, J. M. (1986). *Observing interaction: An introduction to sequential analysis*. New York: Cambridge University Press.

Barnum, C. (2002). *Usability testing and research*. New York: Longman.

Beyer, H., & Holtzblatt, K. (1998). *Contextual design: Defining customer-centered systems*. San Francisco: Morgan Kaufmann.

Berg, S., Taylor, A., Harper, R. (2003). Mobile phones for the next generation: Device designs for teenagers. *Proceedings of CHI '03* (pp. 433–440). New York: ACM Press.

Bias, R. (1994). The pluralistic usability walkthrough: Coordinated empathies. In J. Nielsen & R. Mack (Eds.), *Usability inspection methods* (pp. 63–76). New York: Wiley.

Bias, R., & Mayhew, D. (1994). *Cost-justifying usability*. Boston: Academic Press.

Blythe, M., & Wright, P. (2003). Introduction. In M. Blythe, K. Overbeeke, A. Monk, & P. Wright (Eds.), *Funology: From usability to enjoyment* (pp. xiii–xix). Boston: Kluwer Academic Publishers.

Boren, M., & Ramey, J. (2000). Thinking aloud: Reconciling theory and practice. *IEEE Transactions on Professional Communication, 43*(3), 261–278.

Brooke, J. (1996). SUS: A quick and dirty usability scale. In P. Jordan, B. Thomas, B. Weerdmeester, & I. McClelland (Eds.), *Usability evaluation in industry* (pp. 189–194). London: Taylor & Francis.

Cantani, M. B., & Biers, D. W. (1998). Usability evaluation and prototype fidelity: Users and usability professionals. *Proceedings of the Human Factors and Ergonomics Society 42nd Annual Meeting* (pp. 1331–1335). Santa Monica, CA: Human Factors and Ergonomics Society.

Chapanis, A. (1959). *Research techniques in human engineering*. Baltimore: Johns Hopkins University Press.

Courage, C., & Baxter, K. (2005). *Understanding your users. A practical guide to user requirements. methods, tools, and techniques*. San Francisco: Morgan Kaufmann.

Desmet, P. (2002). *Designing emotions*. Delft, Netherlands: Pieter Desmet.

Desurvire, H., Kondziela, J., & Atwood, M. (1992). What is gained and lost when using evaluation methods other than empirical testing. In A. Monk, D. Diaper, & M. Harrison (Eds.), *People and computers VII* (pp. 89–102). Cambridge, England: Cambridge University Press.

Dumas, J. (1998). Usability testing methods: Using test participants as their own controls. *Common Ground, 8*, 3–5.

Dumas, J. (2001). Usability testing methods: Think aloud protocols. In R. Branaghan (Ed.), *Design by people for people: Essays on usability* (pp. 119–130). Chicago: Usability Professionals' Association.

Dumas, J. (2003). Usability evaluation from your desktop. *Association for Information Systems (AIS) SIGCHI Newsletter, 2*(2), 7–8.

Usability Assessment Methods 137

Dumas, J., Molich, R., & Jeffries, R. (2004). Describing usability problems: Are we sending the right message? *Interactions, 11*, 24–29

Dumas, J., & Redish, G. (1993). *A practical guide to usability testing.* London: Intellect Books.

Dumas, J., & Redish, G. (1999). *A practical guide to usability testing* (rev. ed.). London: Intellect Books.

Dumas, J., Sorce, J., & Virzi, R. (1995). Expert reviews: How many experts is enough? *Proceedings of the Human Factors and Ergonomics Society 39th Annual Meeting,* (pp. 309–312). Santa Monica, CA: Human Factors and Ergonomics Society.

Ebling, M., & John, B. (2000). On the contributions of different empirical data in usability testing. *Proceedings of Designing Interactive Systems,* (pp. 289–296). New York: ACM Press.

Ericsson, K. A., & Simon, H. A. (1993). *Protocol analysis: Verbal reports as data.* Cambridge: MIT Press.

Faulkner, L. (2003). Beyond the five-user assumption: Benefits of increased sample sizes in usability testing. *Behavior Research Methods, Instruments, & Computers, 35*(3), 379–383.

Goodwin, N. C. (1982). Effect of interface design on usability of message handling systems. *Proceedings of the Human Factors Society 26th Annual Meeting* (pp. 69–73). Santa Monica, CA: Human Factors and Ergonomics Society.

Gray, W., & Salzman, M. (1998). Damaged merchandise? A review of experiments that compare usability methods. *Human-Computer Interaction, 13*, 203–335.

Grinter, B., & Eldridge, M. (2003). Wan2tlk?: Everyday text messaging. *Proceedings of CHI '03* (pp. 441–448). New York: ACM Press.

Hackos, J., & Redish, G. (1998). *User and task analysis for interface design.* New York: Wiley.

Hancock, P., Pepe, A., & Murphy, L. (2005). Hedonomics: The power of positive and pleasurable ergonomics. *Ergonomics in Design, 13*(1), 8–14.

Hassenzahl, M., Beu, A., & Burmester, M. (2001). Engineering joy. *IEEE Software,* 70–76.

Hertzum, M., & Jacobsen, N. E. (2001). The evaluator effect: A chilling fact about usability evaluation methods. *International Journal of Human-Computer Interaction, 13*, 421–443.

Hix, D., & Hartson, H. (1993). *Developing user interfaces: Ensuring usability through process and product.* New York: Wiley.

Hong, J., & Landay, J. (2001). WebQuilt: A framework for capturing and visualizing the web experience. *Proceedings of the 10th International Conference on World Wide Web* (pp. 717–724). Southampton, England: University of Southampton.

Hudson, W. (2000). User-Centered Survey Results, e-mail posting to CHI-WEB@ACM.ORG, http://listserv.acm.org/scripts/wa.exe?A1=ind0005a&L=chi-web#19.

Hughes, J., King, V., Rodden, T., & Andersen, H. (1995). The role of ethnography in interactive systems design. *Interactions, 2*(2), 56–65.

Hurst, W. (2003). Indexing, searching, and skimming of multimedia documents containing recorded lectures and live presentations. *Proceedings of 11th Annual Conference of Multimedia Publisher* (pp. 450–451). New York: ACM Press.

International Organization for Standardization. (1998). *Ergonomic requirements for office work with visual display terminals (VDTs)—Part 11: Guidance on usability* (ISO 9241-11:1998[E]). Switzerland: Author.

Jacobsen, N., Hertzum, M., & John, B. E. (1998). The evaluator effect in usability studies: Problem detection and severity judgments. *Proceedings of the Human Factors and Ergonomics Society 42nd Annual Meeting* (pp. 1336–1340). Santa Monica, CA: Human Factors and Ergonomics Society.

Jeffries, R., Miller, J., Wharton, C., & Uyeda, K. (1991). User interface evaluation in the real world: A comparison of four techniques. *Proceedings of CHI '91* (pp. 119–124). New York: ACM Press.

Jordan, P. (2000). *Designing pleasurable products. An introduction to the new human factors.* New York: Taylor & Francis.

Jordan, P. (2002). The personalities of products. In W. Green & P. Jordan (Eds.), *Pleasure with products* (pp.19–48). London: Taylor & Francis.

Jung, Y, Persson, P., & Blom, J. (2005). DeDe: Design and evaluation of a context-enhanced mobile messaging system. *Proceedings of CHI '2005* (pp. 351–360). New York: ACM Press.

Karat, J. (2003). Beyond task completion: Evaluation of affective components of use. In J. Jacko & A. Sears (Eds.), *The human-computer interaction handbook* (pp. 1152–1164). Mahwah, NJ: Erlbaum.

Kirah, A., Fuson, C., Grudin, J., & Feldman, E. (2003). Ethnography for software development. In J. Jacko & A. Sears (Eds.), *The human-computer interaction handbook* (pp. 415–446). Mahwah, NJ: Erlbaum.

Kirakowski, J. (1996). The Software Usability Measurement Inventory (SUMI): Background and usage. In P. Jordan, B. Thomas, B. Weerdmeester, & I. McClelland (Eds.), *Usability evaluation in industry* (pp. 169–177). London: Taylor & Francis.

Kirakowski, J. (2005). Summative usability testing: Measurement and sample size. In R. Bias & D. Mayhew (Eds.), *Cost-justifying usability: An update for the internet age* (2nd ed., pp. 519–554). San Francisco: Morgan Kaufmann.

Kirakowski, J., & Corbett, M. (1988). Measuring user satisfaction. In D. Jones & R. Winder (Eds.), *People and computers, Vol. 4*. Cambridge, England: Cambridge University Press.

Kirakowski, J., & Corbett, M. (1990). *Effective methodology for the study of HCI*. New York: Elsevier.

Kirk, N. S., & Ridgeway, S. (1971). Ergonomics testing of consumer products 2: Techniques. *Applied Ergonomics, 2*, 12–18.

Krahmer, E., & Ummelen, N. (2004). Thinking about thinking aloud: A comparison of two verbal protocols for usability testing. *IEEE Transactions on Professional Communication, 47*(2), 105–117.

Kuniavsky, A. (2003). *Observing the user experience. A practitioner's guide to user research*. San Francisco: Morgan Kaufmann.

Law, C., & Vanderheiden, G. (2000). Reducing sample sizes when user testing with people who have and who are simulating disabilities: Experiences with blindness and public information kiosks. *Proceedings of the XIVth Triennial Congress of the International Ergonomics Association and 44th Annual Meeting of the Human Factors and Ergonomics Society* (pp. 4.157–4.160). Santa Monica, CA: Human Factors and Ergonomics Society.

Lesaigle, E. M., & Biers, D. W. (2000). Effect of type of information on real-time usability evaluation: Implications for remote usability testing. *Proceedings of the XIVth Triennial Congress of the International Ergonomics Association and 44th Annual Meeting of the Human Factors and Ergonomics Society* (pp. 6.585–6.588). Santa Monica, CA: Human Factors and Ergonomics Society.

Lewis, C., & Mack, R. (1982). Learning to use a text processing system: Evidence from "thinking aloud" protocols. *Proceedings of Human Factors in Computer Systems* (pp. 387–392). New York: ACM Press.

Lewis, J. (1991). Psychometric evaluation of an after-scenario questionnaire for computer usability studies: The ASQ. *SICCHI Bulletin, 23,* 78–81.

Lewis, J. (1994). Sample size for usability studies: Additional considerations. *Human Factors, 36*, 368–378.

Lewis, J. (1995). IBM computer usability satisfaction questionnaires: Psychometric evaluation and instructions for use. *International Journal of Human-Computer Interaction, 7*(1), 57–78.

Lewis, J. (2001). Evaluation of procedures of adjusting problem-discovery rates estimated from small samples. *International Journal of Human-Computer Interaction, 13*, 445–480.

Malhotra, A., Gosain, S., & Hars, A. (1997). Evolution of a virtual community: Understanding design issues through a longitudinal study. *Proceedings of the International Conference on Information Systems* (pp. 59–73). Atlanta: Association for Information Systems.

Mao, Ji-Ye, Vredenburg, K., Smith, P. W., & Carey, T. (2005). The state of user-centered design practice. *Communications of the ACM, 48,* 105–109.

McGee, M. (2003). Usability magnitude estimation. *Proceedings of the Human Factors and Ergonomics Society 47th Annual Meeting* (pp. 691–695). Santa Monica, CA: Human Factors and Ergonomics Society.

McGee, M. (2004). Master usability scaling: Magnitude estimation and master scaling applied to usability measurement. *Proceedings of CHI '2004* (pp. 335–342). New York: ACM Press.

Medlock, M., Wixon, D., McGee, M., & Welsh, D. (2005). The rapid iterative test and evaluation method: Better products in less time. In R. Bias & D. Mayhew (Eds.), *Cost-justifying usability: An update for the information age* (pp. 489–517). San Francisco: Morgan Kaufmann.

Medlock, M. C., Wixon, D., Terrano, M., Romero, R., & Fulton, B. (2002). Using the RITE method to improve products: A definition and a case study. *Proceedings of the Usability Professionals Association Annual Meeting* [CD-ROM]. Chicago: Usability Professionals Association.

Meister, D. (1999). *The history of human factors and ergonomics*. Mahwah, NJ: Erlbaum.

Millen, D. (2000). Rapid ethnography: Time deepening strategies for HCI field research. In *DIS 2000, Designing Interactive Systems* (pp. 280–286). New York: ACM Press.

Molich, R., Bevan, N., Curson, I., Butler, S., Kindlund, E., Miller, D., & Kirakowski, J. (1998). Comparative evaluation of usability tests. *Proceedings of the Usability Professionals' Association Annual Meeting* [CD-ROM]. Chicago: Usability Professionals Association.

Molich, R., & Dumas, J. (in press). Comparative usability evaluation (CUE-4). *Behaviour and Information Technology*.

Molich, R, Meghan R., Ede, K., & Karyukin, B. (2004). Comparative usability evaluation. *Behaviour & Information Technology, 23*, 65–74.

Muller, M., Matheson, L., Page, C., & Gallup, R. (1988). Participatory heuristic evaluation. *Interactions, 8*, 13–18.

Murphy, L., Stanney, K., & Hancock, P. (2003). The effect of affect: The hedonic evaluation of human-computer interaction. *Proceedings of the Human Factors and Ergonomics Society 47th Annual Meeting* (pp. 764–767). Santa Monica, CA: Human Factors and Ergonomics Society.

Nielsen, J. (1992). Finding usability problems through heuristic evaluation. *Proceedings of CHI '92* (pp. 373–380). New York: ACM Press.

Nielsen, J. (1993). *Usability engineering*. New York: Academic Press.

Nielsen, J. (1994). Heuristic evaluation. In J. Nielsen & R. Mack (Eds.), *Usability inspection methods* (pp. 25–62). New York: Wiley.

Nielsen, J., & Molich, R. (1990). Heuristic evaluation of user interface. *Proceedings of CHI '90* (pp. 241–256). New York: ACM Press.

Nielson, J., & Mack, R. (1994). *Usabilitiy inspection methods*. New York: Wiley.

Olson, J., & Moran, T. (1995). Mapping the method muddle: Guidance in using methods for user interface design. In M. Rudisill, C. Lewis, P. Polson, & T. McKay (Eds.), *Human-computer interface designs: Success stories, emerging methods, and real world context* (pp. 269–300). San Francisco: Morgan Kaufman.

Pagulayan, R., Keeker, K., Wixon, D., Romero, R., & Fuller, T. (2003). User-centered design in games. In J. Jacko & A. Sears (Eds.), *The human-computer interaction handbook* (pp. 883–906). Mahwah, NJ: Erlbaum.

Palen, L., & Salzman, M. (2002a). Beyond the handset: Designing for wireless communications usability. In *ACM Transactions on Computer Human Interaction, 9*(2), 125–151.

Palen, L., & Salzman, M. (2002b). Voice-mail diary studies for naturalistic data capture under mobile conditions. *Proceedings of Computer Supported Cooperative Work 2002* (pp. 87–95). New York: ACM Press.

Perkins, R. (2001). Remote usability evaluation over the Internet. In R. Branaghan (Ed.), *Design by people for people: Essays on usability* (pp. 153–162). Chicago: Usability Professionals Association.

Polson, P., & Lewis, C. (1990). Theory-based design for easily learned interfaces. *Human-Computer Interaction, 5*, 191–220.

Quesenberry, W. (2004). Balancing the 5Es: Usability. *Cutter IT Journal 17*(2), 4–11.

Quesenberry, W. (2005).The five dimensions of usability. In M. Albers & B. Mazur (Eds.), *Content and complexity: Information design in technical communication* (pp. 81–102). Mahwah, NJ: Erlbaum.

Roberts, T., & Moran, T. (1982). Evaluation of text editors. *Proceedings of Human Factors in Computer Systems* (pp. 136–141). New York: ACM Press.

Rose, A., Shneiderman, B., & Plaisant, C. (1995). An applied ethnographic method for redesigning user interfaces. *DIS '95, Designing Interactive Systems* (pp. 115–122). New York: ACM Press.

Rosenbaum, S. , Cockton, G., Coyne, K., Muller, M., & Rauch, T. (2002). Focus groups in HCI: Wealth of information or waste of resources? *Proceedings of CHI 2002* (pp. 702–703). New York: ACM Press.

Rosenbaum, S., Rohn, J., & Humburg, J. (2000). A toolkit for strategic usability: Results from workshops, panels, and surveys. *Proceedings of CHI 2000* (pp. 337–344). New York: ACM Press.

Rowley, D., & Rhodes, D. (1992). The cognitive jog through: A fast-paced user interface evaluation procedure. *Proceedings of CHI '92* (pp. 389–395). New York: ACM Press.

Rubin, J. (1994). *Handbook of usability testing*. New York: Wiley.

Sawyer, P, Flanders, A., & Wixon, D. (1996). Making a difference—The impact of inspection. *Proceedings of CHI '96* (pp. 378–382). New York: ACM Press.

Sears, A. (1997). Heuristic walkthroughs: Finding the problems without the noise. *International Journal of Human Computer Interaction, 9*, 213–234.

Sheppard, S., Kruesi, E., & Bailey, J. (1982). An empirical evaluation of software documentation formats. *Proceedings of Human Factors in Computer Systems* (pp. 121–124). New York: ACM Press.

Shneiderman, B. (1987). *Designing the user interface: Strategies for effective human computer interaction*. Reading, MA: Addison-Wesley.

Shneiderman, B. (1997). *Designing the user interface: Strategies for effective human computer interaction* (3rd ed.). Reading, MA: Addison-Wesley.

Siegel, D., & Dray, S. (2003). Living on the edges: User-centered design and the dynamics of specialization in organizations. *Interactions, 10*, 19–27

Sklar, A., & Gilmore, D. (2004). Are you positive? *Interactions, 11*, 28–33.
Smith, S., & Mosier, J. (1984). *Design guidelines for user-system interface* software (Report ESD-TR84-190). Bedford, MA: MITRE Corp.
Snyder, C. (2003). *Paper prototyping*. San Francisco: Morgan Kaufmann.
Spencer, R. (2000). The streamlined cognitive walkthrough method, working around social constraints encountered in a software development company. *Proceedings of CHI '92* (pp. 353–359). New York: ACM Press.
Spinuzzi, C. (2000). Investigating the technology-work relationship: A critical comparison of three qualitative field methods. *IEEE 2000*, 419–432.
Sullivan, P. (1991). Multiple methods and the usability of interface prototypes: The complimentarity of laboratory observation and focus groups. *Proceedings of CHI '91* (pp. 106–112). New York: ACM Press.
Teague, R., & Whitney, H. (2002), What's love got to do with it? *User Experience, 1*, 6–13.
Tullis, T., Flieschman, S., McNulty, M., Cianchette, C., & Bergel, M. (2002). An empirical comparison of lab and remote usability testing of web sites. *Proceedings of Usability Professionals Association* (pp. 1–8). Chicago: Usability Professionals Association.
Tullis, T., & Stetson, J. (2004). A comparison of questionnaires for assessing website usability. *Proceedings of Usability Professionals' Association* (pp. 1–12). Chicago: Usability Professionals Association.
Virzi, R. (1990). Streamlining the design process: Running fewer subjects. *Proceedings of the Human Factors Society 34th Annual Meeting* (pp. 224–228). Santa Monica, CA: Human Factors and Ergonomics Society.
Virzi, R. A. (1992). Refining the test phase of usability evaluation: How many subjects is enough? *Human Factors, 34*, 457–468.
Virzi, R. A., Sokolov, J., & Karis, D. (1996). Usability problem identification using both low- and high-fidelity prototypes. *Proceedings of CHI '96* (pp. 236–243). New York: ACM Press.
Virzi, R. A., Sorce, J. F., & Herbert, L. B. (1993). A comparison of three usability evaluation methods: Heuristic, think-aloud, and performance testing. *Proceedings of the Human Factors and Ergonomics Society 37th Annual Meeting* (pp. 309–313). Santa Monica, CA: Human Factors and Ergonomics Society.
Walker, M., Takayama, L., & Landay, J. (2002). High-fidelity, paper or computer? Choosing attributes when testing web prototypes. *Proceedings of the Human Factors and Ergonomics Society 46th Annual Meeting* (pp. 661–665). Santa Monica, CA: Human Factors and Ergonomics Society.
Wharton, C., Bradford, J., Jeffries, R., & Franzke, M. (1992). Applying cognitive walkthroughs to more complex user interfaces: Experiences, issues, and recommendations. *Proceedings of CHI '92* (pp. 381–388). New York: ACM Press.
Whiteside, J., Bennett, J., & Holtblatt, K. (1988). Usability engineering: Our experience and evolution, In M. Helander (Ed.), *Handbook of human-computer interaction* (pp. 791–817). Amsterdam: North-Holland.
Wixon, D., & Ramey, J. (Eds.). (1996). *Field methods casebook for software design*. New York: Wiley.
Wixon, D., & Ramey, J. (2002). Usability in practice: Field methods evolution and revolution. In *Extended Abstracts for CHI '02* (pp. 880–884). New York: ACM Press.
Wood, L. (1996). The ethnographic interview in user-centered work/task analysis. In D. Wixon & J. Ramey (Eds.), *Field methods casebook for software design* (pp. 101–121). New York: Wiley.
Woods, D. (1996). Decomposing automation: Apparent simplicity, real complexity. In R. Parasuraman & M. Mouloua (Eds), *Automation and human performance: Theory and applications* (pp. 3–17). Mahwah, NJ: Erlbaum.
Woolrych, A., & Cockton, G. (2002). Testing a conjecture based on the DR-AR model of UIM effectiveness. *Proceedings of HCI 2002* (pp. 30–33). London: Springer-Verlag.

CHAPTER 5

Satisfying Divergent Needs in User-Centered Computing: Accounting for Varied Levels of Visual Function

By Julie A. Jacko & V. Kathlene Leonard

> The ubiquity of information technologies imposes serious and debilitating limitations on people whose abilities fall outside the range of what is generally considered normal. Those at all ages who experience perceptual, physical, and/or cognitive impairments often have diminished functional abilities, and thus, their technology interaction requirements are often unmet by mainstream technologies. In this chapter, we describe an approach to the advancement of baseline human-centered computing for all users, as well as a summary of advancements championed by research, industry, and government initiatives. These efforts have generated a need for a measured, scientific method to unify accessibility design approaches that can be applied in practice. A user's functional abilities serve as the foundation for this framework on which the design process can evolve. Methods for deriving user needs and establishing thresholds for performance are described in the context of human-computer interactions by users with visual impairment. This framework provides the basis for users who have limited functional abilities to claim or reclaim access to information technologies, sustain or regain independence in their lives, and take full advantage of the opportunities that information technologies afford.

The rapid proliferation of information technology into various facets of daily living has generated a highly variant set of domains, contexts, applications, and, most important, users. The challenge of providing access to the most representative user population and accounting for abilities and needs is not trivial. Although technology provides opportunities to enhance the daily lives of people with limited perceptual, physical, and/or cognitive functioning through assistive technologies, these groups are often overlooked in the release of new mainstream technologies intended to enhance daily functioning and productivity.

In the development of mainstream technologies, diverse users often face barriers when it comes to taking advantage of state-of-the-art information technology. Development and design teams are faced with ethical and, often, legal challenges associated with ensuring that users with divergent needs (those who deviate from the so-called norm) have reasonable access to the activities and information afforded by technology. In addition, design and development teams must incorporate this consideration of divergent user populations while operating under the constraints of project timelines, budgets, and other limited resources.

Although the underlying approaches of mainstream human-computer interaction (HCI) design encompass the assimilation of user needs into the human-centered computing design process, it is more challenging in practice. In situations with divergent user needs, practitioners must command a more practical understanding of user limitations and how they ultimately impose on each user's unique combination of functional abilities. This understanding comes not only with knowledge of user ability but also with experience in design, testing, and implementation of the technology with the actual user population.

In this chapter, we present an overview of concepts, approaches, and methods aimed at meeting the needs of such user populations. In addition, we summarize a discussion about efforts that champion accessible technology by research, industry, and government initiatives. These efforts have generated a need for a systematic scientific paradigm to unify meaningful approaches that can be usefully applied in practice.

We also discuss a framework to advance baseline human-centered computing that leverages users' functional abilities as the foundation from which the rest of the design process can evolve. Methods for deriving user needs and establishing thresholds for performance for divergent user needs are presented. This framework provides the basis for users who have limited functional abilities to claim or reclaim access to information technologies, sustain or regain independence in their lives, and take full advantage of the opportunities that information technologies afford.

DIVERGENT USER NEEDS

The efficacy of design is driven by individual users' functional abilities, including the various aspects of sensation, cognition, processing, and motor skills. That is, each user must be able to adequately perceive, process, and provide information to coordinate actions with the system behind the interface. The challenge is that the human is a complex system, even at its most capable levels of input, output, and processing, the effectiveness of which is highly sensitive to the interactions required, the information available, and extraneous contextual factors.

A review of the literature on human sensation and cognition reveals four major areas of limited functionality (Jacko, Vitense, & Scott, 2002): hearing/speech impairments, cognitive impairments, physical/motor impairments, and visual impairments. Visual impairment is the focus of this chapter, and in the remainder of this section, we briefly introduce some limitations of users' visual functional abilities that affect their interactions with information technology.

Visual Impairment

It is estimated that nearly 20 million Americans have visual impairments that result in reduced vision. This number is expected to rise as aging baby boomers experience typical age-related changes to their functional vision (e.g., reduced visual acuity, presbyopia, contrast sensitivity, color sensitivity, depth perception, glare sensitivity) and ocular diseases associated with aging (e.g., macular degeneration, diabetic retinopathy, glaucoma, and cataracts; see Orr, 1998; Schieber, 1994). Some of these age-related ocular changes and diseases, such as acuity and cataracts, are correctable with lenses and/or surgery, but other conditions, such as age-related macular degeneration (AMD), have no proven remedy.

Visual impairments encompass a range of functional limitations even with the use of corrective lenses.

Visual impairments can result in *severe* limitations, in which people are unable to see words and letters in ordinary print, to mild (or *nonsevere*) visual impairments that cause difficulty in seeing print (Bailey & Hall, 1989). This translates into difficulties performing other near-vision tasks, such as using computers.

It is known that in the presence of visual impairments, user behavior is strongly influenced by the nature and amount of residual vision the user experiences in combination with the characteristics of the computer interface. As an extreme example, a blind user without any functional vision will use fundamentally different coping skills to navigate an interface compared with a person with clouded vision caused by cataracts (Jacko & Sears, 1998). Despite this, although many assistive devices for individuals with visual impairments are on the market, they are typified by three underlying problems: (a) they present one-size-fits-all solutions for a range of visual functionality, (b) they abandon the visual sense entirely or rely only on the visual sensory channel, and (c) they do not accommodate changes in visual functionality over time.

The significance of supporting this population's use of information technologies is two-fold. First, aging baby boomers today are different from older adults of earlier times in terms of their familiarity with and perceived value of information technologies. Second, the anticipated surge in this population segment, when considered with the increased incidence of eye diseases experienced by older adults, generates priority. By the year 2030, the number of Americans age 65 and older will approach 70 million; currently, 1 in 3 people experiences vision-reducing eye disease by the age of 65 (Quillen, 1999).

Assistive technologies are being developed to extend people's ability to live independently. For example, a critical component in obtaining important health-related information often involves using electronic health-monitoring equipment, such as blood glucose monitors (American Foundation for the Blind, 2002). If an individual cannot fully perceive the information presented by these devices, he or she may have to rely on others to do so, sacrificing independence. This population segment and their caregivers are therefore likely to actively seek ways to extend their access to computers (despite visual or other age-related impairments) and extend their independence through computers.

Table 5.1 provides the relative proportion of the U.S. population that experiences disorders related to low vision according to age bracket, as well as the projected growth (based on the 2000 U.S. Census) of this group through the year 2009. By that time, the number of individuals with visual impairment is anticipated to be near 25 million; more than half will be 45 years and older. Without intervention, this rapidly expanding digital divide, caused by the proliferation of technologies and the rising number of aging adults, will have large-scale societal impacts.

AN ORGANIZED APPROACH TO ACCESSIBILITY

Since the 1970s, professionals in the human factors/ergonomics (HF/E) and HCI fields have applied a multidisciplinary, scientifically measured approach to the development of models, theories, and guidelines for designing interactive technologies. Their underlying

Table 5.1. Projected Number (in millions) of Americans with Low Vision through 2009

Population Age Segment	CAGR	2005	2006	2007	2008	2009
18–44-year-olds	0.0%	6.0	6.1	6.1	6.1	6.1
45–64-year-olds	8.1%	7.0	7.6	8.2	9.0	9.8
Age 65+	9.0%	6.3	6.9	7.6	8.4	9.2
Total	6.2%	19.3	20.6	22.0	23.5	25.1

Note. *CAGR: Calculated Annual Growth Rate. Numbers based on the 2000 U.S. Census and the percentage of individuals with low vision (U.S. Census, 2002).

theories and recommendations are continuously attuned to the interaction needs that emerge with the introduction of new input devices, platforms, and contexts in order to communicate their implications to designers and developers. The need to address the accessibility of computers for people with impairments surfaced in the early 1990s with increased awareness that computers could enhance the quality of life for individuals with a variety of impairments—what is best classified as assistive technologies.

By the mid-1990s, in response to the prevalence of computers and technology in integral parts of work, home, and recreation activities, researchers and designers began devoting attention to ensuring that large numbers of individuals could use these technologies *despite* their functional limitations. Terms such as *universal access* and *inclusive design* became more common in this community. Around this time, the Rand Corporation published a report on universal access aimed at achieving a better understanding of the capabilities and limitations of current user-computer interfaces (Anderson, Bikson, Law, & Mitchel, 1995).

Awareness within the HCI and HF/E communities was growing at this time as well, and professionals in those fields were working to meet this growing need. Several groups of researchers and practitioners, who operate under associations with professional organizations such as the Human Factors and Ergonomics Society (HFES, http://hfes.org) and the Association for Computing Machinery (ACM, http://acm.org), have evolved over to address the methods used and research conducted about the needs of diverse users. For example, HFES published a special issue of *Human Factors* as early as 1978 that concerned the needs of "the handicapped" (Nickerson, 1978) and another in 1990, which dealt with the needs of users with functional impairments (Gardner-Bonneau, 1990).

Several organizations are working toward similar agendas for different user groups and technologies and are making important strides toward the intricately complex challenge of technology accessibility. These organizations serve as excellent resources for practitioners as well as arenas for the dissemination of research findings. Examples from professional organizations and U.S. government initiatives addressing divergent user needs and their directives are provided in later sections. In addition, key examples of accessibility in practice are described and general approaches to accessibility are defined.

Professional Organizations

HFES Health Care Technical Group. The Health Care Technical Group supports researchers and practitioners in integrating HF/E principles and processes into medical systems to make them more effective and thus improve the quality of life of people who are functionally impaired. Initially named Medical Systems and Rehabilitation Technical Group, the Health Care TG has evolved to address the needs of a broad base of users, both patients and physicians, who embody a range of functional abilities and needs.

HFES Aging Technical Group. Though not exclusively technology focused, the Aging Technical Group addresses the needs of a group of users who commonly experience limitations in functional ability that affect their daily lives. In addition, the group addresses the needs of individuals who are classified as *special needs populations*. ATG members explore new challenges, assessment techniques, and solutions.

ACM SIG ACCESS. Operating within ACM, the Special Interest Group on Accessible Computing (SIG ACCESS) strives to promote the work of researchers and developers that is intended to help individuals with different types of disabilities. Chartered in 1991 under the name SIGCAPH (Special Interest Group on Computers and the Physically Handicapped), the group aimed to coalesce researchers and practitioners to engage in discussions of ideas and developments related to computer-based systems to help people with diverse needs attributed to auditory, visual, motor, and cognitive impairments. The evolution of this discipline is aptly reflected in the 2004 change in the organization's name to SIGACCESS to reflect a broader consideration of the computing and technology needs of users who experience functional impairment. The group promotes general computer accessibility, including assistive technology and also the agenda of universal accessibility and inclusive design.

Alliance for Technology Access (ATA). ATA coordinates the efforts of community-based developers, vendors, research centers, and associates in a network dedicated to ensuring that children and adults with disabilities have access to information technology. This is achieved through increased use of standard, assistive, and information technologies. ATA serves to connect individuals with disabilities with technologies that enable them to be independent and facilitates use of assistive devices to maintain productivity and recreation with technology. Activities include workshops, seminars, training users in leveraging technology, and consulting on the development of accessible technologies.

W3C Web Accessibility Initiative. The World Wide Web Consortium, or W3C, is an international body dedicated to the development of Web standards to ensure the growth and widespread usability of the Web and associated applications. W3C directs several initiatives, but one of particular interest regarding usability for diverse populations is the Web Accessibility Initiative. This was created to ensure that everyone has access to the Web on every device possible. The group produces guidelines for international standards on Web accessibility, support materials for the implementation of these guidelines, and additional resources that can result from international collaboration on these issues. "The power of the Web is in its universality. Access by everyone regardless of disability is an essential aspect" (*http://www.w3.org/WAI/*). W3C provides resources and guidelines to achieve these goals in the general Web design community.

Refereed journals. In addition to the conferences and meetings of the aforementioned groups, refereed journals such as *Universal Access in the Information Society* are committed to the dissemination of high-quality research and review addressing the needs of accessibility to technology. Published quarterly since 2001, *UAIS* solicits original research on the accessibility, usability, and, ultimately, acceptability of information technology by "anyone, anywhere, at anytime, and through any media device" (from the journal's Web site).

Outcomes are reported that stem from theoretical, methodological, and empirical research that addresses reasonable access and active participation with information technology by users who possess as wide an array of characteristics as possible. The theoretical, often baseline, knowledge derived from these studies generates information that can be considered early in the design process to produce information systems that are usable by a wide range of users.

Government Research Initiatives

Research initiatives funded by government agencies have addressed accessible technology issues for users with limited functional abilities. In the United States, the National Institute on Disability and Rehabilitation Research (NIDRR) was formed in 1978 to support the initiatives of the Department of Education's Rehabilitation Act and the Assistive Technology Act (*http://ed.gov*).

In 1997, NIDRR initiated a grant program focused on the identification of technological solutions for people with different types of functional impairments. Grants are awarded to institutions to conduct research, hold seminars and training sessions, and disseminate the resulting research and development outcomes to better inform the population about both usable and assistive technologies. The Rehabilitation Engineering Research Centers (RERC) that are awarded these grants were established to respond to the needs of several geographical areas to address the true diversity of their populations. The National Rehabilitation Information Center Web site lists 22 active RERCs in the following areas (NARIC, 2005):

- Cognitive Technologies
- Accessible Medical Instrumentation
- Accessible Public Transportation
- Communication Enhancement
- Hearing Enhancement
- Technology Access for Land Mine Survivors
- Mobile Wireless Technologies for Persons with Disabilities
- Prosthetics and Orthotics
- Recreational Technologies and Exercise Physiology Benefiting Persons with Disabilities
- Rehabilitation Robotics and Tele-manipulation: Machines Assisting Recovery from Stroke
- Spinal Cord Injury
- Technologies for Children with Orthopedic Disabilities
- Technology for Successful Aging
- Technology Transfer
- Telecommunication Access
- Tele-Rehabilitation

- Universal Design and the Built Environment at Buffalo
- Information Technology Access
- Wheelchair Transportation Safety
- Wheeled Mobility
- Workplace Accommodations

In addition, the Division of Information and Intelligent Systems within the Directorate for Computer Information Science and Engineering at the National Science Foundation established a long-term initiative, called "Universal Access," in 1998. The goal of this initiative was to support fundamental research that provides information on ways to empower people with disabilities to use computers. Specifically, funding was directed at programs investigating how users with diverse needs could use computer interfaces to find, manipulate, process, and use information with efficiency and efficacy (Glinert, 2001). This effort to promote research in universal access served to support investigations that could generate outcomes that affect relatively broader segments of the population.

Accessibility in Practice

Outside the research arena, a key milestone that marks the integration of accessibility research in mainstream consumer software was the release of Windows 95. Prior to 1995, Microsoft Corporation offered accessibility add-ons to enhance the usability of their products for individuals who experience limitations to their sensory and/or motor systems. Access-Pac, created for Windows 2.0, was developed with funding from NIDRR and the IBM Corporation and in collaboration with the Trace Research & Development Center at the University of Wisconsin-Madison. The add-on included the original release of such common utilities as StickyKeys, FilterKeys, MouseKeys, SerialKeys, and ShowSounds.

Windows 95 was the first operating system to integrate these accessibility options into the standard installation package. Microsoft has since continued the pursuit and support of research and development on the topic of accessibility. In 2005, the company offered a royalty-free cross-platform license for user interface automation to make available the information needed to create the most complete and effective interactions with accessibility products that run on the Windows Vista platform (to be released in 2007).

Apple Computer has also incorporated an evolving set of accessibility features into its operating systems since 1985. Features that have been integrated to address user needs include Zoom, full keyboard navigation, Sticky Keys/Slow Keys, Mouse Keys, QuickTime (closed captioning), visual alert, spoken items, talking alerts, speech recognition, and display adjustments. Apple promotes these as universal accessibility features to its system. This approach to inclusive technology is typified in the company's integration of Zoom magnification and VoiceOver, a screen reader, into every operating system, regardless of whether or not the user has visual/sensory limitations.

Although this is by no means an exhaustive list of the research and development activities under way that aim to meet diverse user needs, it paints a portrait of the depth and breadth of ongoing work in research and practice. The range of activities demonstrates the growing importance of this area and the vastness of the problem space for computer accessibility. In practice, the accessibility solutions applied in the rapidly evolving information technology landscape tend to be derived and applied in a very ad hoc way and typically are not founded on evidence-based outcomes from actual users (Keates & Clarkson, 2003). Much remains to be learned about how functional limitations manifest in the interaction between humans and computers—interaction that is further complicated by new and emergent hardware and usage contexts.

Approaches to Accessibility

The range of activity in the area of technology accessibility has led to disparities in the use of terminology and concepts. How abilities and interactions are classified and described can have a direct impact on the efficacy of the solutions. In addition, the language used to describe users' individual conditions and capabilities must be carefully considered. In this section, we introduce readers to the contemporary terminology employed in the field of accessible technology.

Though functional limitation(s) experienced by an individual can affect many areas of life, people bring a rich set of memories, experiences, knowledge, and skills to the fore. This is important when considering the impact of functional ability on human-computer interaction, because other aspects of the individual can influence performance levels. Researchers must consider the user in a more holistic way, not only in terms of the impairment experienced. For example, when understanding the needs of users with motor impairments during human-computer interaction, the researcher must account for (statistically, or in controlled recruitment) aspects other than simply those related to motor control. In particular, they must take into account visual perception and hand-eye coordination, in addition to personal experience using computers. The type and level of user dysfunction can inhibit performance and direct the interaction, but this is also influenced by other personal characteristics and abilities. In characterizing the needs of the diverse user population, one must account for comprehensive user traits.

Four approaches that are commonly leveraged to account for divergent user needs include *accessible design, assistive technologies, universal access*, and *universal usability*. The terms are sometimes used interchangeably in popular culture, but they are based on distinct philosophies and direct different approaches to addressing the needs of users with limited functional abilities. The concepts are useful in reference to any type of system (e.g., accessibility of a physical building or parking space); in the context of this chapter, we define and use them relative to information technology.

Accessible design. *Accessibility* is defined by Webster's online dictionary as "the quality of being at hand when needed or the attribute of being easy to meet or deal with." It refers to the extent to which a design affords use by different users with a variety of characteristics under various scenarios. Accessibility also refers to the ease with which a person can use a technology or the ease with which a group of individuals, linked by a common profile, can leverage the technology. The term is an appropriate qualifier in terms of mainstream user populations and technology interaction; however, it is most commonly used in reference to users with limited functional abilities.

Accessibility is the goal of both assistive and universal design, but the extent of accessibility afforded (e.g., which populations and which usage scenarios) differs between the two ideologies. It is important to note that an interface may be considered usable according to mainstream usability heuristics but can contain features that are not accessible by individuals with specific disabilities. In fact, several Web pages that are adequately designed from a usability standpoint for a specific set of users do not meet the interaction requirements to make them accessible by individuals with disabilities (Hanson, 2004; Keates & Clarkson, 2003). Simply put, the techniques and tools used to achieve usability in mainstream populations do not always transfer to the interaction of diverse populations, in terms of both design and evaluation.

Information technology may be immediately accessible to a user (without any changes) with adaptation to the interface and/or input mechanism (e.g., changing screen resolution), via a third-party device (e.g., a screen reader), if the user employs customized adaptations, or if the user requires a completely separate device in order to retain the information or accomplish the activity. In light of the rapid proliferation of new technologies and applications in the marketplace, the best solution for diverse user needs is technology that enables users to "walk up and use" the system without adaptations, specialized equipment, or alternative technology. This is not entirely feasible given real-world design constraints and conflicting user needs. Designers should make informed decisions based on the user population; needs that cannot immediately be met by the application should be addressed through adaptations to the interface settings and be compatible with common third-party devices. If this is not possible, the user should be directed to an alternative means of access. In the handbook *Universal Principles of Design,* the authors elaborated on the principle of accessibility by stating, "Objects and environments should be designed to be usable, without modification, by as many people as possible" (Lidwell, Holden, & Butler, 2003, p. 14).

Assistive technology. The majority of solutions for information technology access are classified as assistive or adaptive technologies. The Assistive Technology Act of 2004 aims to empower individuals with functional limitations to maintain greater control over their lives and more fully participate in their home, school, and work environments and in their communities by enhancing access to assistive technology devices and services. The legislation strives "to maintain, increase, or improve functional capabilities of individuals with disabilities" (Assistive Technology Act, 2004). In these cases, typically, the technology is designed to accommodate the needs of a specific profile of users; it is not a device that is marketed or commonly found in mainstream computing or technology markets.

Table 5.2 provides a classification of different types of assistive technologies and their underlying purpose. In the area of computing, a collection of assistive devices and adaptations focus on improving the accessibility of technologies. It is truly a catch-22, because computers can serve in unique capacities to maintain, increase, and improve functional capabilities (such as providing a handheld computer as a cognitive aid for an individual with memory problems attributed to brain injury). However, as computers become an increasing necessity for daily living and quality of life, assistive design needs to be available to ensure access for a variety of individuals across all ages. Assistive technologies are typically specific to a limited consumer group, sometimes existing as completely customized devices. They are not commonly sold in the mainstream consumer marketplace, and those who can benefit usually need to be made aware of their existence by a vendor, through word of mouth, or on the recommendation of a physician or occupational therapist.

Table 5.2. Taxonomy of Assistive Technologies

Computer Access Aids	**Daily Living Aids**
Alternative input devices	Clothing and dressing aids
Alternative output devices	Eating and cooking aids
Accessible software	Home maintenance aids
Adaptive interfaces	Toileting and bathing aids
Affective computing	Appointment reminders
Education and Learning Aids	**Environmental Aids**
Cognitive aids	Environmental controls and switches
Early intervention aids	Home-workplace adaptations
Recreation and Leisure Aids	**Mobility and Transportation Aids**
Sports aids	Ambulation aids
Toys and games	Scooters and power chairs
Travel aids	Wheelchairs
Communication	Vehicle conversions
Hearing and Listening Aids	**Vision and Reading Aids**
Hearing aids	Magnifiers
Captioning	Auditory interfaces
Sign language interpreters	Braille interfaces
Health Care	**Communication Aids**
Home health monitoring	Speech and augmentative communication
Physician record systems	Writing and typing aids
Pharmaceutical	
Life monitoring devices	
Prosthetics and orthotics	

One contemporary example of an assistive technology that aims to improve interactions for a very specific population is Jaws®. This popular screen reader program developed by Freedom Scientific (*http://www.freedomscientific.com/fs_products/software_jaws.asp*) provides Braille and synthesized speech output from computer displays. In addition, catalogue companies such as EnableMart (*http://www.enablemart.com*) and Gold Violin (*http://www.goldviolin.com*) provide both high- and low-tech solutions to improve various activities of daily living, including reading, writing checks, and mobility and navigation support.

Alternatively, efforts can be made to retrofit technologies to be compatible with accessible features. For example, a Web site can be designed in a format to enable a person using a screen reader to more easily scan page content. Because such devices are not available in the mainstream marketplace, this approach often means high costs for the users with the specialized need. Thus, the assistive technology approach at times is slow in assimilating technological change, incorporating some of the solutions, and addressing novel difficulties posed by new technologies (Stephanidis & Savidis, 2001).

Universal access. Universal access has been interpreted many different ways. In the most basic sense, universal access relates to the architectural design that facilitates ease of entry into and mobility around building space (via elevators, wheelchair ramps, corridors designed to accommodate wheelchair access). In the realm of information access, however, universal access implies much more; it accounts for the global need to cope with the diversity of user characteristics, aspects of the task, and varied contexts of use, including the proliferation of information technology systems into various aspects of daily life. Consequently, universal access assumes that the increased expansion of technology into new interaction scenarios can trigger an increase in the number of users who will likely face accessibility issues—and not just users with disabilities (Stephanidis & Savidis, 2001). In that sense, the design solutions that are targeted at meeting the needs of people with disabilities may enhance the interactions of users without impairment in certain limiting situations.

One of the most tangible examples of this phenomenon is the widespread popularity of closed-caption television shows. Intended to be a useful tool for individuals with auditory impairments, the transcription of television shows is a commonly viewed feature on television in noisy or crowded public places such as airports, restaurants, bars, and fitness studios.

The universal design approach entails meeting the needs of as many users as possible and limiting the need for special features or third-party accessibility tools. This inclusive design approach fits the design to as many user needs as possible. Looking into the future, some assert that the best way to achieve universal design is through design that fosters automatic individualization of the interface and interaction to meet the user's needs at a specific time and context. Ultimately, interfaces that can automatically adapt and anticipate user needs are the best solution for attaining universal—or at least highly inclusive—accessibility (Brown & Robinson, 2004; Stephanidis & Savidis, 2001). The keys to achieving the necessary knowledge for interface customization to achieve truly universal accessibility include (a) the user profile, (b) the technology hardware or platform, (c) the task, and (d) the usage context.

The user profile includes the various aspects of user functional abilities, including sensory, cognitive, and motor abilities, as well as background, experience, and user values. To accommodate the broadest possible range of users, the design must take into account attitudes, capabilities, skills, values, and experiences from the potential user population. Modeling the user population's needs in terms of interface requirements also poses a substantial challenge and yet is a necessary step in the process.

Regarding the approach to user modeling, Stary (2001) identified a variety of appropriate tactics for the deployment of universal design methods, depending on the constraint of the user group and the study. These methods include (a) calculating predictions of user behavior based on the operators from cognitive science; (b) specifying basic interaction tasks, such as elementary menu operations, and combining these with cognitive procedures; and (c) integrating existing techniques and models used in HCI.

Relevant to all three approaches is the need to establish the user's capacity for input and output dialog with the interface to understand the baseline ability when the technology is used. Diversity among users can be depicted with respect to their functional roles and personal characteristics. This serves as the basis for representations of the actual user characteristics juxtaposed against the minimum requirements needed to accomplish the interaction (Stary, 2001). In mainstream HCI studies, usability techniques typically profile users based on skill level, experience, and, often, basic demographics (e.g., age, gender, education). To provide accessibility to a more variant population, designers must extend their understanding of the users beyond merely the data on their skill and experience to allow a much more detailed and broader user profile, including how the functional capability of the user maps to the so-called normal population (Keates & Clarkson, 2003).

Universal usability. A term coined by Shneiderman (2000), *universal usability* is achieved by "having more than 90% of all households as successful users of information and communications at least once a week." The biggest challenges in achieving universal usability include (a) the ever-changing portfolio of technologies that need support, including hardware, software, and network accessibility; (b) the accommodation of users with different backgrounds (socioeconomic, literacy, skill, experience, knowledge), demographics, disabilities, and disabling conditions (environmental/contextual constraints); and (c) providing users with the knowledge required to succeed in the interactions. According to Shneiderman, universal usability ensures both access and usability, which are two separate constructs. This approach is a more comprehensive technique and addresses the shortcomings of providing only access, or only usability, as discussed by Hanson (2004) and by Keats and Clarkson (2003).

In summary, there are several areas of overlap between the four approaches of accessibility, assistive design, universal access, and universal usability. Although it is apparent that they are discrete constructs, they all have the same goal: to provide reasonable access to technology for users with diverse needs. There is also agreement about the need for a complete understanding of the user profile, interaction requirements, and the performance capabilities of the mainstream versus diverse user groups with functional limitations.

A FRAMEWORK OF INTERACTION THRESHOLDS FROM FUNCTIONAL ABILITIES

In this section, we present an approach to accessible, inclusive design that is partly validated by research on users with visual impairments. It is based on the establishment of critical performance thresholds for HCI involving users with disabilities. We introduce a classification system for functional profiles relative to user performance in interacting with computers. The integration of a user's profile of functional ability and imitations can be used to identify the optimal design solution for users of different sensory, motor, and cognitive impairments.

Here we summarize an HCI approach introduced by Jacko and Vitense (2001; further clarified by Jacko, Vitense, & Scott, 2002) that derives the interaction thresholds of users based on their profiles of functional ability. We present a detailed overview of this approach as it is used in research to understand and clarify the needs of users with visual impairments. This demonstrates the power of gaining a fundamental understanding of users' ability when combined with the skills and tools of the HCI discipline for the area of visual impairments.

Framework Description

As stated previously, limitations to human functioning can be categorized into four areas: visual impairment, speech/hearing impairments, cognitive impairments, and physical/motor impairments. In their work aimed at deriving a framework of functional ability, Jacko and Vitense (2001) illustrated that each of these categories involves a unique set of clinical diagnoses. The clinical diagnoses determine the ability of the user to provide and receive information critical to the efficacy and efficiency of interacting with technology. Figure 5.1, which is an adaptation of the work of Jacko et al (2002), illustrates the connection between clinical diagnoses, user functional ability, and the ability to achieve the requirements of the interactions critical to manipulating the computer interface. In the figure, diagnoses are depicted in by $A_1,...,A_n$, $B_1,...,B_n$, $C_1,...,C_n$, $D_1,...,D_n$ and $E_1,...,E_n$. The functional capabilities that are critical to the technology access are depicted by $(Y_1,...,Y_n)$.

DISABILITIES RESULTING FROM:

A	B	C	D	E
Speech Impairment	Hearing Impairment	Mental Impairment	Physical Impairment	Visual Impairment

CLINICAL DIAGNOSES

$A_1, A_2, A_3,.....A_n$ $B_1, B_2, B_3,.....B_n$ $C_1, C_2, C_3,.....C_n$ $D_1, D_2, D_3,.....D_n$ $E_1, E_2, E_3,.....E_n$

FUNCTIONAL CAPABILITIES

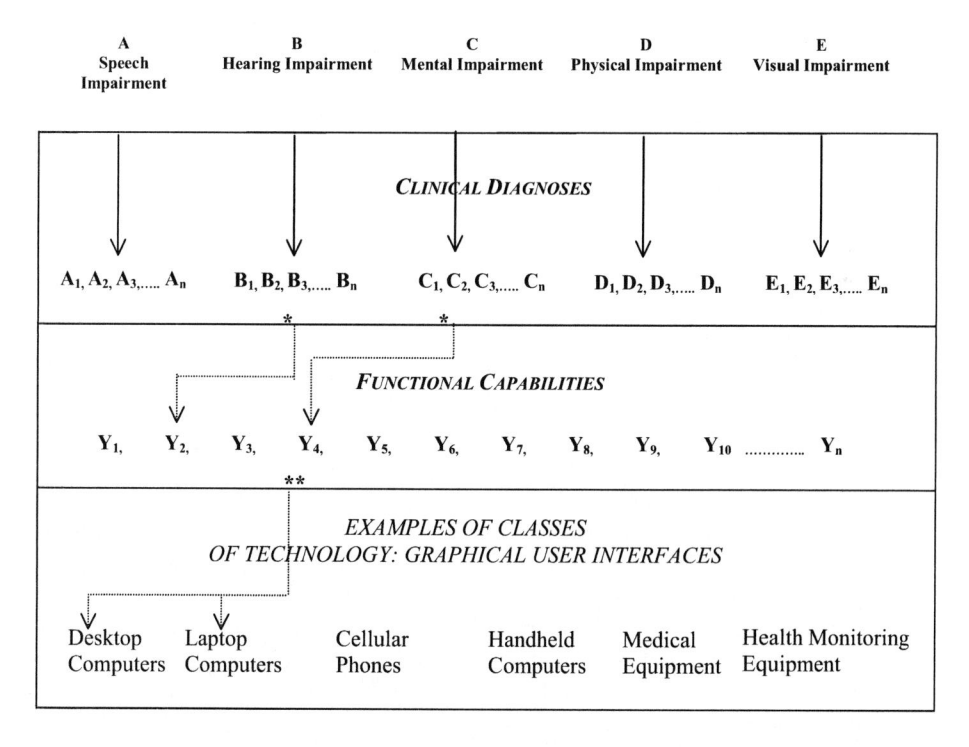

$Y_1,$ $Y_2,$ $Y_3,$ $Y_4,$ $Y_5,$ $Y_6,$ $Y_7,$ $Y_8,$ $Y_9,$ Y_{10} .... Y_n

EXAMPLES OF CLASSES OF TECHNOLOGY: GRAPHICAL USER INTERFACES

Desktop Computers   Laptop Computers   Cellular Phones   Handheld Computers   Medical Equipment   Health Monitoring Equipment

Figure 5.1. *Framework for the integration of clinical diagnoses, functional capabilities, and access to classes of technologies. Adapted from Jacko, Vitense, and Scott (2002). Reproduced with permission.*

As discussed by Jacko and Vitense, a subset of these functional capabilities affects interactions that are dependent on the specific classes of technology employed. To illustrate, consider a person who is diagnosed with a specific type of hearing disorder (represented as B_3 in Figure 5.1); this disorder results in the decline of the functional capability Y_2. In this instance, the capability Y_2 is a function not required to complete interactions with the listed technologies in Figure 5.1. However, other functional impairments, such as the mental impairment represented by C_2, reduce the user's functional capacity in terms of Y_4 and ultimately impinge on his or her successful interactions to manipulate information using a personal computer with communication devices.

A requisite to this framework is research that bridges ocular clinical diagnoses, functional abilities, and interactions relative to the specific platforms. To validate this framework, it has been necessary to reverse-engineer critical HCI scenarios for users exhibiting a range of visual function. This ongoing validation enables someone using the framework to direct attention to those components of the interaction that can be augmented or modified to compensate for users' decrements in functional vision.

To this end, the authors have undertaken a research agenda to validate this framework. This agenda includes three major elements:

- Establish links between clinical diagnoses and sets of functional capabilities (depicted in Figure 5.1 by *).
- Define the set of functional capabilities required to access information technologies (depicted in Figure 5.1 by Y_n).
- Derive empirical bases for the influence of specific functional capabilities on access to specific classes of technologies (depicted in Figure 5.1 by **).

A comprehensive discussion of the basic mechanics and physiological mechanisms that drive functional abilities for human-computer interactions can be found in Jacko et al. (2002), in addition to fundamental human factors/ergonomics texts such as Sanders and McCormick (1993). The taxonomy presents each type of diagnosis and subsequent impairment manifested in terms of functional aspects of human ability and its relevance to HCI.

We next provide an overview of the relative success of this approach in the development of an understanding of the impact of visual impairment on HCI. This represents a major research endeavor aimed at identifying user performance thresholds across a range of visual abilities, clinical diagnoses, and classes of technologies. Although the approach we outline details the work done in the area of vision, note that similar agendas are being pursued in the areas of cognitive and motor impairments (Davis, Moore, & Story, 2003; Hwang, Keates, Langdon, Clarkson, & Robinson, 2001; Steriadis & Constantinou, 2003).

The collection of studies presented on vision, though they are not yet fully translated into actionable design recommendations, is one of the largest-scale collections of investigations that, using quantitative HCI methods and clinical visual profiles, systematically derive performance thresholds for enabling this population to leverage the advantages that information technology affords. This application to visual function presents an example of how visual impairment can be analyzed in a four-phase process:

- classification of clinical diagnoses;
- extrapolation of visual functional abilities;

- derivation of interactions critical to graphical user interface (GUI) interaction across classes of technology; and
- evaluation of performance thresholds as well as the lessons gleaned from the emergent patterns of these quantitative approaches to understanding user functional ability and vision.

DERIVING PERFORMANCE THRESHOLDS ON GUI INTERACTION FOR VISUAL IMPAIRMENTS

This work is part of a substantial research agenda aimed at the empirical characterization of the manipulations critical to GUI interaction for a variety of users with a variety of visual profiles, including visual function and ocular diagnoses. Jacko and colleagues have completed numerous systematically derived, empirically based studies demonstrating how characteristics of users' functional vision and ocular diagnoses can influence performance on a variety of direct manipulation tasks for effective GUI interaction.

In this section, we present an in-depth review of ocular function, followed by an overview of the approaches being used to broaden knowledge about abilities relative to the range in visual function of users with visual impairments.

Classifying visual impairment. Visual impairment is defined by the American Foundation for the Blind (AFB) as encompassing all degrees of vision loss, from slight visual field loss to total blindness (Bailey & Hall, 1989). It is useful to differentiate among several terms that describe visual functioning beyond visual impairment because it may be indicative of the specific interaction strategies an individual is likely to employ to use GUIs. In addition to *visual impairment,* these terms include *blindness* and *low vision.*

The term *blindness* has both legal and functional definitions. *Legal blindness* is a level of visual function, as defined by the U.S. government, that identifies individuals whose vision affords them certain benefits and services but also restrictions (such as driving). In the United States, legal blindness is defined by visual acuity (the ability to resolve fine visual detail) and visual field (the physical space that is visible through the eyes).

Total blindness refers to the complete loss of visual function, and *functional blindness* comprises individuals who have the ability to perceive light but cannot resolve its shape or source. *Low vision* is visual loss that impedes tasks of daily living, though the ability to discriminate visual detail is still possible. Individuals classified as having low vision possess visual capabilities below what is considered normal, even with the use of corrective lenses (Biglan, Van Hasselt, & Simon, 1988; Jacko et al., 2002).

The AFB estimates that of the 1.3 million Americans who report legal blindness, 80% retain some residual vision (American Foundation for the Blind, 2002). It is also estimated that individuals with low vision outnumber completely blind individuals 3:1 (Newell & McGregor, 1997). Despite these statistics, research and development on assistive and universal technology solutions for individuals who are blind are more prevalent. Given the sheer number of people with low vision and projected growth rates, there is a clear need to investigate the nature of their interactions with information technologies. However, the disproportionate nature of solutions under development may be attributable in part to an unanticipated paradox: *Providing access to users who have limited vision can be more challenging than providing access to users who do not see at all* (Arditi, 2002, 2003a, 2003b).

Visual function is most commonly assessed in terms of visual acuity (e.g., 20/20). In addition to visual acuity, other visual functions that may be used to characterize low vision include color perception, contrast sensitivity, eye movements, and visual fields (Jacko & Sears, 1998; Jacko, Dixon, Rosa, Scott, & Pappas, 1999; Jacko, Rosa, Scott, Pappas, & Dixon, 1999; Jacko & Vitense, 2001). Table 5.3 provides a definition of each component of visual function, the impact it can have on computer use, and common, clinically based assessment techniques. (See Orr, 1998, and Schieber, 1994, for a complete review of the biological functioning and components of the ocular sensory system.)

Schieber (1994) reported a number of additional age-related functional visual characteristics that affect driving, which are also relevant to the use of GUIs and direct manipulation. A useful analogy is made between driving and using computers; both require a significant amount of visual attention in order to orient and navigate to a final destination. Specifically, Schieber described the age-related changing attributes of eye movements, attention, and visual search as functions of the visual sensory system. The detection of and orientation to events is a critical component of driving and other visually intensive tasks, such as interaction with GUIs. Although driving and GUI use differ in terms of near and far vision requirements (driving requires both near and far vision, but GUIs typically require only near vision), they are analogous in requiring visual attention and motor skill. In both tasks, for efficient task performance, an individual must attend to task-relevant stimuli while rejecting the extraneous.

Eye movements are a necessary component of processing information, especially in dynamically changing visual environments such as driving or GUIs. According to Schieber, "Optimal spatial resolution depends on the ocular motor system's capacity to acquire, track, and image a visual target at or near the fovea" (1994, p. 3). The fovea is the central portion of the retina and has a high concentration of photoreceptors. Eye movements are therefore composed of saccades and pursuit movements. Saccades are short, rapid movements of the eye with the purpose of centering visual information on the fovea. Pursuit eye movements are larger in nature, and their purpose is to track moving targets (keeping the target in range of the fovea). An additional component of eye movements are stationary periods, known as fixations, that occur in between saccades, during which information may be processed. Furthermore, deficits in the central, paracentral, and peripheral visual fields can pose different demands on vision, resulting in different search strategies and subsequent eye movements.

These studies have addressed the relative performance on several desktop computer tasks of a cohort of users with visual impairments caused by ocular disease and a cohort of age-matched controls without ocular dysfunction. Assessments of the interactions are achieved not only via traditional time and accuracy measures of performance but also using physiological methods such as electroencephalogram and eye tracking.

Identifying critical interaction scenarios. To date, the aforementioned studies have focused on direct manipulation tasks using the graphical user interface, primarily on desktop computers with the mouse (e.g., Jacko, Moloney, Kongnakorn, Barnard, Edwards, Emery, et al., 2005; Jacko, Rosa, Scott, Pappas, & Dixon, 1999; Jacko, Salvendy, & Koubek,

1995; Jacko, Scott, Sainfort, Moloney, Kongnakorn, Zorich, et al., 2003). More recently, the work has been extended to new classes of technology to address the needs of handheld human-computer interaction (Leonard, Jacko, & Pizzimenti, 2005). Perhaps the most challenging question is where to begin when trying to derive the needs of the population: Which tasks should be used to understand the most representative performance thresholds?

In response to this need, in 2001, Emery and colleagues identified the set of interaction scenarios that are essential for effective GUI interaction. Using hierarchical task analyses, the authors identified critical interactions relative to visual search, object manipulation, and text entry and manipulation.

Linking clinical functions to HCI performance thresholds. The specific critical direct manipulations identified by Emery et al. (2001) are the focus of the empirical research on visual impairment and GUI interaction. These studies addressed the visually rigorous task of icon search and selection in the presence of distractors, cursor movement, the direct manipulation of drag and drop in the absence of distractors, and the identification and selection of targets in a drop-down menu with distractors.

The emergent theme, which links these research studies, is the importance of understanding the specific details of users' impairment in terms of their functional ability to achieve reasonable levels of performance in HCI tasks. Empirical investigations by Jacko and her colleagues completed to date used the interaction framework approach for visual impairments and human-computer performance thresholds. (See Jacko, Leonard, & Scott, in press, for references on this series of studies.) They explored the problem space by considering the following: populations (types of visual impairment, age of users), task phases (i.e., critical interactions such as visual search, icon selection, cursor movement, drag and drop, and menu selection), complexity of the task (e.g., the presence of distractors), and various aspects of the interfaces taken that the user took into consideration and/or manipulated (e.g., icon size, background, visual profiles, tactile cues, auditory cues, text size, and desktop or handheld computer).

The outcomes of this collection of investigations by Jacko and her colleagues provided a scientific basis to enhance computer interaction for users with visual impairments and supported the derivation of design guidelines for such users. The concept of interaction threshold (also referred to as *performance threshold*) should be taken into consideration when profiling users in terms of ability and providing the baseline framework to direct future research in this area. These generalized findings on functional interaction include the following:

- IT solutions for individuals who are blind are typically inappropriate for those who maintain useful residual vision.
- The efficacy of design interventions depends on the nature and amount of a user's residual vision.
- Increasing text size and image size can be more problematic than assistive, especially considering the nature of the visual impairment.

Clearly, a user's clinical profile of visual function is one key determinant in interface accessibility and defines the limits of the interaction threshold. This supports the need

Table 5.3. Age-Related Visual Functioning Assessment Method and Impact on Computer Use

Ocular Function	Definition	HCI Impact	Assessment Method
Visual acuity	The ability to resolve fine detail in high contrast (Bailey & Hall, 1989).	The inability to resolve fine detail on a computer display has been shown to increase both the rate of errors and the time to complete direct manipulation tasks on GUIs (Jacko, Dixon, Rosa, Scott, Pappas, 1999; Jacko, 2000; Jacko, Scott, Sainfort, Moloney, Kongnakorn, Zorich, & Emery, 2003).	Snellen Acuity is assessed via the ability to correctly identify visual characters at 20 feet compared with what a "normal" individual can see at 20 feet (e.g. 20/20, 20/80, etc.; Ferris, Kassoff, & Bresnick, 1982).
Accommodation	The ability of the lens of the eye to focus light rays onto the retina, or the focusing power as viewing distance changes (Orr, 1998; Sanders & McCormick, 1993).	If not corrected with lenses (bifocals, trifocals), eye fatigue or headaches may occur when reading or working on computer display terminals (Fraser & Gutwin, 2000; Orr, 1998).	Measured with visual acuity scores (Sanders & McCormick, 1993)
Contrast Sensitivity	The measure of how visible an object is before it is indistinguishable from the environment—a person's ability to detect small changes in brightness (Bailey & Hall, 1989).	To discern objects from their background, older adults benefit from higher contrast and sharper edges around objects and texts (Jacko, Rosa, Scott, Pappas, & Dixon, 1999; Orr, 1998).	Assessed using the Pelli-Robson chart. This test assesses contrast sensitivity at different spatial frequencies (Pelli, Robson, & Wilkins, 1988).
Visual field	The area of physical space that is visible through the eyes; the sensitivity of both the central area and periphery of the retina. For normally sighted people, visual field is about 90° on either side of the nose when person is looking straight ahead (Bailey & Hall, 1989).	Blind spots in the visual field create barriers to a user's ability to systematically inspect a display (Bailey & Hall, 1989). Visual field loss has been shown to increase the time needed for visual search and target selection in a drop-down menu selection task. Restricted fields also make it impossible to survey the entire display at once (Fraser & Gutwin, 2000).	Commonly measured with the Humphrey Visual Field Analyzer, which generates maps of a person's visual field and measures his/her ability to detect small spots of lights on a constantly illuminated background. Two key measures of the Humphrey Visual Field Analyzer are mean pattern deviation and pattern standard deviation (Nelson-Quigg, Cello, & Johnson, 2000).

Color perception	The ability of an individual to discriminate between colors (Sanders & McCormick, 1993).	Color-coding of screen elements, especially blue-green coding, without effective illumination could lead to confusion for users with color deficits. Studies have shown that color perception is a predictor of performance with GUIs for individuals with visual impairments (Jacko, Rosa, Scott, Pappas, & Dixon, 1999).	Assessed using the Farnsworth-Munsell 100 Hues Test. Participants arrange colored caps in an ordinal series extending between two anchored colored caps. The order of caps as arranged by the individual is assessed to determine the presence and type of color confusion. This classification includes protan, deutan, tritan, and noncongenital color deficiency (Farnsworth, 1947).
Dark/light adaptation	How well the eye adjusts between dark and light illumination conditions so the optical system can resolve small differences in luminance (Orr, 1998; Schieber, 1994).	The dynamic nature of displays with respect to brightness and colors can impose performance decrements with diminished visibility attributable to adaptation.	Measured by pupillary response, contrast sensitivity, and visual acuity at different levels of luminance (Schieber, 1994).
Depth perception	The ability to determine the distance to objects in the environment (Orr, 1998; Schieber, 1994).	With respect to HCI, deficiencies in depth perception will have a significant impact in virtual reality applications.	Measured with the Worth-4 dot test. A patient wears glasses with one green lens and one red and looks at a target with two green dots, one red dot, and one white dot. Depending on the number and color of dots the patient sees, the examiner can determine whether the vision of one eye is being suppressed.

for universally accessible technologies to employ both strategies common to usability (such as determining user skills and abilities) and also basic aspects of the user's ability to sense and perceive the information presented on the interface. In addition, the visual profile and resulting interaction thresholds provide additional evidence that the most universally accessible and usable technologies are those that can dynamically adapt to the functional needs of each user as these needs change over time.

Framework extension. The foundation for this framework and methods used in its application across the field of visual function can be leveraged to account for other categories of functional limitation. These other nonvisual functions can be scoped and categorized in similar ways relative to clinical diagnoses, and their impact on performance thresholds for various interactions can be quantified. For example, one ocular condition that often results in motor impairment is tremors. This diagnosis can be associated with different types of tremors, such as action, resting, intention, essential, senile, and familial. These tremor types all possess varying levels of severity and frequencies of occurrence (Law, Sears, & Price, 2005). Hence, what is needed is an empirical link between the clinical assessment of tremor and functional IT capability. This will, in turn, translate into accurate assessments of motor capabilities necessary for IT use to enable the categorization of user performance when evaluating the merit of HCI solutions. In this way, the framework presented in this chapter is applicable to motor impairments as researchers endeavor to use functional assessment to improve information systems research, design, and technology matching (Price, 2006).

CONCLUSIONS AND FUTURE DIRECTIONS

The potential for individuals who experience perceptual, physical, and/or cognitive impairments to overcome their limited ability and lead more productive, independent lives has perhaps never been greater. This is a result of increased social awareness and legislation, and the emergence of new technologies has created substantial possibilities for enhancing the quality of people's daily lives. Yet, rapidly proliferating technology platforms and their applications to a greater number of usage scenarios intended to enrich and increase productivity often lack the design consideration necessary to accommodate the interaction capacities of the representative user population. This is largely because the current development and design cycles—though they promote the facets of usability and usefulness, incorporating user skill level, experience, and information needs—often do not address issues related to accessibility.

The rapid emergence of new technology has not only created an increased global demand for information and productivity but has also increased the likely set of users and the variance in their fundamental perceptual, sensory, cognitive, and motor capabilities. Attempts to meet these needs are often last minute, ad hoc, and based on a designer's and developer's personal experience and heuristics, a handful of field observations, limited user feedback, and/or legal responsibility. What have emerged are not the solutions that are typically incorporated into the mainstream product but a separate design to address each very specific group of users' needs (e.g., assistive technologies). This patchwork approach, though admittedly critical to enabling subsets of users, is highly inefficient and ineffective in light of the rapid proliferation of new technologies and platforms into various aspects life.

It is well established that the efficacy of design solutions is largely driven by the functional ability and limitations of each user, but the specific implications of different functional impairments for computer interactions are not necessarily well documented or understood by designers and developers. In this chapter, we have presented an integrated, scientific approach to derive interaction thresholds based on user functional impairment. The challenge to provide access to technology is not to be taken lightly and warrants a well-planned empirical approach to build quantitative- and qualitative-driven methods that can be applied in practice. An integrated research framework provides the potential for a heightened awareness of the restrictions that functional limitations impose on HCI interaction thresholds across interface conditions and contexts of use.

Achieving these goals is not trivial. For example, although the work accomplished to date on users with visual impairments is extensive, additional knowledge is needed regarding other aspects of visual function in connection with several interaction scenarios and technologies. Each new functional constraint and technology platform has the potential to be unique in its demands on the users. Leonard, Jacko, and Pizzimenti (2005) demonstrated that for users with macular degeneration, the design guidelines derived for desktop interaction with a mouse are not wholly applicable to interaction with a handheld computer and stylus. The handheld interaction was more highly sensitive to a greater number of aspects of visual function and thus was affected uniquely when different interface features were changed. Furthermore, this framework has yet to be applied across functional categories (e.g., cognitive, hearing, speech, or motor function) and in consideration of multiple impairments.

Despite heightened awareness of the unmet needs of the user populations discussed in this chapter, a considerable number of obstacles remain in order to facilitate accessible technologies for the widest possible range of users. Developers and designers need to not only routinely integrate functional profiles early in the design process of state-of-the-art technologies but also retrofit substantial numbers of technologies and systems to accommodate a greater variance in user ability. In addition, to meet the needs of the growing numbers of older adults, there is a need to consider the implications of multiple functional limitations across several sensory, cognitive, and motor abilities.

Furthermore, research needs to expand to additional interaction paradigms, such as voice- or gesture-based interfaces. The optimal solution, no doubt, is a customizable interface that can adapt to a user's unique set of functional abilities, even as those abilities change over time. Through technological advancements in HCI research, the concepts of perceptual and adaptive interfaces provide vast opportunities for individuals with limited functional ability to fully access electronic information. More specifically, multimedia and multimodal systems have substantial untapped possibilities for supporting a range of user needs.

In order to achieve universal access across classes of computing technologies, researchers must realize several specific objectives:

- establish empirical links between clinical diagnoses and sets of functional capabilities,

- define the set of functional capabilities required to access information technologies, and
- establish empirical bases for the influence of specific functional capabilities on access to specific classes of technologies.

In this chapter, we have aimed to establish a basis on which one can address these issues and have provided an example of one approach being used to understand the needs of GUI computer users with visual impairments. The efficacy and quality of the solutions derived for universal access depend on the quality and validity of the methods generated for HCI performance associated with functional impairment, clinical diagnoses, and the facets of direct manipulation and interface design technology.

Aside from working to meet productivity needs and legal requirements, technology designers, developers, and evaluators have a social and moral obligation to ensure that no user is unnecessarily burdened with the interactions required for access. The systematic approach outlined in this chapter provides a basis for computer users with limited functional abilities to claim or reclaim access to information technologies, sustain or regain independence in their lives, and take full advantage of the opportunities that information technologies afford.

ACKNOWLEDGMENTS

The authors gratefully acknowledge Erin Kinzel, who contributed to the final preparation of this manuscript.

REFERENCES

American Foundation for the Blind (2002). *Statistics and sources for professionals*. Retrieved August 11, 2006, from http://www.afb.org/info_document_view.asp?documentid=1367.

Anderson, R. H., Bikson, T. K., Law, S. A., & Mitchel, B. M. (1995). *Universal access to e-mail. Feasibility and societal implications*. Santa Monica, CA: Rand Corporation.

Arditi, A. (2002). Web accessibility and low vision. *Aging & Vision, 14*(2), 2–3.

Arditi, A. (2003a). Low vision web browsing and allocation of screen space resources. *Investigative Ophthalmology and Visual Science, 44*(5), 2767.

Arditi, A. (2003b). *Making text legible: Designing for people with partial sight*. Retrieved August 11, 2006, from http://www.lighthouse.org/print_leg.htm.

Assistive Technology Act (2004). Assistive Technology Act of 2004 *Putting Technology into the Hands of Individuals with Disabilities*. Retrieved August 30, 2006, from http://www.house.gov/ed_workforce/issues/108th/education/at/billsummary.htm

Bailey, I. L., & Hall, A. (1989). *Visual impairment: An overview*. New York: American Foundation for the Blind.

Biglan, A. W., Van Hasselt, V. B., & Simon, J. (1988). *Handbook of developmental and physical disabilities* New York: Pergamon Press.

Brown, S. S., & Robinson, P. (2004). Transformation frameworks and their relevance in universal design. *Universal Access in the Information Society, 3*, 209–223.

Davis, A. B., Moore, M. M., & Story, V. C. (2003). Context-aware communication for severely disabled users. In *ACM Conference on Universal Usability* (CUU'03; pp. 106–111). Vancouver, British Columbia New York: ACM Press.

Emery, V. K., Jacko, J. A., Kongnakorn, T., Kuruchittham, V., Landry, S., Nickles, G. M., Sears, A., and Whittle, J. (2001). Identifying critical interaction scenarios for innovative user modeling. *UAIS*, New Orleans, USA: 481–485, Mahwah: Lawrence Erlbaum Associates.

Farnsworth, D. (1947). *The Farnsworth Dichotomous Test for Color Blindness: Manual.* New York: Psychological Corporation.

Ferris, F. L., III, Kassoff, A., & Bresnick, G. H. (1982). New visual acuity charts for clinical research. *American Journal of Ophthalmology, 94,* 91–96.

Fraser, J., & Gutwin, C. (2000). A framework of assistive pointers for low vision users. In *Fourth Annual International ACM/SIGCAPH Conference on Assistive Technologies* (ASSETS 2000; pp. 9–16). Arlington, VA. New York: ACM Press.

Gardner-Bonneau, D. J. (Ed.). (1990). Assisting people with functional impairments [Special Issue]. *Human Factors, 32*(4).

Glinert, E. P. (2001). NSF funding for research on universal access. In *Proceedings of the Workshop on Universal Accessibility of Ubiquitous Computing (WUAUC'01),* Alcácer do Sal, Portugal, May 22–25, 2001 *(WUAUC '01)* (pp. 11–13). New York: ACM Press.

Hanson, V. L. (2004). The user experience: Designs and adaptations. In *Proceedings of the Workshop on Web Accessibiltiy at the World Wide Web Conference 2004 , New York, NY, May 18, 2004, (W4A at www2004),* (pp. 1–11). New York: ACM Press.

Hwang, F., Keates, S., Langdon, P., Clarkson, P. J., & Robinson, P. (2001). Perception and haptics: Towards more accessible computers for motion-impaired users. *In the Proceeding of the 2001 workshop on Perceptive User Interfaces, Orlando, Florida* (pp. 1–9). New York: ACM Press.

Jacko, J. A. (2000). Visual impairment: The use of visual profiles in evaluations of icon use in computer-based tasks. *International Journal of Human-Computer Interaction, 12*(1): 151–164.

Jacko, J. A., Dixon, M. A., Rosa, R. H., Jr., Scott, I. U., & Pappas, C. J. (1999). Visual profiles: A critical component of universal access. *Human factors in computing systems, Pittsburgh, PA* (pp. 330–337). New York: ACM Press.

Jacko, J. A., Leonard, V. K., & Scott, I. U. (in press). Perceptual impairments: New advancements promoting technological access. In A. Sears & J. A. Jacko (Eds.), *Human-computer interaction handbook* (2nd ed.). Mahwah, NJ: Erlbaum.

Jacko, J. A., Moloney, K. P., Kongnakorn, T., Barnard, L., Edwards, P. J., Emery, V. K., Sainfort, F., & Scott, I. U. (2005). Multimodal feedback as a solution to ocular disease-based user performance decrements in the absence of functional visual loss. *International Journal of Human-Computer Interaction, 18*(2), 183–218.

Jacko, J. A., Rosa, R. H., Jr., Scott, I. U., Pappas, C. J., & Dixon, M. A. (1999). Linking visual capabilities of partially sighted computer users to psychomotor task performance. In *9th International Conference on Human-Computer Interaction* (pp. 975–979). Mahwah: Lawrence Erlbaum Associates.

Jacko, J. A., Salvendy, G., & Koubek, R. J. (1995). Modeling of menu design in computerized work. *Interacting With Computers, 7*(3), 304–330.

Jacko, J. A., Scott, I. U., Sainfort, F., Moloney, K., Kongnakorn, T., Zorich, et al. (2003). Effects of multimodal feedback on the performance of older adults with normal and impaired vision. *Lecture Notes in Computer Science, 2615,* 3–22.

Jacko, J. A., & Sears, A. (1998). Designing interfaces for an overlooked user group: Considering the visual profiles of partially sighted users. In *ACM Conference on Assistive Technologies* (pp. 75–77). Marina del Rey, CA. New York: ACM Press.

Jacko, J. A., & Vitense, H. S. (2001). A review and reappraisal of information technologies within a conceptual framework for individuals with disabilities. *Universal Access in the Information Society, 1,* 56–76.

Jacko, J. A., H. S. Vitense, & Scott, I. U. (2002). Perceptual impairments and computing technologies. In *The human-computer interaction handbook: Fundamentals, evolving technologies and emerging applications* (pp. 504–522). Mahwah, NJ: Erlbaum.

Keates, S., & Clarkson, P. J. (2003). Countering design exclusion: Bridging the gap between usability and accessibility. *Universal Access in the Information Society, 2,* 215–225.

Law, C. M., Sears, A., & Price, K. J. (2005). Issues in the categorization of disabilities for user testing. In *Proceedings of HCII 2005* (CD Version). Mahwah, NJ: Erlbaum.

Leonard, V. K., Jacko, J. A., & Pizzimenti, J. J. (2005). An exploratory investigation of handheld computer interaction for older adults with visual impairments. In *Seventh International ACM SIGACCESS Conference on Computers and Accessibility (ASSETS '05) Baltimore, MD,* (pp. 12–19). New York: ACM Press.

Lidwell, W., Holden, K., & Butler, J. (2003). *Universal principles of design: 100 ways to enhance usability, influence perception, increase appeal, make better design decisions, and teach through design.* Gloucester, MA: Rockport Publishers.

NARIC. (2005). Rehabilitation Engineering Research Centers (RERCs), Retreived August 30, 2006. from http://www.naric.com/research/pd/results.cfm?type=type&display=detailed&criteria=Rehabilitation%20Engineering%20Research%20Centers%20(RERCs.

Nelson-Quigg, J. M., Cello, K., & Johnson, C. A. (2000). Predicting binocular visual field sensitivity form monocular visual field results. *Investigative Ophthalmology and Visual Science, 41,* 2212–2221.

Newell, A. F., & McGregor, P. (1997). Human-computer interfaces for people with disabilities. In M. G. Helander, T. K. Landauer, & P. V. Prabhu (Eds.), *Handbook of human-computer interaction* (pp. 813–824). New York: Elsevier Science.

Nickerson, R. S. (Ed.). (1978). Human factors and the handicapped [Special Issue]. *Human Factors, 20*(3).

Orr, A. L. (1998). *Issues in aging and vision.* New York: American Foundation for the Blind.

Pelli, D. G., Robson, J. G., & Wilkins, A. J. (1988). The design of a new letter chart for measuring contrast sensitivity. *Clinical Visual Science, 2,* 187–199.

Price, K. J. (2006). *Using functional assessment to improve information systems research, design, and technology matching.* Unpublished doctoral dissertation, University of Maryland Baltimore County, Baltimore, MD.

Quillen, D. A. (1999). Common causes of vision loss in elderly patients. *American Family Physician, 60*(1), 99–108.

Sanders, M. S., & McCormick, E. J. (1993). *Human factors in engineering and design,* 7th ed. New York: McGraw-Hill.

Schieber, F. (1994). *Recent developments in vision, aging and driving: 1988–1994.* Washington, DC: Transportation Research Board.

Shneiderman, B. (2000). Universal usability: Pushing human-computer interaction research to empower every citizen. *Communications of the ACM, 43*(5): 84–91.

Stary, C. (2001). User diversity and design representation: Towards increased effectiveness in design for all. *Universal Access in the Information Society, 1,* 16–30.

Stephanidis, C., & Savidis, A. (2001). Universal access in the information society: Methods, tools, and interaction technologies. *Universal Access in the Information Society, 1,* 40–55.

Steriadis, C. E., & Constantinou, P. (2003). Designing human computer interfaces for quadriplegic people. *ACM Transactions on Computer Human Interaction, 10*(2), 89–110.

U.S. Census. (2002). *Facts and figures: U.S. Census Bureau.* Retrieved January 31, 2006, from http://www.census.gov/Press-Release/www/2002/cb02ff11.html.

CHAPTER 6

Multidimensional Aspects of Slips and Falls

By Krystyna Gielo-Perczak, Wayne S. Maynard,
& Angela DiDomenico

The study of slips and falls has traditionally focused on body kinematics and tribology. However, this strictly mechanical approach does not allow scientists to assess the importance of each component in relation to the complete system, and thus it lacks integration. The purpose of this chapter is to present and demonstrate the components of a broad analysis for in-depth understanding of slips and falls while walking on level surfaces. In most slip-and-fall studies, balance analysis is simplified and attributed to the point of heel contact. To determine sufficient fall prevention strategies, however, one must analyze balance before the critical moment of lost control. Such an approach requires the sciences of biomechanics, mechanics, anatomy, and neuromuscular control, as well as tribology. Causes of slips and falls are complex, and prevention approaches are often reactive, driven by high-injury trends and lawsuits. Prevention strategies need to be more proactive: Understanding the causes of accidents can help in identifying and correcting hazards *before* they cause problems. Examples include reporting incidents, selecting the right flooring, selecting footwear, and implementing proper floor-cleaning procedures. A combined effort among all members of the organization, including communication across the entire work system, is critical to the success of slip-and-fall prevention efforts.

Slips and falls represent either the highest or second-highest type of workers' compensation claim for most industry groups in the United States (Leamon & Murphy, 1995). Further, 11% of claim cases and 12% of claim costs related to low back pain are attributed to slips and falls (Murphy & Courtney, 2000). The direct cost of same-level falls is more than $4 billion, second only to overexertion. But according to a survey of workplace executives, most executives perceive same-level falls to be much less of a problem—the seventh most important cause overall (Liberty Mutual Group, 2004).

It is not fully understood why there is a discrepancy between reality and the perceived importance of slips and falls, and why slips and falls continue to represent one of the most costly safety problems today. The reason might lie in a lack of understanding as to how slips and falls occur and industries' failure to implement a managed safety process that targets those complex causes.

To address this problem, we describe the multidimensional aspects of slips and falls. The first part of the chapter provides a multifaceted perspective on how slips and falls occur through an exploration of how existing biomechanical, tribological, and operational concepts relate to stability control during slips. Some slips occur because of a general disregard for industry standards and guidelines, inappropriate floor design, poor housekeeping, or insufficient management involvement. In the second part, we present considerations that should be taken into account to reduce slips and falls.

SYSTEM COMPONENTS OF SLIPS AND FALLS

A person's stability control, dictated by physical constraints, represents his or her ability to maintain and recover balance after a slip. Some of these constraints are attributable to environmental limitations (e.g., the magnitude and direction of the ground reaction force, surface friction coefficient, and compliance). Others may be attributable to anatomical dimensions (e.g., foot geometry, body mass and its distribution, or segment length and height), physiological parameters (e.g., strength, rate of muscle force rise, or gains and delays of feedback control), or perhaps cognitive and behavior origins (e.g., reaction time, attention, or fear of falling; Redfern, Cham, Gielo-Perczak, Grönqvist, Hirvonen, Lanshammar, et al., 2001). Each of these constraints has a different impact on a person's ability to maintain and recover balance, which can be assessed in terms of performance kinematics (Pai & Iqbal, 1999). Systems analysis provides a convenient quantitative condition in which the biomechanics, neuromuscular control, and tribology systems and their components can be compared.

The systems approach also integrates all components necessary for a successful slip-and-fall prevention strategy (Grönqvist et al., 2003). Many slips and falls occur as a result of distractions or failure to detect hazards. They are embarrassing events and, unless injury is involved, are rarely reported. Hazards such as water, grease, soil, and debris change continually and are also underreported. Such hazards often go unabated until there has been a fall. A slip-and-fall prevention process must consider injury and hazard surveillance methods in addition to technical interventions such as selection of flooring, floor treatments, cleaning chemicals, slip-resistant footwear, and matting.

Biomechanical Aspect

The systems approach is applied to identify functional balance limitations based on biomechanical analysis of joint loading (Gielo-Perczak, 2001). In predicting balance performance during slips and falls, it is important to consider the ways in which the nervous and musculoskeletal systems control and constrain movement. The dynamic model of a person's stability or ability to recover balance after a slip has not yet been formulated. This would require a mathematical model that expresses the systems variables in input and output form that may be observed, manipulated, or learned.

Human body movement. In the study of biomechanics, human movement is described as a result of the constant influence of several systems—the most influential being the musculoskeletal, nervous, and sensory systems (see Figure 6.1). During a state of dynamic balance, there are continual fluctuations of the system components. This implies that the task of controlling movement is complex and requires a multidisciplinary approach. Several hypotheses applied to recovery of balance after a slip have been suggested, not only for analyzing balance of the musculoskeletal system but also for integrating central commands of the nervous system and sensory feedback to produce coordinated recovery movements.

Musculoskeletal system. The musculoskeletal system is composed of the three subsystems: bones, muscles, and ligaments. The components of human body movement shown

Multidimensional Aspects of Slips and Falls

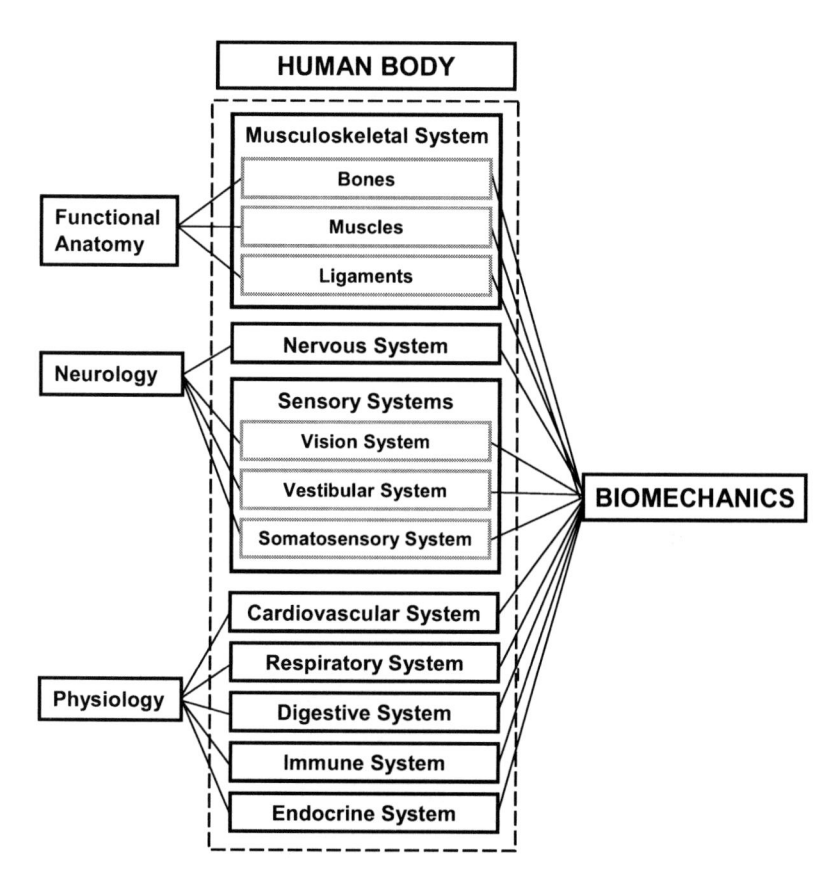

Figure 6.1. Biomechanics is the science that examines forces acting on and within the human body that are the result of the musculoskeletal, nervous, and sensory systems.

in Figure 6.1 are under unique musculoskeletal, physiological, and environmental constraints. This implies three different types of analyses: muscular, movement, and behavioral, at the quantitative or qualitative level. Movements are the result of a complex set of interacting inputs (external loadings), mediating processes (e.g., internal responses of the musculoskeletal system), and outputs (kinematic parameters of chosen body parts). They demonstrate an almost infinite variety of interacting factors that support the idea that there are no fundamental patterns of human movement, and human movement is a self-regulating system. How the human organism learns to move, and how body movements adapt to simple and complex environmental situations, are the questions that should be addressed in resolving problems of slips and falls.

Human body control during slips and falls is a component of a complex system involving the interaction between the physical world and the organism itself. However, in most slips-and-falls studies, the musculoskeletal system is not considered as a prime element influencing human body balance.

Nervous system. Recovery from a slip is a result not only of the mechanical body response but also of neuromuscular coordination. The nervous system controls body movement and, by sending the proper series of impulses along nerves to activate the appropriate muscles, decides which joints need to be moved, exactly when they should be moved, and by how much.

Sensorimotor system. Balance control is a complex motor skill that involves integrating many types of sensory information and performing movement in order to achieve balance during unpredictable slips (Horak, 1997). The structure of the body, the geometry of muscular actions, and the locations and characteristics of orientation and motion receptors within the body all contribute to alternative sensory and motor strategies for a postural task (Nashner, 1985).

Balance is not static but consists of a dynamic interplay between the sensory systems, which makes the whole body flexible and open to change. It is believed that using visual, vestibular, and somatosensory information appropriately under various sensory contexts maintains stability.

Vision is the single most important source of sensory information during goal-directed locomotion. It has been shown that subjects employ recovery patterns that are different if visual information is available (Grabiner, Koh, Lundin, & Jahnigen, 1993).

Biomechanical attribution of stability. The importance of stability during slips and falls is well known, but stability has a complex nature and is best represented by a multi-level approach: biomechanical and neuromechanical. It integrates different points of view that interact and overlap. The most important task within the biomechanical approach is to assess balance by predicting the risk of a fall and to define the range of stability before losing balance. A large number of individual components in the human body affect balance, but effects also arise from mutual interactions of those components in both physiological and neurological systems (Figure 6.1).

The neurological system must be capable of predicting the emergent properties of the human body system, those properties that are possessed by the body as a physiological and musculoskeletal system. The neuromechanical approach relates to controlling the nervous system and provides information on the feedback properties in combination with the musculoskeletal system dynamics. There is a need for an integration of physical principles and neurophysiology in the study of human movement. Understanding the nature of stability control during slips and falls compels one to analyze the controlling role of the nervous system.

In human body motion, one can distinguish between motion stability, which focuses on the response of a dynamic system to disturbances in an unlimited time, and balance control, which is the stability of a dynamic system restricted within certain time limits. Stability during slips and falls is not static but consists of a dynamic interplay between two complementary tendencies that make a body flexible and open to change. The internal flexibility of a body that is controlled by dynamic relations rather than rigid mechanical structures gives rise to a number of characteristic properties that can be seen as different aspects of the same dynamic principle of self-organization. If one considers a body as a system, then the equilibrium state is termed *stable* when a system tends to return to its initial equilibrium position after perturbation. The stable system implies the capacity to develop the forces that are necessary to oppose the perturbing forces. A fall is usually caused by multiple factors that may amplify one another through interdependent feedback loops. This leads scientists to wonder why it is sometimes so difficult to avoid a fall.

In postural control, three important aspects can be distinguished (Jacobs & Burleigh-Jacobs, 2000):

1. The central nervous system continuously receives afferent information from all peripheral sensory systems.
2. The central nervous system continuously adapts and learns based on all sensory inputs to achieve postural stability.
3. The central nervous system adaptation and (re-)learning of postural control depends on intent, experience, instruction, and environment.

Biomechanical studies of balance point out the large number of fluctuations of the center of pressure (COP) as the point location of the vertical ground reaction force for the left and right foot in both the anterior/posterior (A/P) and medial/lateral (M/L) directions. The location of the COP reflects the neural control of the ankle muscles (Winter, Patla, Prince, Ishac, & Gielo-Perczak, 1998).

Body equilibrium and posture are different phenomena. Body equilibrium must take into account the relation of the body to gravity, which means the center of gravity projected onto the ground must be within the support perimeter. The greater this perimeter, the more stable is the body. During normal human locomotion, the center of mass (COM) falls outside the boundaries of the base of support for 80% of the stride, which, from a purely static viewpoint, represents an unstable condition. The traditional goal of postural control—holding the body's center of mass over the base of support—is limited and represents only a small subset of the many postural capabilities of the central nervous system. The nervous system does not calculate body center of mass. However, researchers use center of mass in an effort to have a self-limited, understandable variable.

Posture describes the overall body attitude and is a composite of segmental postures that are under neurophysiological control. Postural stability is a dynamic phenomenon that depends on several factors. Most studies by Bouisset and Zattara (1987, 1990) have been concerned with *reaction equilibration* (from a perturbation unknown by the subject) and *action equilibration* (which results from an anticipated perturbation). These include the movements controlled by the autonomic nervous system, such as thoracic movements, and voluntary movements controlled by the sensorimotor system. As Bouisset and Zattara stated further, the relations between posture and voluntary movement are complex and apparently conflicting: Each voluntary movement perturbs balance, but unbalance must be controlled—and, to a certain extent, limited—so as to perform the movement efficiently. This confirms the existence of anticipatory postural adjustment (APA), which depends on movement and the support conditions. Each body part must be balanced with respect to the adjacent ones.

Bouisset and Zattara also assumed that intentional movement is a perturbation to balance and that a counter-perturbation must be developed for movement to be performed efficiently. They proposed to call this ability to react a *posturo-kinetic capacity*. Posture and balance are not mutually independent; Do and his collaborators asked whether posture and balance might have the same control mechanism (Do, Brenière, & Brenguier, 1982).

Among the different theoretical insights on slips and falls is the intuitive recognition by Leamon and Li (1990) that the likelihood of a fall arising from a slip is considerably greater at a point of discontinuity in the frictional characteristics of the walking surface. Thus, to analyze a full spectrum of stability, one should integrate the tribological aspect.

Tribological Aspect

Tribology, the study of the interaction between sliding surfaces, is derived from the Greek *tribos*, meaning "rubbing." The field of tribology includes the analysis of friction, wear, and lubrication and the application of these principles to mechanical design, manufacturing processes, and machine operation. One of the goals in tribology is to develop a method for estimating the probability of slips and falls based on measurements of available friction and required friction. Thus, tribology helps describe causes and support preventative strategies. Friction is the resistance to movement of one body over another. In this case, it is a function of the interface between the floor and shoe sole, the wear of the shoe sole and floor surface material, and lubrication attributable to contaminants such as grease, water, and dirt.

The coefficient of friction (a dimensionless quantity, μ) between two contacting surfaces is defined as the ratio of friction (horizontal) force, F, to normal (vertical) force, F_N, at the interface, which is mathematically expressed as follows:

$$\mu = \frac{F}{F_N} \qquad (1)$$

The higher the coefficient of friction (COF) value, the higher the slip resistance. Guidelines on COF and safety will be discussed later. Other factors that may increase the likelihood of slips and falls include poor lighting, unexpected changes in environment (including transitions from slip-resistant to nonslip-resistant surfaces), and transitions in surface height.

Tribological attribution of stability. When the frictional requirements of walking exceed the frictional capabilities of the environment, a slip and fall may occur. Recent studies of the *required* or *utilized* coefficient of friction during the gait cycle and its relationship with the *available* friction between the shoe and floor surface help provide an increased understanding of slip-and-fall probability. Required coefficient of friction—sometimes called *required friction, friction usage,* or *utilized friction* (Grönqvist et al., 2001)—represents the minimum friction that can be supported during ambulation to prevent a slip and resulting fall. Redfern et al. (2001) described *required* friction as the baseline friction demand for safe gait, typically measured on dry, nonslip surfaces. Burnfield and Powers (2003) described *utilized* coefficient of friction during walking as the ratio between the shear (resultant of the fore-aft and medial lateral forces) and vertical components of the ground reaction forces. Ground reaction forces generated as a person walks across a given surface can be measured by a force plate. When the required COF is greater than the available or measured COF, then, theoretically, the probability of a slip and resulting fall is increased.

Multidimensional Aspects of Slips and Falls 171

Measuring slip resistance and limitations of the slip resistance value. Static and dynamic coefficients of friction have dynamic natures. They are dependent on many factors, including angle (e.g., Hanson, Redfern, & Mazumdar, 1999), temperature, viscosity, contamination, and velocity (Grönqvist, 1995). Tribometers, or slipmeters, allow one to measure the relative slipperiness of a floor by determining a slip index (Irvine, 1970, 1976) or static coefficient of friction (SCOF) between the shoe sole and the floor surface. Traditional drag-type devices are suitable only for static friction measurements on dry and clean surfaces (Chang, Grönqvist, Leclercq, Brungraber, et al., 2001; Chang, Grönqvist, Leclercq, Myung, et al., 2001). For this reason, a horizontal pull slipmeter (HPS) shown in Figure 6.2 is not recommended for use on wet floors.

In tribometry, stiction (pronounced *stickshon,* also spelled sticktion) is the tendency of two surfaces in forceful contact, in the presence of a lubricating interface or contaminant, to bond together if there is a period of time between initial contact and initiation of relative motion (ANSI A1264.2-2006). It is a function of the dwell time between the instant of surface contact and the application of horizontal force. This phenomenon has also been referred to as *surface-tension adhesion* (Smith, 2002). When the water is squeezed away, the two surfaces have a tendency to stick to each other. An example of stiction is when water is placed between two panes of glass. Notice that the two surfaces essentially stick to each other.

Some experts believe a similar but less obvious phenomenon occurs when the test foot of a drag-sled slipmeter (such as the HPS) is in contact with a floor in the presence of water. Water is squeezed away from the test pad surfaces by the weight of the slipmeter. In other words, a period of time goes by from when the device is placed on the wet floor to when the horizontal pulling force is applied to obtain the COF or slip index. When the horizontal forces are applied *after* the normal force F_N, stiction can occur.

Figure 6.2. Horizontal Pull Slipmeter (HPS).

Stiction can artificially increase the COF and produce an erroneous measure of slip resistance. In other words, the floor appears to be more slip resistant than it really is. The Brungraber Mark II (PIAST) shown in Figure 6.3 and English XL shown in Figure 6.4 are two slipmeters that eliminate stiction by applying the horizontal and vertical components *at the same time*; thus, they are usable on both wet and dry floors.

The floor traction results measured by the various slipmeters are a great way to show floor conditions before and after a recommendation is implemented. For example, the horizontal pull slipmeter (Figure 6.2) might be used to determine the slipperiness of a vinyl composition tile (VCT) before and after a wax or polish is applied, or before and after the floor is cleaned. The Brungraber Mark II (PIAST) or English XL testers might be used in a restaurant kitchen to determine the effectiveness of a floor-cleaning protocol or the differences between various floor surface materials. Either way, the output of a slip tester is generally used by safety professionals to justify solutions. It is difficult to use tribometer results to show compliance with federal, state, or local regulations.

Other popular slipmeters available for walkway surfaces include the Universal Walkway Tester (UWT), now called the BOT-3000 tribometer. In the United Kingdom, popular portable slipmeters include the Tortus and the pendulum slip tester.

Tribometers or slipmeters report results as either SCOF or slip index; however, measuring SCOF precisely is difficult. It is not known, however, what method and which tribometer or slipmeter produces the most accurate measurements. Several portable walkway surface slipmeters are available (Chang, Grönqvist, Leclercq, Brungraber, et al., 2001). Measurements obtained from slipmeters provide a relative measure of slipperiness. There can be wide variability in slip-resistance readings between types of slipmeters, sensor pad

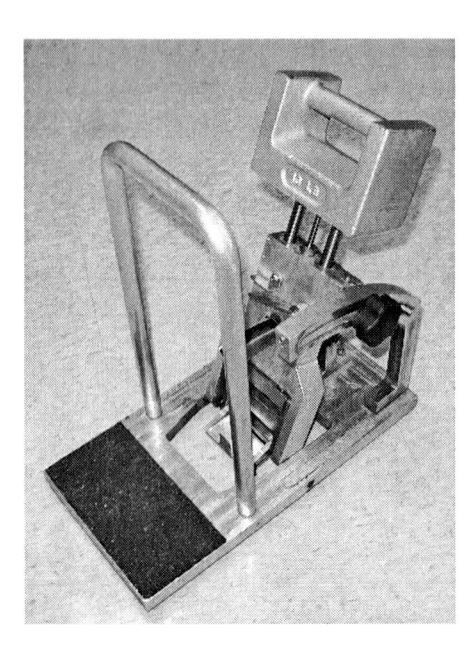

Figure 6.3. Brungraber Mark II Portable Inclinable Articulated Strut Slip Tester (PIAST).

Multidimensional Aspects of Slips and Falls

Figure 6.4. English XL Variable Incidence Tribometer (VIT).

materials, and whether the floor tested is wet or dry. Slipmeter readings can also be interpreted differently. Consider some viewpoints below of floor tile manufacturers and safety professionals:

Floor tile manufacturers. Many floor tile manufacturers market the slip resistance of their products by providing lab test results that certify the tile to a tested SCOF of 0.5 or more. The 0.5 SCOF level is known to offer some degree of safety, but these measurements are taken only when the tile is new and not in real-world conditions such as when dirty or worn.

The American Society for Testing and Materials (ASTM) C1028 is the test method often cited by tile manufacturers to support the slip resistance of their products. ASTM C1028 specifies a Horizontal Dynamometer Pull-Meter Method with a Neolite™ heel assembly. The Horizontal Dynamometer Pull-Meter is a drag-sled slipmeter with a 50-pound drag-weight. Wet tests are acceptable under ASTM C1028, though some experts recommend disregarding wet floor results when using drag-sled type slipmeters because of the stiction effect (described earlier).

Safety professionals. Most floor surfaces are slip resistant when clean and dry. The performance of that same floor in the presence of contaminants and wear over time determines the slip-resistance performance of that floor. It is amazing how many companies choose flooring on the basis of aesthetics and/or cost rather than slip-resistance performance and durability over time. Installing the right floor the first time can potentially save millions of dollars in costly floor treatments, repairs, or even replacement.

Safety professionals deal with real-world conditions and accept floor surface SCOF as only a small part of the slip-and-fall exposure. Slipperiness depends on the floor surface material, the presence of contaminants, water, the type of footwear used, and wear of the floor and footwear over time. A dry, clean floor may have an SCOF over 0.5, but add a wet spot and/or a buildup of grease, and the SCOF can be reduced to a much lower level.

For this reason, floor-cleaning protocols are essential to maintaining good traction. SCOF can tell us only what the static coefficient of friction or slip index (depending on device) of the floor was at the exact moment of testing, but not that the floor is always safe.

Operational Aspect

The operational aspect offers a framework to conceptualize work system elements and is concerned with the design of the overall work system. It must involve the technological and personnel subsystems as well as the external environment and the interrelationships between them (Hendrick & Kleiner, 2001). It should involve employees at all organizational levels. Work system design that has congruent subsystems and alignment and that considers the sociotechnical characteristics of the organization can lead to substantial benefits, such as increased productivity, health, and safety.

Technological subsystem. The technology subsystem includes proactive facility design to minimize hazards. This includes installing the right floor surface for the environment, selecting appropriate floor surface treatments, selecting front- and back-of-the-house matting, using appropriate floor-cleaning chemicals for likely contaminants, and selecting appropriate footwear. Aesthetics and cost often drive the selection of most of these interventions, but all are involved in a successful slip-and-fall prevention process.

Personal subsystem. Frontline employees play an important role in the process. Their involvement can be promoted by offering flexibility of different footwear and styles. Employees should feel free to report hazards and incidences to their managers or supervisors without reprisal or penalties. Action should be taken to correct hazards as soon as possible.

A sociotechnical approach can be used to illustrate why many slip-and-fall prevention processes are less effective than they otherwise could be (Maynard & Curry, 2005). To conceptualize macroergonomic issues as related to slips and falls, a process can be described using three levels of a work system: organizational, group, and individual. Within each of these levels, the sociotechnical elements (technological and personnel subsystems) and physical work environment factors can be related to potential outcomes that could be used to measure the success of a safety and health process including the falls prevention program (Robertson, Maynard, Huang, & McDevitt, 2002).

Organizational. Senior managers must take the lead and set an example for slip-and-fall prevention. This includes being proactive in prevention, holding managers and supervisors responsible and accountable for program implementation, and providing funding for interventions. Proper flooring and floor treatments, lighting improvements, slip-resistant footwear, and appropriate matting are examples of crucial interventions.

Group. Everyone in the organization must work together to prevent slips and falls. Stakeholder groups in a slip-and-fall prevention process include facilities managers, operations managers, risk managers, and safety, purchasing, occupational health, engineering, maintenance, and housekeeping personnel. Designing facilities to reduce risk by selecting the right flooring, matting systems, cleaning chemicals, footwear, and more must be done right the first time.

Individual. Injury and hazard surveillance, or worksite analysis, is essential and must involve the worker. This is usually the weakest point of a slip-and-fall prevention process, and no easy solution for dealing with the issue exists. Three safety and health surveillance approaches—employee reports, review of existing records, and job surveys—are recommended to assist managers and safety and health professionals in managing risks associated with slips and falls.

Employee reports. Prompt reporting of slips and/or falls with or without injury is important. Employees are often reluctant to do this because of embarrassment, and for noninjurious falls, there is rarely a system in place to accomplish this. The American Society for Testing and Materials addresses this issue in ASTM F1694, *Standard Guide for Composing Walkway Surface Evaluation and Incident Report Forms for Slips, Stumbles, Trips, and Falls* (ASTM International, 2004a). The standard includes incident-reporting guidelines, investigation approaches, information to collect, and sample forms.

Review of existing records. Records such as workers' compensation claims reports and Occupational Safety and Health Administration (OSHA) logs can provide valuable information. Focus on these data, however, is a reactionary approach. Unfortunately, loss trends, serious injuries, and expensive lawsuits (rather than proactive prevention) seem to drive most interventions.

Job surveys. Job surveys include facility audits, supervisor interviews with workers, and surveys dealing with hazards. The challenge in dealing with slips and falls is that physical hazards are dynamic and variable from minute to minute. Employee involvement is thus critical. "Clean as you go" policies in the restaurant industry and sweep logs in the retail grocery industry are two examples of programs that recognize the dynamic nature of hazards, and they can be an important part of a successful slip-and-fall prevention program.

Participation by key stakeholders inside and outside the organization is critical to the success of the program. A process that addresses prevention (preinjury) and return-to-work (postinjury) is important. Key individuals who need to be involved include human resources personnel, leaders, safety and health professionals, engineering and maintenance staff, the health care provider, the rehabilitation provider, the workers' compensation insurer, and others.

The worker is in the middle of this process. Communication between the worker and all other parties is essential for promoting safe behavior on the job and enabling return to work after an injury occurs (Robertson et al., 2002).

CONNECTIVITY CONSIDERATIONS OF THE SYSTEM COMPONENTS

The key to a comprehensive multidimensional approach to the reduction of slips and falls lies in the synthesis of the three different components, as illustrated in Figure 6.5: the study of human body movement, or *biomechanics*; the study of the design, friction, wear, and lubrication of interacting surfaces, or *tribology*; and an examination of the corporate *operational* environment. To understand the linkage, one must map a configuration of these approaches.

The *biomechanics component* contributes research information about the whole body movement, its body part velocities, accelerations, symmetry, forces, mutual joint activity, multijoint timing, and stability measures (COP variability) during voluntary activities requiring balance control. The whole dynamics of a human body during slips and falls includes this group of parameters.

During initiation of a slip or fall, the first observable variable is most often the position of the center of mass (COM) linked to the base of support. If the COM position is over the base support, then the body is in balance. The positions of the COM, as well its velocities and accelerations, are linked with the reaction time of the human body, which is critical during recovery from a stumble. Researchers studied the dynamics of postural control in human locomotion with a model using experimentally applied impulsive force disturbances. The simulation results suggested that early responses (within 80 ms) can be effective in compensating for disturbances in this early period (Yang, Winter, & Wells, 1990).

The results of Grabiner et al. (1993) suggest that recovery from a stumble is dependent on lower-extremity muscular power and the ability to restore control of the flexing trunk. Why a body loses balance and how balance can be recovered to prevent a fall are important study areas adding to the overall knowledge of prevention.

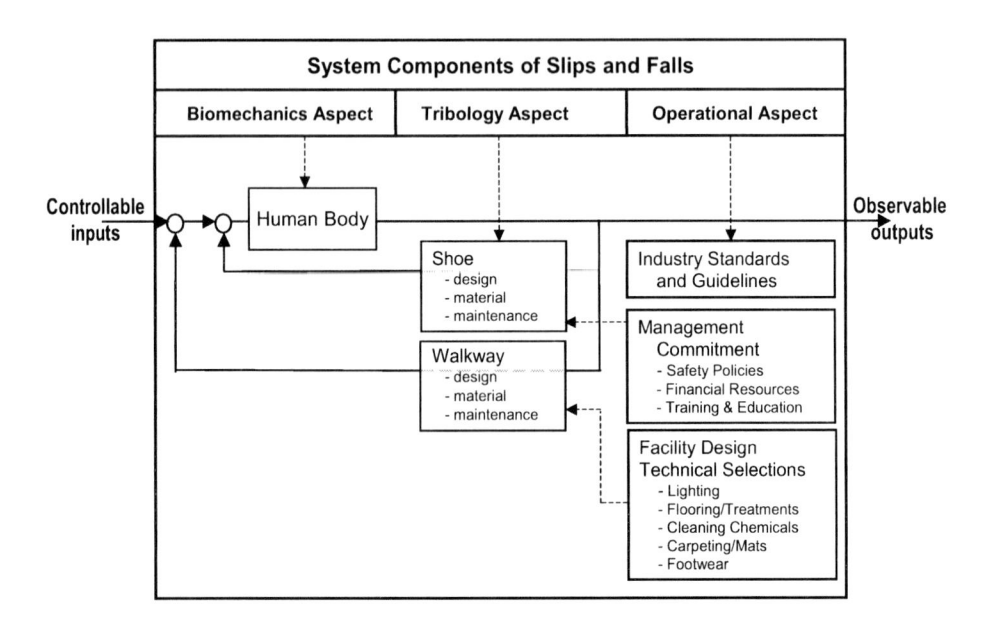

Figure 6.5. Modeling dynamic system of slips and falls; possible dependencies and interactions among biomechanical, tribological, and operational components during slips-and-falls accidents.

The second condition that should be considered in parallel with the COM position over the base support relates to the *tribology component*. This condition is expressed by the equilibrium of forces at the contact area of a heel with a surface. During slips and falls accidents, there are several dependencies and interactions among considered biomechanical and tribological elements. Tribology is linked with the human body by the shoe and walkway surfaces; both provide feedback to the body control system (see Figure 6.5). The mechanical characteristics of shoes and walkways and their designs and maintenance influence the coefficients of friction. Environmental disturbances can be created by contamination and by changes of temperature and humidity. Without the information on floor slipperiness, optimum floor surfaces, adequate shoe materials, contamination consequences, and design elements provided by tribological studies and testing, one would have no scientific guidance for solving the practical problem of preventing slips and falls.

Outside the laboratory, slips and falls occur in real environments and must be addressed by the *operational component* of an organization. Industry guidelines and scientific testing can point the way, but many slips and falls are caused by poor planning, poor design, or poor management. When a slip and fall occurs, several behavioral factors are often involved, including inattention of the person who slipped and fell or of the one who failed to clean up a spill. Mistakes include installing the wrong flooring for the expected environment, unforeseen transition issues, and not keeping the floor clear of obstacles. They may also include inadequate cleaning or repairing of defective floors, inadequate reporting of spills leading to lack of cleanup, limited training provided to employees, inadequate footwear, and inadequate lighting. These mistakes are examples of the poor application of tribology or failure to consider the characteristics of the shoe/floor interface and slips and falls.

Managing slip-and-fall exposures requires participation from everyone in the organization. The operational aspect of slips and falls prevention is a dynamic component that stimulates changes and poses challenges for continued study in the fields of biomechanics and tribology.

An understanding of the connectivity of the biomechanical, tribological, and operational systems components will help to determine an optimal control strategy for avoiding slips and falls.

CONSIDERATIONS IN REDUCING SLIPS AND FALLS

An effective slip-and-fall prevention strategy incorporates the information learned from research and applied in the real environment of an organization. Management commitment and facility design are important for the proactive prevention of slips and falls. Baseline knowledge of floor slipperiness assessment measures and the implementation of industry safety guidelines and standards are essential.

Industry Safety Guidelines and Standards

A performance definition for a reasonable COF for slippery work surfaces was proposed by Miller in 1983 as 0.5. Most studies show that people can walk comfortably and safely on surfaces with a coefficient of friction greater than 0.4, but 0.5 offers an additional safety factor (Miller, 1983). One U.S. guideline mentions that the 0.5 COF applies only to dry floors (ANSI/ASSE, 2006). Although no guideline specific to wet floors has been determined, most experts agree there is no rationale for having different traction thresholds for dry and wet conditions. It is important, however, not to overrely on this measurement as an indicator of slip-and-fall safety performance.

Several agencies issue standards and guidelines that can be applied to floor products, public buildings, and workplaces. A comprehensive list of supporting references is provided next.

American National Standards Institute. ANSI A1264.2, *Standard for the Provision of Slip Resistance on Walking-Working Surfaces* cites a slip-resistance guideline for working and walking floor surfaces of 0.5 or more for dry floor conditions only. ANSI A1264.2 specifies four ASTM slipmeters to collect this measurement: the HPS, NBS Brungraber Mark I, Brungraber Mark II (PIAST), and English XL Variable Incidence Tribometer (VIT). The ANSI A1264.2 subcommittee mentions no known rationale for having different traction thresholds for dry and wet conditions.

ANSI A117.1, *Accessible and Usable Buildings and Facilities* is a voluntary consensus standard that sets forth specifications for making buildings and facilities accessible to and usable by people with physical disabilities. The 1980 version of the A117.1 standard was adopted in part in the body of the Americans with Disabilities Act (ADA), which is now in effect. The term *slip resistant* is used in the standard but without a definition.

American Society for Testing and Materials. ASTM C1028, *Test Method for Determining the Static Coefficient of Friction of Ceramic Tile and Other Like Surfaces by the Horizontal Dynamometer Pull Meter* is a test method for determining the COF on ceramic tiles and similar surfaces under wet and dry conditions. Floor tile manufacturers occasionally use this test method to certify the slip resistance of their products. Test results can be found in marketing materials or on display tiles or samples.

ASTM D2047, *Test Method for Static Coefficient of Friction of Polish-Coated Surfaces as Measured by the James Machine* is one of the few test methods prescribing a minimum criterion of 0.5 SCOF as the threshold of safety, but for dry floors only. The James Machine is a laboratory-only slipmeter and is not used in the field.

ASTM F609, *Standard Test Method for Using a Horizontal Pull Slipmeter (HPS)* is the test method for Liberty Mutual's HPS slipmeter. Coefficient of friction results are stated by multiplying the measured slip index by 10. Categorizations of surfaces are as follows: relatively nonslippery (more than 6), generally acceptable (5–6), relatively slippery (less than 5). The HPS is used for dry floor surfaces only (ASTM International, 2005a).

ASTM F1679, *Standard Test Method for Using a Variable Incidence Tribometer (VIT)* is a test method for the English XL Tribometer. No coefficient of friction is stated, but the aforementioned guidelines are generally followed by users—that is, a 0.5 slip index or higher for wet and dry surfaces (ASTM International, 2004b).

ASTM F1677, *Standard Test Method for Using a Portable Inclineable Articulated Strut Slip Tester (PIAST)* is a test method for the Brungraber Mark II slipmeter. No coefficient of friction is stated, but the earlier guidelines are generally followed by users (i.e., a 0.5 SCOF or higher for wet and dry surfaces; ASTM International, 2005b).

A new standard entitled *Standard Practice for Validation and Calibration of Walkway Surface Tribometers Using Reference Surfaces* is in development at ASTM. This standard provides validation and calibration requirements for all slipmeters using standardized reference flooring surfaces.

Underwriters Laboratories. UL 410, *Standard for Slip Resistance of Floor Surface Materials* is the UL standard for listing of materials as slip resistant using the James Machine. The performance portion of the standard requires an average SCOF of 0.5 for floor treatment materials, floor covering materials of wood or composite materials, and walkway construction materials, including metal (Underwriters Laboratories Inc., 1996).

Some floor wax manufacturers cite research that their products are UL listed with a 0.5 COF. This number is for dry, clean-floor conditions only. Adding water or other contaminants to the floor will invalidate this number.

OSHA and ADA. The 1990 Americans with Disabilities Act Accessibility Guidelines (ADAAG) formally cited a minimum SCOF of 0.6 for level surfaces and 0.8 for ramps. The U.S. Access Board recently revised this requirement, and now floor surfaces must simply be slip resistant. This guideline was found in the ADAAG Appendix section A4.5.1, "Ground and Floor Surfaces (General)." The guideline does not mention what slipmeter(s) to use, so determining compliance is very difficult.

In 1990, OSHA published a proposed general industry rule in the Federal Register in *Subpart D. Walking-Working Surfaces* on floor slip resistance. This rule included a nonmandatory appendix specifying a minimum static coefficient of friction of 0.5. This proposal was reintroduced for public comment in May 2003. No update is available as of 2006.

OSHA 1926 *Subpart R - Steel Erection, 1926.754 (c)(3)* mentions that after July 18, 2006, workers will not be able to walk on any structural steel member coated with paint or similar material unless the coating has been certified to a minimum slip resistance rating of 0.5 wet. Appendix B to Subpart R specifies the Brungraber Mark II and English XL slipmeters for collecting these measurements and determining compliance. On January 18, 2006, OSHA revoked this portion of the standard, citing that the test methods for compliance could not be validated by July 2006.

Management Commitment

Management commitment can be depicted as a work system continuum shown in Figure 6.6. The circular arrangement (continuum) represents an extensive palette of the various elements influencing the slips-and-falls management process. Managing slip-and-fall exposures requires participation from everyone in the organization. It demands taking action at individual and group levels.

All the elements of the continuum must be accessed, including surveillance of hazards, assessment of floor slipperiness, selection of surface treatments, and identification of those areas that need improvement. It is necessary to make organizational decisions and determine the right housekeeping and maintenance procedures. Slip-resistant footwear, warning signs, and instructions are useless if floor-cleaning protocols are omitted.

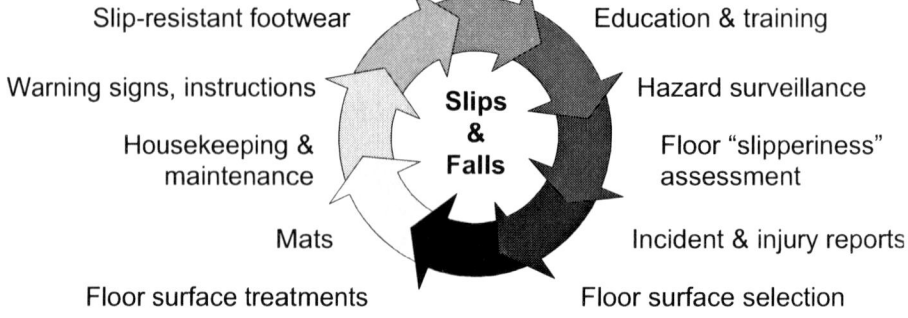

Figure 6.6. The circular arrangement, or continuum, represents the various elements influencing the slips-and-falls management process.

A closer look at each element of the continuum can help reduce same-level slip-and-fall incidents. If only one link in this chain is weak, the probability of an unexpected accident grows.

Injury and Hazard Surveillance

Surveillance is defined as the ongoing systematic collection, analysis, and interpretation of health and exposure information (National Safety Council, 2002). For slips and falls, that means analysis of (a) proactive or preloss hazard and incident data obtained through inspections, surveys, employee interviews, self-reports, and so on; and (b) reactive or postloss data, such as past accidents and injuries from insurance workers' compensation and/or general liability claim reports and other accident reports. Both are essential in a managed safety process and for establishing safety priorities. Unfortunately, it is usually the reactive data, such as a big claim or high overall claim costs, that determine whether or not resources will be expended for slip-and-fall prevention.

Although a proactive approach is often underutilized, a valuable source of information is unreported incidents—that is, slips without a fall and slips with a fall but without injury. These incidents need to be reported but are often disregarded, leaving the hazard uncorrected. Limited data are available concerning the number of unreported slips and/or falls, but it is estimated to be many times more than reported. For this reason, an essential element of a proactive managed slip-and-fall prevention program is a system to report close-call incidents that did not cause injury, in combination with hazard information, and using those data along with claims and injury cost data to establish safety priorities.

Facility Design

Preventing slips and falls requires a strategy that contains goals and objectives and is managed like other safety hazards and exposures. Slip-and-fall prevention needs to be included in safety preplanning for new facilities or renovations. Rarely is selection and maintenance of floors included in a strategy for the prevention of slips and falls. It is often an afterthought; decisions are based mostly on cost and aesthetics rather than on safety. The result is costly repairs or floor replacement sooner than expected.

The determining factor for floor selection should be to install floor surface materials for the expected traffic load and environment. If the floor is expected to be wet with contaminants present, then that factor should drive the decision of what floor surface to install. There is more flexibility in the choice for a floor used in mostly dry conditions, given that most dry, clean floors are slip resistant. However, when the floor surface is not quite right for whatever reason, floor surface treatments might be an option.

Floor Design and Selection

There are many different types of flooring, including a variety of tiles, carpeting, epoxy floors, terrazzo, and concrete. The selection of flooring should consider contaminants expected and transition areas. A transition from a carpeted floor or nonslippery floor to a glazed tile or more slippery walking surface could increase the likelihood of a slip and fall because of lack of detection of the transition (change in slip resistance) and appropriate gait adjustments. In general, flooring should have similar slip-resistance properties when a person is transitioning between different types of flooring, especially when liquid contaminants may be present.

Slip-resistant qualities of a new floor may be altered by high traffic, especially if the floor is not durable. What might seem inexpensive today could be more expensive in the long run if the floor has to be replaced sooner than expected. Most floors are slip resistant when dry. Wet or contaminated conditions determine whether the floor offers the best slip-resistant qualities. The following is a list of various floor surface materials and their slip-resistant benefits.

Quarry tile is an extruded natural clay tile with a porous surface. These tiles are very common in restaurant kitchen floors and some dining areas. Quarry tiles are unglazed, but polyurethane sealers are available to protect the surface and prevent staining. As a general rule, the lower the water absorption, the better the stain resistance. Most are vitreous (absorption of more than 0.5% but not more than 3.0%), although some are semivitreous (absorption more than 3.0% but not more than 7.0%).

Quarry tiles offer good slip-resistant qualities when clean and wet but poor slip qualities when soiled and wet, especially when polymerized grease is present. Some quarry tiles come with a texture or abrasive surface material of aluminum oxide grit. This grit material is sprinkled on the surface of the tile and baked in place, offering improved surface roughness. However, this material does not last long (one or two years) in heavy foot traffic areas and is removed or worn away over time. Some manufacturers offer a so-called double abrasive product that actually imbeds the grit material into the clay during the extrusion process. The manufacturer cites improved durability and slip-resistance performance.

Ceramic tile is a versatile clay product mixed with ceramic materials and baked at a higher temperature than quarry tile. Glazed and unglazed surfaces are available, and some tiles have abrasive granules imbedded in the glaze to enhance traction. An unglazed tile is visibly detected. The colors in unglazed tiles extend through the full thickness of the tile and will not wear off. For glazed tiles, the glaze can be seen on the surface and can vary from a high-gloss glaze to a low-luster glaze. Smooth glazed tiles offer poor slip-resistant qualities when wet unless treated. Glazed tiles are impervious (almost zero absorption) and do not stain. Unglazed ceramic tiles are not popular in kitchens because of their tendency to stain.

Glazed tiles for floor surfaces usually are rated for hardness using a Mohs scratch hardness scale. The Mohs scale of mineral hardness was devised by the German mineralogist Frederich Mohs (1773–1839), who selected an arbitrary list of 10 commonly available minerals, the softest being talc (1) and the hardest being diamond (10). Each mineral higher in the list can scratch the minerals lower on the list. Glazed tiles for floors usually have a hardness rating of about 6 using a case-hardened 1/4–3/8 inch diameter drill bit.

Porcelain paver tile is a clay product baked in a kiln at high temperatures. Porcelain pavers are very dense, hard, and impervious to water (i.e., water absorption less than 0.5%) and wear. They are commonly found in vestibules, lobbies, and restaurant dining areas and are available glazed or unglazed. Decorative porcelain tiles simulate mineral floors, such as slate or stone. Some textured styles offer good slip-resistant properties when wet.

Vinyl composition tile is a very common and inexpensive floor surface generally found in schools, hospitals, and offices and usually waxed to a high shine. Vinyl composition tile is often slippery when wet. Floor polishes and treatments are commonly available and applied to improve the slip resistance of floors made with this type of tile. The downside is that these products must be continually reapplied to be effective.

Rubber tile is a common flooring material used in airport lobbies, elevators and elevator lobbies, parking garages, and similar locations. Rubber flooring comes in a variety of styles and colors. It offers excellent slip resistance when dry and, overall, is a good floor in wet environments as well.

Marble and granite offer good slip resistance when dry, but when wet, it can be problematic unless treated.

Terrazzo is a poured-in-place decorative flooring commonly seen in airports and retail stores. Terrazzo flooring is made of chips of glass, marble, and other decorative aggregates suspended in a urethane or epoxy resin with a bonding agent. Dividers separate each floor section or decorations within the floor. Terrazzo has similar slip-resistance issues as marble and granite floors.

Concrete is also a poured-in-place flooring material. Overall, concrete makes for a very good slip-resistant floor finish as long as it is clean. Sealed concrete floors, however, may lose some of their slip-resistant qualities. Although sealed concrete floors are easier to clean and more attractive, trade-offs must be considered for slip resistance. Broom-finished concrete is considered the gold standard for slip-resistant floors and is very durable as well.

Surface roughness also affects friction; floor surfaces with adequate roughness characteristics may potentially reduce slip-and-fall accidents. A floor that will be used under mostly dry conditions offers more flexibility in terms of both selection and use, as most dry, clean floors are slip resistant by design. If liquid contaminants are expected on the floor, potential interventions could include molded surface patterns or profiled surfaces at the macroscale or surface roughness on nominally flat surfaces at the microscale. One of the selection criteria should be floors with high values in particular surface roughness parameters. These surface roughness parameters have been shown to relate to increased friction under such conditions.

Greater friction can be achieved by increasing the overall surface roughness level, although such an increase is not always desirable because of cleaning and other requirements. Alternatively, friction can be increased without an increase in the overall roughness level through the proper selection of floor surfaces that incorporate a range of microscopic geometric features on nominally flat surfaces.

In the United Kingdom (Health and Safety Executive, 1999) and Australia (Bowman, 1997), surface roughness and friction are recommended as two discrete parameters, both of which should be evaluated when making a detailed assessment of slip resistance. Although surface roughness is important in determining slipperiness, there is insufficient information to establish a safety criterion based solely on this single parameter. It can, however, readily provide a relative comparison between alternative floor surfaces or treatments.

Although other surface roughness parameters exist that are good indicators of friction, the parameter R_{pm}, which represents the allowable volume of contaminant before the surface is fully covered, is the easiest to measure using a relatively inexpensive profilometer, an instrument designed to measure the degree of surface roughness in micrometers (Chang, Kim, Manning, & Bunterngchit, 2001). A surface with a larger void volume can contribute to higher friction by allowing direct contact between the shoe and floor surfaces covered with liquid contaminants. An increase in surface irregularities at the peaks of the surfaces resulting from a large R_{pm} value also makes it more difficult to establish lubrication because of liquid contaminants at the shoe-floor interface. Even under conditions where a floor is completely covered with liquid contaminants, it is easier for the footwear surface to penetrate the contaminants and establish a direct solid-to-solid contact when the floor surface has a larger R_{pm} value.

Background information regarding surface roughness measurements and the desired settings of a profilometer or a surface roughness meter are available in the research literature. A surface with a higher R_{pm} value, the lower drawing in Figure 6.7, is preferred to a surface that has a lower R_{pm} value, the top drawing in Figure 6.7 (Chang, 2004).

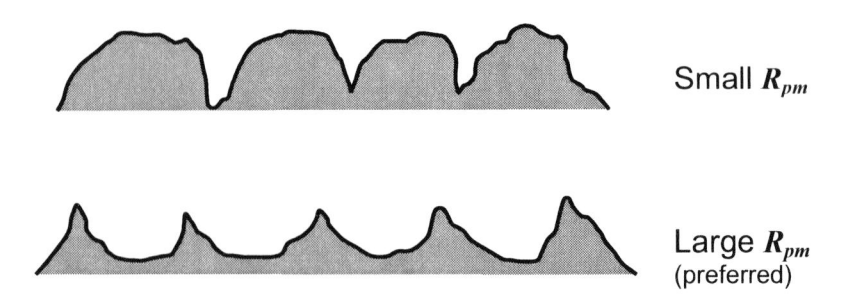

Figure 6.7. An illustration of surface roughness profiles with different R_{pm} values. Reprinted from Chang (2004), with permission from Elsevier.

Floor Mats

Mats can be an integral component of slip-and-fall prevention programs. Mats may be installed when a walking surface does not meet slip-resistance requirements, such as when wet or contaminated floors are present. Mats, when designed and installed properly, clean off the bottom of the footwear and help improve the slip resistance of floors, especially when contaminants are present. Mats with antifatigue properties can improve comfort and prevent injuries when working on hard surfaces over prolonged periods.

There are two types of matting systems: (a) entrance mats, also called "front-of-the-house" mats in the restaurant industry, and (b) multipurpose mats, or "back-of-the-house" mats. Examples of back-of-the-house mats might be those used around machinery process areas or water fountains, near food counters, and where spills may occur. Whether mats are in the front or the back of the house, a strategy needs to be employed in selecting the correct mat for the right environment and the expected contaminant. Often, matting systems are chosen based on cost or simply subcontracted to a vendor for cleaning and replacement. Mats that are dirty, worn, and old offer little slip prevention benefit.

Entrance matting can improve overall floor maintenance by scraping and absorbing soil particles and moisture from footwear and keeping the floor in a clean, dry condition. A rule of safe practice is that footprints or water prints should not be seen beyond the last mat of an entrance. Mats can protect a floor from unnecessary wear and remove water between the shoe and the floor.

The depth, or length, of the mat is important. The number of steps required to effectively scrape and wipe feet depends on climate conditions. As weather improves, the demands on floor matting become less intense. In snowy conditions, a mat length allowing a minimum of 10–12 walking steps for removing moisture from the footwear outsole is a good guide for the depth of the floor mat needed. When rain is present, the mat should gauge about 8–10 steps, and dry conditions require about 6–8 steps (Wolf, 1998).

In addition, the size of the mat depends on expected foot traffic, moisture, and debris. For example, a store that has 1,000 customers a day in a snowy climate needs a larger mat than the same store in a dry climate or an area typically with little rain or snow.

There are four types of entrance mats:

Multidimensional Aspects of Slips and Falls

1. Well and grate systems require a structural commitment. Mats funnel and drain moisture down and are a permanent fixture at entrances.
2. Glue-down mats can be installed at any time. The floor surface may be damaged by the adhesive. Some types require a metal strip and rubber reducer that is screwed into the floor as the finish edge. Replacement is time consuming.
3. Recessed mats are permanent and inserted into a well or recessed surface, becoming the finished floor. The finished height of the mat should be at least flush with the lip of the well and not represent a trip hazard.
4. Loose-lay matting stays in place without the use of adhesives, frames, screws, or duct tape. Be aware of the type of backing. Guard against mold and mildew damage to the underlying floor surface. Air should circulate through the mat.

Back-of-the-house mats are multipurpose mats that can absorb liquids, elevate workers above standing water, provide a slip-resistant working or standing surface, and/or provide antifatigue properties. Absorbing or retaining spills is common for mats used in grocery produce areas, around water fountains and sinks, to name a few. Liquid absorption characteristics, containment of spills and debris, and durability under grocery cart and foot traffic need to be considered when selecting a mat for this purpose.

Mats with slip-resistant surfaces are useful in standing work areas where grease and oil are common, such as in restaurant kitchens and manufacturing and food-processing areas. These mats have durable, slip-resistant surfaces that are also easy to clean.

When the surface is wet, some mats can be more slippery than the surface they rest on. Test by kicking the wet mat and then kick the adjacent floor surface to determine which is more slippery. Mats can also interfere with wheeled equipment, and moisture and debris can be tracked onto other areas if the wrong mat is selected. It can be difficult to practice "clean as you go" when mats are in place.

Back-of-the-house mats have varying surfaces, including some that are slip resistant. Use the following guidelines when selecting mats:

- Select mats with edges that are designed not to curl. These mats often have a beveled or flat edge to reduce tripping exposure. Mats more than 0.64 cm (¼ inch) thick should have tapered edges to reduce the potential for creating a tripping hazard.
- Select mats with nonslip backing that resists movement.
- Select mats that guard against damage to the underlying floor surface caused by mold and mildew.
- Routinely inspect mats for damage and excess wear and replace as necessary.
- Store mats or runners in a way to prevent curling of edges.
- Do not place mats or runners against objects that do not allow the mat to lie flat (e.g., machinery and process areas, doors, and furniture).

Antifatigue mats are common in work areas where long-term standing work is performed (e.g., retail cashiering, machine operation, and packing). Studies have shown that hard flooring may contribute to lower leg and low back fatigue or discomfort after as little as three hours of standing work (Cham & Redfern, 2001). This effect is believed to be the result of venous pooling rather than muscle fatigue. Subjective ratings of perceived fatigue have been shown to be better than more quantitative measures when comparing the physical benefits of antifatigue mats against possible alternatives. Studies have also shown that in general, floor mats characterized by increased elasticity, decreased energy absorption, and increased stiffness result in lower levels of both discomfort and fatigue (Cham & Redfern).

No study has yet recommended a specific material for antifatigue matting, so during selection, some experimentation with different types should be performed. It is usually beneficial to involve the affected workers in the final mat selection.

Floor Surfaces and Treatments

A variety of conditions may affect the coefficient of friction on floor surfaces, one of the most common being the presence of contaminants (often liquid). Lubrication of the contact area between the shoe and floor surface will, in most cases, substantially reduce its slip resistance. The reduction in friction between dry and wet surfaces may range from as little as 2% to as much as 81%, depending on the combination of shoe and walking surface materials involved (Grandjean, 1973). Almost all combinations of shoes and dry, clean walking surfaces exhibit a coefficient of friction in excess of 0.5, whereas only limited combinations of wet shoes and floor surfaces result in a friction coefficient greater than 0.4.

No estimate of the reduction in slip resistance caused by walking on wet surfaces can be made globally, but the following values excerpted from a report by the British Standards Institution (see Table 6.1) provide some perspective (British Standards Institution, 1984). It should be noted, however, that for known slippery or wet conditions, walkers commonly alter their gait by decreasing stride length. Altering gait results in a reduced angle of incidence between the heel and walking surface and thus a reduction in the forward directed force imparted when the heel strikes the ground. This in turn reduces the required level of slip resistance to prevent slipping.

Table 6.1. Selected Wet and Dry COFs for Various Surface Materials

Surface Material	COF (dry and unpolished)	COF (wet)
Clay tiles (Carborundum finish)	>0.75	>0.75
Clay tiles (textured)	>0.75	0.40 to <0.75
Clay tiles (smooth)	0.40 to <0.75	0.2 to <0.40
Carpet	>0.75	0.40 to <0.75
PVC (with nonslip granules)	>0.75	0.40 to <0.75
PVC	>0.75	0.2 to <0.40
Mastic asphalt	0.40 to <0.75	0.40 to <0.75
Vinyl asbestos tiles	0.40 to <0.75	<0.40
Linoleum (untextured)	0.40 to <0.75	0.2 to <0.40
Concrete (untextured)	0.40 to <0.75	0.2 to <0.40
Terrazzo	0.40 to <0.75	0.2 to <0.40
Rubber	>0.75	<0.2
Cast Iron	0.40 to <0.75	0.2 to <0.40

Note. Excerpted from British Standards Institution (1984).

Multidimensional Aspects of Slips and Falls 187

Floor treatments might be utilized for two reasons: to attempt to correct a problem because the wrong floor was installed in the first place and a hard lesson is learned (i.e., slips and falls are occurring), or a surface application is desired to improve an existing floor's slip resistance. Examples of slip-resistant treatments include abrasive floor treatments, chemical etches and cleaners, carpeting and mats, floor waxes, and new slip-resistant floor treatments that are applied daily.

Abrasive treatments. Abrasive floor applications provide a rough surface treatment to enhance surface traction and impart greater slip resistance. Grit material can be sand or silica quartz, garnets, and silicon carbide. Cleaning, durability, and cost must be considered when selecting grit material and treatments. Binder materials include paints, epoxies, and inexpensive abrasive tapes or strips. Some inexpensive floor applications can deteriorate or wear away with time and need to be reapplied.

Chemical etches and cleaners. Chemical etching or hydrofluoric acid and other chemicals professionally applied to marble, granite, ceramic and porcelain tiles, or concrete floors produce microscopic ridges and valleys in the floor and increase surface roughness. Etching produces a higher coefficient of friction with most shoe sole materials and with bare feet. Etching is commonly employed to improve the slip resistance of tiles used in showers and bath areas. Some new products are available that deep-clean quarry tile floors in restaurant kitchens and compete with some products offered by cleaning vendors. An etched floor can lose its effectiveness if not cleaned thoroughly and frequently.

Carpeting and mats. Carpeting offers inherent slip-resistant qualities but can be difficult to keep clean. It must be replaced often in high-traffic areas. Mats with slip-resistant surfaces are available that can be used in areas with slippery floor conditions, such as around restaurant fryers and in production operations.

Waxes and slip-resistant applications. Floor cleaners, polishes, and waxes are available, and some have been tested for slip resistance using ASTM D2047. A limitation of the ASTM D2047 test method is that it is for dry floors only. Some new slip-resistant floor surface treatment additives are available that must be applied daily when the floor is washed. These treatments must be reapplied anytime the floor is waxed and can wear away with heavy pedestrian traffic. The advantage of these treatments is that they can be applied over vinyl composition tile floors, and there are few treatments available for such floors. The disadvantage is that they must be continually applied to be effective.

Technical Selections

General floor-cleaning procedures. Floor-cleaning procedures may vary by the type of product; check the manufacturer's guidelines or protocol for the product used. Recommendations for removing grease include the following: Use the proper amount of cleaning product with hot, softened tap water; apply the cleaning product evenly on the floor surface with a clean mop; temporarily block any floor drains to permit the chemical sufficient time to penetrate built-up contaminants; allow sufficient time for the cleaning product to loosen contaminants on the floor (usually 5 to 10 minutes); scrub the floor briskly using a deck brush; open floor drains; wet-vacuum or squeegee before rinsing; and rinse floor with clear, hot, softened tap water to avoid leaving a residue on floors after drying.

Floor cleaners. Major categories of cleaners include alkaline, acidic, neutral pH, and microbial. Most solutions used to clean floors are intended to act chemically on contaminants and emulsify or break down the contaminant so it can be removed easily by rinsing. Surfactants and water conditioning additives are common.

Alkaline cleaners (pH greater than 7) react with fats and oils, converting them into soap (saponification), and must be thoroughly rinsed with clean, hot water to prevent polymerization. Therefore, they are often used to remove grease and can remove sealers, finishes, and waxes.

Typically, alkaline cleaners are used on restaurant kitchen and dining area tile floors. In restaurant kitchens, animal and vegetable fats (fatty acids and triglycerides) used in cooking oils hydrolyze and fall to the floor as a grease contamination. Grease, in the presence of water, can produce a very slippery floor. Over time, triglyceride molecules can unite to form a long-chain polymer (called *polymerization*) and form a hard grease film on floors that is resistant to most detergents. Cleaning a restaurant kitchen floor with a mop and pail using detergent and hot water only partially cleans the floor. A restaurant kitchen floor is not clean until the polymerized grease film is also removed. This requires the use of a suitable amount of detergent applied to the floor in very hot water with a dwell period followed by vigorous deck brushing. Greasy residue must be picked up using a squeegee or wet vacuum or rinsed away using hot, softened tap water. A hose rinse is best, assuming good drainage is available (English, 2003; Filiaggi & Courtney, 2003).

Acidic cleaners (pH less than 7) use a process known as *oxide reduction* to remove rust, scale, and oxides from floors. They are commonly used for cleaning porcelain tiles, ceramic tiles, and grout, but if they are too strong, they can etch the floor surface. Bathroom toilet bowl cleaners and household cleaners that remove calcium, rust, and lime deposits are examples of acid cleaners.

Neutral cleaners (pH around 7) are typically used on floors with glossy finishes or those that can be dulled by the abrasive qualities of acidic or alkaline cleaners (Di Pilla, 2003). Retail stores and malls, offices, airports, or any establishment with resilient flooring, terrazzo, marble, or granite floors would benefit from using neutral cleaners.

Microbial cleaners are also becoming popular. These cleaners use a nonpathogenic form of bacillus bacteria that can consume and digest oil, fat, grease, and petroleum hydrocarbons. Microbial cleaners are multipurpose cleaners and have been used to clear drains and clean concrete floors, tiles, and grout areas.

The wrong soap or detergent or incorrect application can add to the slipperiness of floors. Some commercial cleaners may also leave a residue on the floor. Floors should be rinsed only with clear water to avoid leaving such residue on floors after drying.

Housekeeping. Dirty floors are a common cause of slips and falls. Contaminants may accumulate on floor surfaces because of inadequate cleaning processes, resulting in the reduction of surface roughness as soil, grease, or other contaminants fill in the pores or

Multidimensional Aspects of Slips and Falls 189

valleys in the floor surface. The accumulation of contaminants alters these surface features and, consequently, reduces the floor's original uncontaminated friction characteristics. Therefore, it is very important to keep floors clean in order to maintain these desirable features of the walking surface and improve slip resistance. A floor-cleaning protocol must consider the floor type, the contaminants involved, and the cleaning solvent most suitable for both. A good housekeeping program should include written instructions regarding floor maintenance. Basic elements of an effective program should include the following:

- Identification of the specific contaminants and selection of a cleaner or chemical that effectively breaks each down.
- Establishment of a floor-cleaning protocol for the removal of contaminants.
- Provision of appropriate tools to clean the floor (e.g., mops, buckets, deck brushes, and squeegees). Designation of dedicated tools for specific areas is necessary in order to avoid cross-contamination (e.g., mops used in areas with grease should not be used in nongreasy areas).
- Implementation of a floor-cleaning schedule that is consistently followed, including the identification of responsible employees and the time of day during which cleaning should take place.
- Establishment of a training program for persons responsible for inspection, maintenance, and cleaning. This includes definition of cleaning requirements, cleaning procedures, safe handling and disposal of chemicals and solutions, emergency conditions and operations, and recordkeeping or reporting related to housekeeping and maintenance.
- Routine inspection of all floor surfaces for wear, damage, debris, and contaminants. Clear communication of any needed repairs to the facilities or maintenance department is critical.
- Occasional testing of floor surfaces to monitor slip-resistance levels and determine effectiveness of the floor-cleaning protocol.

In addition, the housekeeping safety program should address the following procedural questions: How are potential hazards identified and reported to appropriate supervision? Are "sweep logs" maintained? Are routine inspections performed, including unannounced inspections, and results recorded? Are first-line supervisors held accountable for hazards in their departments? Are warnings or signage provided whenever a slip-and-fall hazard has been identified and left in place until appropriate action is taken? Warning signs should use symbols that follow ANSI Z535.3, "1991 Criteria for Safety Symbols." Are enough trash containers provided, and are they located close to points of waste generation?

Training and Education

The best cleaning protocols will be ineffective without proper employee training or reinforcement of best practices. Many managers in the organization do not understand their role in the slip-and-fall prevention process or the benefits of one intervention over another. Education and training on the causes of slips and falls, their roles in the process, and the importance of recognizing, evaluating, and controlling hazards before they become expensive lawsuits/claims are critical to the success of the program.

Training should also include technical training about types of floors, types of treatments or coatings, types of abrasive or grit material, the chemistry of cleaning chemicals, the design of slip-resistant footwear, and matting systems. The training program should be widely promoted to persons responsible for inspection, maintenance, and cleaning.

Slip-Resistant Footwear

Utilizing slip-resistant footwear is an additional intervention strategy for reducing the likelihood of a fall. A slip-resistant footwear program needs to consider the working environment (e.g., wet/greasy floors, etc.) and, when selecting footwear, whether it is worn indoors or outdoors.

Slip-resistant is a specific term given to footwear that reduces the likelihood of slipping. Any shoe with that label should have corresponding test data to support the claim. In the United States, there are no standardized test methods for slip-resistance testing of shoes. However, in Europe, as of March 2004, the European committee for standardization (CEN) required slip testing of personal protective equipment (PPE) footwear under ISO EN 20344. CEN published EN13287 as the test method for slip-resistance testing using the STM 603 SATRA (Shoe and Allied Trade Research Association) slip-resistance tester (SATRA Technology Centre, 2005). PPE footwear must be tested to receive the CE mark. Terms such as oil-, fat-, acid-, alkaline-, or skid-resistant do not mean slip-resistant.

Manufacturers of slip-resistant footwear perform laboratory slip testing of their products under so-called realistic conditions mostly using wet and greasy quarry tile. The slipmeter commonly used is the Brungraber Mark II. The Mark II has a 3 × 3-inch test pad surface, a good size for attaching a sample of outsole material. The English XL Variable Incidence Tribometer (VIT) has a smaller test foot surface but is also a suitable shoe tester.

There are two major components to consider when selecting an appropriate slip-resistant footwear product: tread design and tread material.

Tread design. There are no specific recommended tread patterns, but there are recommendations for tread design. SATRA has produced guidelines for slip-resistant sole designs as follows (SATRA Technology Centre, 2005): The sole should have (a) a raised tread pattern on heel and sole with a leading edge in many directions—in other words, a crosshatch or similar design; (b) cleat width between 3 and 20 mm; (c) channel width at least 2 mm; (d) tread pattern extending over the whole sole and heel area; (e) a flat, flexible bottom construction—consider a low-density midsole that conforms to the ground and maximizes contact area; and (f) a square heel breast (acts as leading edge) as opposed to a rounded edge.

Tread material. In addition to tread design, tread material is very important for slip resistance and is the aspect of slip-resistant footwear about which manufacturers cite competitive differences between their products. The tread material is usually a proprietary, softer rubber material with slip-resistance benefits arising from the fact that the heel and shoe sole conform to the surface of the floor. Styrene butadiene rubber, nitrile-butadiene rubber, and polyurethanes are some of the more commonly used footwear soling materials. If a tread material is too hard, it will not conform to the floor surface and may not provide maximum slip-resistance protection (Di Pilla, 2003).

The softer, rubber soling typical of slip-resistant shoes offers limited durability in demanding outdoor work environments, such as construction work. Tread durability can be an issue with slip-resistant shoes worn in such environments. The softer, rubber materials can wear quickly and the tread can be damaged over time. A good work boot with a harder polyurethane sole with raised tread pattern will usually suffice for outdoor work. The tread material, regardless of use, should also be oil resistant.

In winter weather, softer heel and sole materials of thermoplastic rubber are often recommended, but they are not adequate for wet ice. Footwear worn in wet, icy conditions should consist of harder materials, preferably studded heals and soles (Grönqvist & Hirvonen, 1995).

Implementing a slip-resistant footwear program. A slip-resistant footwear program should be in writing and include a policy for the selection, purchase, reimbursement, and replacement of footwear. A slip-resistant footwear policy needs to be customized to meet the needs of an organization. Before implementing such a program, legal counsel should review the policy for potential legal exposures.

FUTURE DIRECTIONS

Prevention strategies need to be guided by scientific findings. Research in slips and falls, however, can never provide a complete and definitive description of reality. There will always be approximations of the true nature of slip-and-fall events because of limited descriptions of reality. A new way of looking at slips-and-falls control is needed in which a new definition of the goal of postural control and modeling using soft computing techniques is stated.

Future research on the mechanics of slips and falls should recognize the following guiding points:

1. Balance control of the human body during slips and falls relates to anatomy, mechanics, and neuromuscular control.
2. Experiments conducted under natural movement conditions and functional postural perturbations instead of static conditions can offer insight into real-life situations.
3. Equations of dynamic analysis at heel contact include a control aspect of the human body.
4. The human body structure in modeling includes the moment of inertia, accelerations, velocities, and displacements of the body segments.
5. The stiffness control mechanism is a crucial explanation of the body's response to unexpected perturbations during walking on slippery surfaces.
6. Theories and models can take into account central nervous system development, adaptation, and learning over a lifetime of experiences, as well as changes in biomechanics (Jacobs & Burleigh-Jacobs, 2000).
7. The adaptability of gait and the central nervous system control mechanisms on slippery surfaces should be taken into account.
8. A different order of events can follow the perturbation underlying frictional mechanisms in the shoe-floor interface, such as which comes first—foot slip or loss of balance.
9. Fluctuations in vertical ground reaction forces for both the left and right feet should be considered during experiments with the next generation of slip-testing devices.

Safety programs are an ever-evolving system. As research advances are made into the nature and causes of slips and falls, the information can be incorporated by managers into more effective prevention strategy plans. Prevention approaches can be more proactive rather than reactively triggered by undesirable loss trends or expensive injuries. Flooring selection, floor treatments, housekeeping programs, slip-resistant footwear, and mats are all valid interventions, but only when they are part of an overall prevention strategy that combines the advantages of each. In combination, interventions can form the basis for a comprehensive plan of action for slips-and-falls prevention. Continued research and collaborations will improve existing safety programs and present challenges for future study.

REFERENCES

ANSI/ASSE. (2006). A1264.2-2006 *Standard for the provision of slip-resistance on walking/working surfaces*. New York: American National Standards Institute.

ASTM International. (2004a). *Standard guide for composing walkway surface evaluation and incident report forms for slips, stumbles, trips and falls* (No. ASTM F1694). West Conshohocken, PA: Author.

ASTM International. (2004b). *Standard test method for using a variable incidence tribometer (VIT)* (No. F1679). West Conshohocken, PA: Author.

ASTM International. (2005a). *Standard test method for using a horizontal pull slipmeter (HPS)* (F605-05). West Conshohocken, PA: Author.

ASTM International. (2005b). *Standard test method for using a portable inclinable articulated strut slip tester (PIAST;* No. F1677-05). West Conshohocken, PA: Author.

Bouisset, S., & Zattara, M. (1987). Biomechanical study of the programming of anticipatory postural adjustments associated with voluntary movement. *Journal of Biomechanics, 20*, 735–742.

Bouisset, S., & Zattara, M. (1990). Segmental movement as a perturbation to balance? Facts and concepts. In J. M. Winters & S. L.-Y. Woo (Eds.), *Multiple muscle systems: Biomechanics and movement organization* (pp. 498–506). New York: Springer-Verlag.

Bowman, R. (1997). *Overstepping the mark—Practical difficulties in maintaining a slip resistance standard*. Paper presented at the International Ergonomics Association 13th Triennial Congress, Tampere, Finland.

British Standards Institution. (1984). *Stairs, ladders and walkways* (No. BSI 5395). Milton Keynes, England: Author.

Burnfield, J. M., & Powers, C. M. (2003). Influence of age and gender on utilized coefficient of friction while walking at different speeds. In M. I. Marpet & M. A. Sapienza (Eds.), *Metrology of pedestrian locomotion and slip resistance* (ASTM STP 1424; pp. 3–16). Conshohocken, PA: ASTM International.

Cham, R., & Redfern, M. S. (2001). Effect of flooring on standing comfort and fatigue. *Human Factors, 43*, 381–391.

Chang, W. R. (2004). Preferred surface microscopic geometric features on floors as potential interventions for slip and fall accidents. *Journal of Safety Research, 35*(1), 71–79.

Chang, W. R., Grönqvist, R., Leclercq, S., Brungraber, R. J., Mattke, U., Strandberg, L., et al. (2001). The role of friction in the measurement of slipperiness, Part 2: Survey of friction measurement devices. *Ergonomics, 44*, 1233–1261.

Chang, W. R., Grönqvist, R., Leclercq, S., Myung, R., Makkonen, L., Strandberg, L., et al. (2001). The role of friction in the measurement of slipperiness, Part 1: Friction mechanisms and definition of test conditions. *Ergonomics, 44*, 1217–1232.

Chang, W. R., Kim, I. J., Manning, D. P., & Bunterngchit, Y. (2001). The role of surface roughness in the measurement of slipperiness. *Ergonomics, 44*, 1200–1216.

Di Pilla, S. (2003). *Slip and fall prevention: A practical handbook*. Boca Raton, FL: Lewis Publishers.

Do, M. C., Brenière, Y., & Brenguier, P. (1982). A biomechanical study of balance recovery during the fall forward. *Journal of Biomechanics, 15*, 933–939.

English, W. (2003). *Pedestrian slip resistance: How to measure it and how to improve it*. Alva, FL: William English, Inc.

Filiaggi, A. J., & Courtney, T. K. (2003). Restaurant hazards: Practice-based approaches to disabling occupational injuries. *Professional Safety, 48*(5), 18–23.

Gielo-Perczak, K. (2001). Systems approach to slips and falls research. *Theoretical Issues in Ergonomics Science, 2*, 124–141.

Grabiner, M. D., Koh, T. J., Lundin, T. M., & Jahnigen, D. W. (1993). Kinematics of recovery from a stumble. *Journal of Gerontology: Medical Sciences, 48*(3), M97–M102.

Grandjean, E. (1973). *Ergonomics of the home*. London: Taylor & Francis.

Grönqvist, R. (1995). *A dynamic method for assessing pedestrian slip resistance* (People and Work Research Report No. 2). Helsinki: Finnish Institute of Occupational Health.

Grönqvist, R., Abeysekera, J., Gard, G., Hsiang, S. M., Leamon, T. B., Newman, D. J., et al. (2003). Human-centred approaches in slipperiness measurement. In W.-R. Chang, T. K., Courtney, R. Grönqvist, & M. S. Redfern (Eds.), *Measuring slipperiness: Human locomotion and surface factors* (pp. 67–99). London: Taylor & Francis.

Grönqvist, R., Chang, W.-R., Courtney, T. K., Leamon, T. B., Redfern, M. S., & Strandberg, L. (2001). Measurement of slipperiness: Fundamental concepts and definitions. *Ergonomics, 44*, 1102–1117.

Grönqvist, R., & Hirvonen, M. (1995). Slipperiness of footwear and mechanisms of walking friction on icy surfaces. *International Journal of Industrial Ergonomics, 16*, 191–200.

Hanson, J. P., Redfern, M. S., & Mazumdar, M. (1999). Predicting slips and falls considering required and available friction. *Ergonomics, 42*, 1619–1633.

Health and Safety Executive. (1999). *Preventing slips in the food and drink industries—Technical update on floor specifications* (HSE No. Food Sheet No 22). Sudbury, England: HSE Books.

Hendrick, H. W., & Kleiner, B. M. (2001). *Macroergonomics: An introduction to work system design*. Santa Monica, CA: Human Factors and Ergonomics Society.

Horak, F. B. (1997). Clinical assessment of balance disorders. *Gait and Posture, 6*, 76–84.

Irvine, C. H. (1970). Shoe sole slipperiness on structured steel. *Materials Research and Standards, 10*(4), 21–23.

Irvine, C. H. (1976). Evaluation of some factors affecting measurements of slip resistance of shoe sole materials on floor surfaces. *Journal of Testing and Evaluation, 4*(4), 133–138.

Jacobs, R., & Burleigh-Jacobs, A. (2000). Neuro-muscular control strategies in postural coordination. In J. M. Winters & P. E. Cargo (Eds.), *Biomechanics and neural control of movement*. New York: Springer-Verlag.

Leamon, T. B., & Li, K.-W. (1990). *Microslip length and the perception of slipping*. Paper presented at the 23rd International Congress on Occupational Health, Montreal.

Leamon, T. B., & Murphy, P. L. (1995). Occupational slips and falls: More than a trivial problem. *Ergonomics, 38*, 487–498.

Liberty Mutual Group. (2004). *Liberty Mutual workplace safety index*. Boston: Author.

Maynard, W. S., & Curry, D. G. (2005, June). *A macroergonomic approach to managing slips and falls in the workplace*. Paper presented at the 8th International Symposium on Human Factors in Organizational Design and Management, Maui, HI.

Miller, J. M. (1983). Slippery work surfaces: Towards a performance definition and quantitative coefficient of friction criteria. *Journal of Safety Research, 14*, 145–158.

Murphy, P. L., & Courtney, T. K. (2000). Low back pain disability: Relative costs by antecedent and industry group. *American Journal of Industrial Medicine, 37*, 558–571.

Nashner, L. M. (1985). Strategies for organization of human posture. In M. Igarashi & F. O. Black (Eds.), *7th International Symposium International Society of Posturography: Vestibular and Visual Control on Posture and Locomotor Equilibrium* (pp. 1–8). Houston, TX: Karger, Basel.

National Safety Council. (2002). *ASC/ANSI Z365 Management of work-related musculoskeletal disorders* (Final Draft December 2002). Retrieved August 18, 2006, from http://www.nsc.org/ehc/z365/newdrft.htm.

Pai, Y.-C., & Iqbal, K. (1999). Simulated movement termination for balance recovery: Can movement strategies be sought to maintain stability in the presence of slipping or forced sliding? *Journal of Biomechanics, 32*, 779–786.

Redfern, M. S., Cham, R., Gielo-Perczak, K., Grönqvist, R., Hirvonen, M., Lanshammar, H., et al. (2001). Biomechanics of slips. *Ergonomics, 44*, 1138–1166.

Robertson, M. M., Maynard, W., Huang, Y. H. E., & McDevitt, J. R. (2002). *Telecommuting: An overview of macroergonomic issues*. Paper presented at the Human Factors and Ergonomics Society 46th Annual Meeting, Baltimore, MD.

SATRA Technology Centre. (2005). Retrieved August 18, 2006, from *http://www.satra.co.uk/*.
Smith, R. H. (2002). Examination of sticktion in wet-walkway slip-resistant testing. In M. I. Marpet & M. A. Sapienza (Eds.), *Metrology of pedestrian locomotion and slip-resistance*. West Conshohocken, PA: ASTM International.
Underwriters Laboratories Inc. (1996). *Standard for slip resistance of floor surface materials, UL 410*. Northbrook, IL: Author.
Winter, D. A., Patla, A. E., Prince, F., Ishac, M., & Gielo-Perczak, K. (1998). Stiffness control of balance in quiet standing. *Journal of Neurophysiology, 80,* 1211–1221.
Wolf, D. (1998). Keep it clean by being up front. *Professional Retail Store Maintenance,* September, 84–86.
Yang, J. F., Winter, D. A., & Wells, R. P. (1990). Postural dynamics of walking in humans. *Biological Cybernetics, 62,* 321–330.

CHAPTER 7

Protection and Enhancement of Hearing in Noise

By John G. Casali & Samir N. Y. Gerges

Humans are subjected to noise in many occupational, military, transportation, recreational, and other settings, sometimes with attendant ill effects. Some noises may pose the threat of inducing hearing loss; most noises compromise one's ability to hear signals and other desirable sounds, as well as to communicate via speech. In this chapter, we review the role of noise in human hearing impairment and its effects on signal audibility and speech intelligibility, task performance, and nonauditory responses. Particular emphasis is paid to the needs of the human factors/ergonomics practitioner who is confronted with the objective of protecting individuals' hearing in noise, designing signals so that they are audible and recognizable, and in general, optimizing the use of the auditory sense under poor acoustic conditions. In addition to providing guidance and consensus standards for auditory danger signal design, we provide a review of new technologies in augmented hearing protection. These technologies are aimed at protecting the user's hearing in noise from injury while affording improved audibility for certain applications.

In this chapter, we concentrate on three major challenges, all amenable to the benefits of human factors engineering: (a) the deleterious effects of noise on humans, especially noise-induced hearing loss (NIHL); (b) signal detection and speech communications in noise; and (c) contemporary technologies for protecting hearing and improving the audibility of signals and speech in noise.

Before embarking on these subjects, we believe it is important that readers gain an appreciation for the extent of the noise problem in society, including noise-related annoyance, performance effects, and noise-imposed risks to hearing. For the reader who is unfamiliar with the acoustical terms used in this chapter, the following references are recommended: ANSI S1.1-1994 (R2004), entitled "Acoustical Terminology," and Crocker (1998).

WHAT CONSTITUTES NOISE?

Noise, which is often defined as undesired sound and also as "an erratic, intermittent, or statistically random oscillation," pervades people's very existence (ANSI S1.1-1994 [R2004]). In the occupational sectors of manufacturing, construction, and mining, noise is typically an expected phenomenon that surrounds daily work life, and laws govern workers' exposure to it (OSHA, 1971a, 1971b, 1983; MSHA, 1999, respectively). In the military sector, noise is characteristic of weapons, airplanes, ships, and ground vehicles such as tanks and mobile command units, and its presence often affects mission performance. On a civilian flight deck or inside a passenger vehicle, noise affects operations when it interferes with the hearing of desirable warning signals and speech communications. In the residential community, noise is a by-product of a host of sources, such as traffic, aircraft, trains, barking dogs, construction work, outdoor events, and local industry, and it often results in vigorous complaints. In leisure life, people regularly subject themselves to noise from myriad sources, such as power tools, all-terrain vehicles and snowmobiles, target shooting and hunting, gaming arcades, dance clubs, personal music headsets, and spectator events such as automobile racing and athletic competitions. In fact, in many walks of life, noise is inescapable, and our ears seem never to have a respite from it.

Noise as Annoyance

There are large differences among individuals as to the desirability or offensiveness with which they view a particular noise. There is an element of truth to the old adage that one person's music is another one's noise, as evidenced, for instance, by the fact that although the driver of a "boom" car equipped with a high-powered stereo system may be enjoyably immersed in the music, at the same time that the occupants of nearby vehicles or residences may consider the persistent booming bass sounds to be very offensive.

In any case, the aforementioned examples represent only a few well-known sources of noise to which people are exposed, and there are many others. One must remember that since the hearing sense is always "on"—that is, people cannot shut their ears as they can their eyes, and, short of wearing an earplug or an earmuff, they simply cannot escape noise unless they leave its vicinity. Furthermore, unlike vision, wherein an object must be in the viewer's line of sight to be seen, the hearing sense is omnidirectional; one need not be oriented to a sound source to hear it. Consequently, though turning the eyes away from an offensive image renders it unseen, simply turning one's head away from an offensive noise source does not render it unheard.

In terms of its effects on humans, at the very least, noise constitutes an *annoyance,* such as the disruption of conversation in a noisy restaurant, the inability to hear quiet musical passages from a symphony because of the background noise produced by a ventilation system, or the disruption of sleep because of traffic outside. As aptly described by Berger (2003a), as population density increases, there are fewer opportunities to experience "complete, absolute quiet." A natural example might include an isolated tundra in the Canadian northwest (that is, at least between aircraft flyovers), and an artificial example is a sound-isolated laboratory chamber used for acoustical experiments.

The World Health Organization (WHO, 1995) reported that in Europe, transportation noise alone exposes 40% of the population during daytime hours to A-weighted equivalent sound levels ($L_{Aeq, T}$) above 55 dBA, and 20% are exposed to levels above 65 dBA, both of which are perceived as annoying or intrusive (Berger, 2003a). In many communities in the United States, it is not uncommon to find residential zoning restrictions requiring that noise not exceed 55 dBA during the day. However, in some of these communities, 55 dBA is no longer feasible because of increasing levels of traffic or other sources (Casali, 1999).

Noise as a Risk to Hearing

Perhaps at its worst, noise creates a *hazard* to human hearing, which can result in a temporary threshold shift (TTS), or can permanently and irrecoverably injure the neural structures of the ear, resulting in noise-induced hearing loss (NIHL). The common result is impairment of the victim's ability to hear desirable sounds and speech, and in many cases it results in tragic isolation of the individual in family, social, and occupational settings. Hearing aids can compensate auditorially for the effects of NIHL in some individuals, but the benefits of hearing aids are limited to amplification, filtering, and compression of certain sounds, and those benefits are minimized if the neural receptors that receive specific sounds have been destroyed.

Reliable data on the number of people exposed to dangerous levels of noise outside the workplace are scant, but Berger (2003a) provided a review of exposure risk in occupational work settings. For instance, in the U.S. workplace, the National Institute for Occupational Safety and Health (NIOSH) estimated that as many as 30 million workers are exposed to hazardous sound levels above 85 dBA (NIOSH, 1996). In 1998, using 8-hour time-weighted full workday exposures rather than simple sound levels, NIOSH estimated that more than 5 million workers were overexposed.

Other industrialized countries face similar trends. For example, in British Columbia, workers' compensation records from 1987–1996 demonstrate that NIHL claims exceeded, by a factor of three, claims for each of the other physical agents that were recorded, the only exception being cumulative trauma disorders (WCB, 1998). In Chile, the number of industrial employers who combat noise in the workplace is five times larger than those who are dealing with the next most common hazard, the use of solvents (Dummer, 1997). The National Institutes of Health (NIH) convened a consensus conference in 1990 and concluded that of the hearing loss that afflicted some 28 million Americans, 10 million cases were at least partly attributable to NIHL (NIH, 1990).

The Role of Human Factors Engineering in Addressing Noise

Noise—and, more broadly, sound—is a phenomenon that confronts the human factors/ergonomics (HF/E) professional in many applications that involve human operators, consumers, and workers. By combining a basic knowledge of acoustics with in-depth training in ergonomics, the human factors engineer or ergonomist can make strong contributions to acoustical application issues such as (a) designing auditory informational and warning displays for vehicles; (b) designing auditory alarms and warnings for personal protection equipment (such as firefighter localization alarms) and buildings (e.g., directional evacuation enunciators); (c) developing communications systems that provide sufficient intelligibility to achieve mission success; (d) developing hearing protection devices and communications headsets; (e) developing in-ear devices for music rendition or surreptitious communications; (f) preparing and promulgating noise ordinances; (g) supporting litigation involving hearing loss, noise annoyance, warning signal detection, and intellectual property matters involving acoustical products intended for human use; (h) designing architectural environments to achieve desired sound quality and appropriate room acoustics (e.g., reverberation time, diffusivity); (i) designing, developing, selecting, and/or locating auditory signals to convey localization and velocity cues and other three-dimensional information; (j) designing products to convey aspects of quality via their acoustical emission, such as manual control detents, vehicle door closures, and engine exhausts; (k) developing psychophysical metrics for outdoor noise annoyance sources such as aircraft flyovers and stadium noise; and (l) developing standards and regulations to govern the acoustical design of various products (such as the output of power tools) and signals (e.g., vehicle reverse alarms and sirens).

By no means could a single chapter prepare the HF/E professional to effectively deal with all these applications in a comprehensive fashion. However, in this chapter, we introduce certain methods, standards, and technologies that will support the HF/E professional in efforts aimed at prevention of NIHL, in the selection of appropriate hearing protection and communications devices for use in specific noise problems, and in the evaluation of noise environments for which speech or auditory warning signals are necessary. (Consult Chapter 8 in this volume [Laughery & Wogalter] for details about visual warnings.)

Before he or she can become a well-rounded designer or consultant, the HF/E professional must also become knowledgeable in the basics of sound physics, sound measurement instrumentation, techniques, and standards for its measurement and quantification, how to interpret those measurements, and other areas of acoustical expertise. At the outset, it should be recognized that the science of acoustics comprises a vast body of reference, research, and standards literature. A broader book chapter concerning sound and noise, particularly as it affects humans, appears in Casali (2006). For even more in-depth information, three broad-coverage texts are Berger, Royster, Royster, Driscoll, and Layne (2003); Crocker (1998); and Kryter (1994).

Sound and Noise: Parameters and Measurements

The remainder of this chapter relies on standard terminology relating to the parameters of sound and noise and the measurement thereof. Acoustical measurement is a subject of considerable depth, beyond the scope of this application-oriented chapter. Earshen (2003) and Harris (1991) provided fundamental information on sound measurement indices and related instrumentation.

Very briefly, *sound* is a disturbance in a medium (in industry, home, or recreational settings, most commonly air or a conductive structure such as a wall) that has mass and elasticity. For example, a helicopter rotor has blades that rotate in the air, creating sound waves that propagate in all directions, including to the ground below. Because the blades are coupled to the air medium, they produce pressure waves that consist of alternating compressions (above ambient air pressure) and rarefactions (below ambient pressure) of air molecules, the *frequency* (f) of which is the number of above or below ambient pressure cycles per second, or *hertz* (Hz).

The reciprocal of frequency, *1/f,* is the *period*. The linear distance traversed by the sound wave in one complete cycle of vibration is the *wavelength*. As shown in equation 1, wavelength (λ in meters or feet) depends on the sound frequency (f in Hz) and velocity (c in m/s or ft/s; in air at 68° F and pressure of 1 atmosphere [atm], 344 m/s or 1127 ft/s) in the medium.

The speed of sound is influenced by the medium's temperature and, in air, increases about 1.1 ft/s for each increase of 1° F. Also, the attenuation of sound over distance decreases with increased temperature and increases as frequency increases (Harris, 1991).

Vibrations are oscillations in solid media and are often associated with the production of sound waves.

$$\lambda = c/f \tag{1}$$

The unit of *decibel*, or 1/10 of a *bel*, is the most common metric applied to the quantification of noise amplitude. The decibel, hereafter abbreviated as dB, as a measure of *level*, is defined as the logarithm of the ratio of a quantity to a reference quantity of the same type. In acoustics, it is applied to sound level, of which there are three types: *sound power level*, *sound intensity level*, and *sound pressure level* (Earshen, 2003). Sound pressure level is by far the most commonly measured and cited quantity, and it constitutes the basic measurement for most standards and regulations involving sound and noise. Thus, only it will be discussed here.

Sound pressure level, often abbreviated as SPL, or in formulae as L_P, is typically expressed in dB. SPL is defined as

$$\text{Sound pressure level (SPL or } L_P\text{) in dB} = 10 \log_{10} P_1^2/P_r^2 = 20 \log_{10} P_1/P_r \tag{2}$$

where P_1 is the pressure level of the sound in microPascals (µPa), or other pressure unit; and P_r is the pressure level of a reference sound in µPa, usually taken to be the pressure at hearing threshold, or the quantity 20 µPa, or 0.00002 Pa. Other equivalent reference quantities are 0.0002 dynes/cm^2 and 20 µbars.

The application of the dB scale to acoustical measurements yields a convenient means of collapsing the vast range of sound pressures that would be required to accommodate sounds that can be encountered into a more manageable, compact range. As shown in Figure 7.1, using the logarithmic compression produced by the dB scale, the range of typical sounds is 120 dB, and the linear pressure scale applied to the same sounds produces a range of 1,000,000 Pa. Sounds do occur that are higher than 120 dB (for instance, artillery fire or fighter jet takeoff) or lower than 0 dB (below-normal threshold on an audiometer). A comparison of dB values of example sounds with their pressure values (in Pa) is also depicted in Figure 7.1.

In considering changes in sound level measured in dB, a few numerical relationships emanating from the aforementioned dB formulae are often helpful in practice. An increase (decrease) in sound pressure level by 6 dB is equivalent to a doubling (halving) of the sound pressure. Similarly, on the power or intensity scales, an increase (decrease) of 3 dB is equivalent to a doubling (halving) of the sound power or intensity. This latter relationship gives rise to what is known as the *equal energy rule* or *trading relationship*. Because sound represents energy that is itself a product of intensity and duration, an original sound that increases (decreases) by 3 dB is equivalent in total energy to the same original sound that does not change in dB value but decreases (increases) in its duration by half (twice).

Measurement of noise levels provides fundamental data for assessing hearing damage risk, speech and signal masking effects, hearing conservation program needs, and

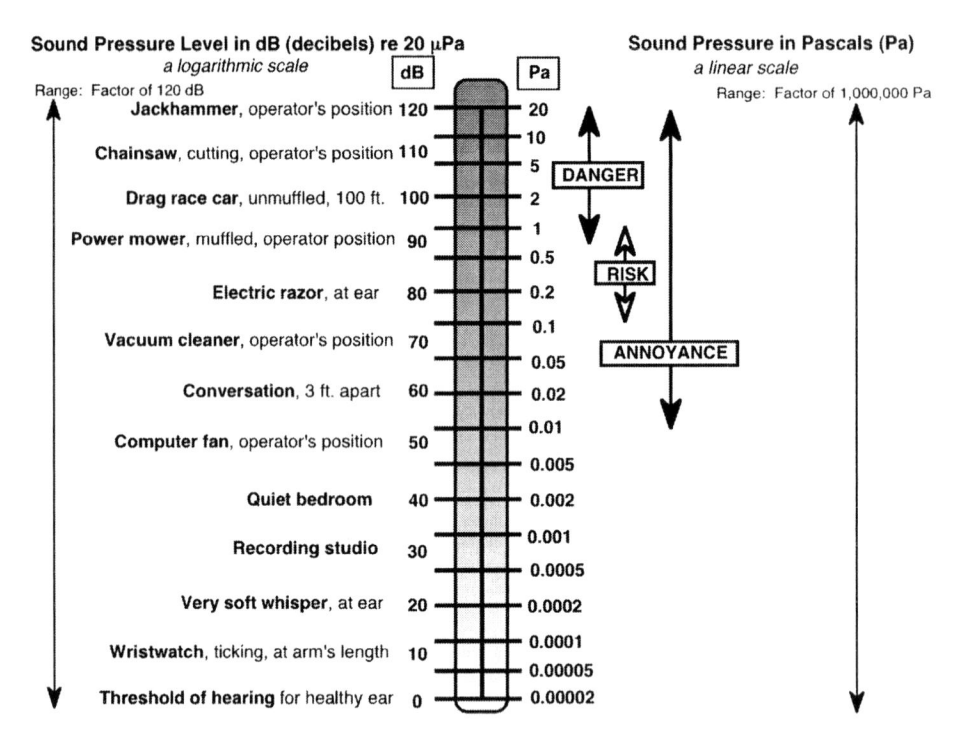

Sound Pressure Level in dB (decibels) re 20 µPa
a logarithmic scale
Range: Factor of 120 dB

	dB		Pa	
Jackhammer, operator's position	120		20	
			10	
Chainsaw, cutting, operator's position	110		5	DANGER
Drag race car, unmuffled, 100 ft.	100		2	
			1	
Power mower, muffled, operator position	90		0.5	RISK
Electric razor, at ear	80		0.2	
			0.1	
Vacuum cleaner, operator's position	70		0.05	ANNOYANCE
Conversation, 3 ft. apart	60		0.02	
			0.01	
Computer fan, operator's position	50		0.005	
Quiet bedroom	40		0.002	
Recording studio	30		0.001	
			0.0005	
Very soft whisper, at ear	20		0.0002	
Wristwatch, ticking, at arm's length	10		0.0001	
			0.00005	
Threshold of hearing for healthy ear	0		0.00002	

Sound Pressure in Pascals (Pa)
a linear scale
Range: Factor of 1,000,000 Pa

Figure 7.1. *Sound pressure level (SPL) values in dB and sound pressure values in Pascals for typical sounds.*

engineering noise control strategies. A vast array of instrumentation is available (see Earshen, 2003); however, most devices derive from the basic sound level meter (SLM), which is covered by a standard (ANSI S1.4-1983 [R2001]). Among other things, this standard specifies several ratings of SLM quality, (e.g., Type I, Precision, Type II, General Purpose). SLMs include frequency-response weighting networks that most commonly include A- and C-weighting functions (as depicted in Figure 3.2 of Earshen, 2003).

For OSHA noise-monitoring measurements and for many community noise applications, the A-scale is used; this scale de-emphasizes the low frequencies and, to a smaller extent, de-emphasizes the high frequencies. The A-weighting, or dBA, is a rough approximation of the human ear's frequency response to moderate (i.e., 40 phon) sound levels (Earshen, 2003). The C-scale is more linear (flat) in its response and thus is used when there is a desire for minimal spectral bias in the measurement.

SLMs also have dynamic response settings that define the short-time average display of the sound waveform. The two most common settings are FAST, which has a time constant of 0.125 s, and SLOW, which has a time constant of 1.0 s. These time constants were established decades ago to give analog needle indicators a rather sluggish response so they could be read by the human eye even when highly fluctuating sound pressures were measured. For OSHA or MSHA measurements, the SLOW setting is required, and this setting is best when the average value (as it is changing over time) is desired. The FAST setting is

Protection and Enhancement of Hearing in Noise

more appropriate when the variability or range of fluctuations of a time-varying sound is desired, or when the maximum value reached is desired. On certain SLMs, a third time constant, IMPULSE (having an exponential rise time constant of 35 ms and a decay time of 1.5 s), may also be included for measurement of sounds that have sharp transient characteristics over time and are generally less than 1.0 s in duration; examples are gun shots or impact machinery such as pile drivers.

Earshen (2003) and Harris (1991) addressed issues of SLM microphone selection and other usage guidance. One issue of paramount importance is that any microphone-SLM combination should be calibrated both pre- and postmeasurement to a known calibration reference source.

A *spectrum analyzer* is an advanced type of SLM that incorporates selective frequency-filtering capabilities, such as octaves or 1/3 octaves, to provide an analysis of the noise dB level as a function of its frequency (see Figure 7.2). Although occupational noise is usually measured in dBA with an integrating/averaging SLM (known as a *dosimeter*) for the purpose of noise exposure compliance, a spectrum analyzer can distinguish noises as to their hazard potential even though they may have similar sound pressure levels in dBA. This is illustrated in Figure 7.2. As shown in the figure, both noises would be considered to be of equal hazard by the OSHA-required dBA measurements (as they both are 90 dBA on a SLM or dosimeter). However, the 1/3-octave analysis suggests that the lowermost noise may in fact be more hazardous, as evidenced by the heavy concentration of energy in the more harmful midrange and high frequencies. A spectrum analyzer is necessary

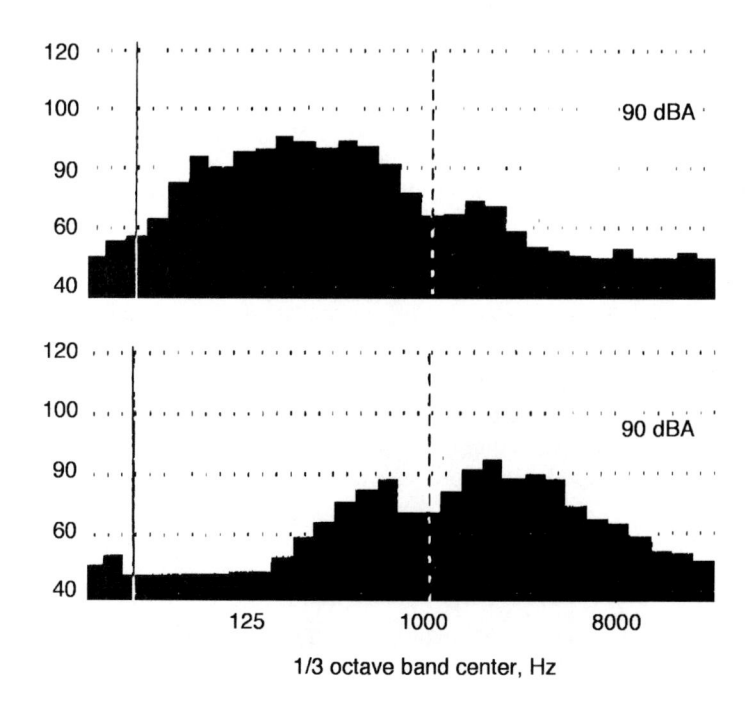

Figure 7.2. Spectral differences for two different noises with the same dBA value.

to obtain data that will provide the basis for engineering noise control solutions. Furthermore, spectrum analyzers are also required for performing the frequency-specific measurements needed to predict either signal audibility or speech intelligibility in noisy situations, according to the techniques and standards discussed later in this chapter.

NOISE REGULATION AND HEARING CONSERVATION

The Need for Attention to Noise

In the brief discussion in this section, we concentrate on the control of noise exposures in the occupational setting because that is the major hearing hazard risk for most individuals. Many of the techniques also apply to other exposures, such as those encountered in recreational or military settings.

The need for attention to noise is indicated when (a) noise creates sufficient intrusion and operator distraction such that job performance (and even job satisfaction) are compromised; (b) noise creates interference with important communications and signals, such as interoperator communications, machine- or process-related aural cues, and alerting/emergency signals; and/or (c) noise constitutes a hazard of noise-induced hearing loss (NIHL) in workers.

OSHA Noise Exposure Limits

With regard to combating the hearing loss problem for *general industry,* in OSHA terms, if the noise dose equals or exceeds the OSHA *action level* of 50%, which corresponds to an 85 dBA time-weighted average (TWA), the employer must institute a *hearing conservation program,* which consists of several facets (OSHA, 1983). If the *criterion level* of 100% dose is exceeded (which corresponds to the permissible exposure limit [PEL] of 90 dBA TWA for an 8-hour day), the regulations specify that steps must be taken to reduce the employee's exposure to the permissible exposure level or below through administrative work scheduling and/or the use of engineering controls. The regulation specifically states that hearing protection devices (HPDs) shall be provided if administrative and/or engineering controls fail to reduce the noise to the PEL. Therefore, in applying the letter of the law, HPDs are intended to be relied on only when administrative or engineering controls are infeasible or ineffective.

Another OSHA noise level requirement pertains to impulsive or impact noise, which is not to exceed a *true peak* sound pressure level limit of 140 dB(Z; where Z denotes linear).

Hearing Conservation Program Components

Hearing conservation programs are mandated by federal law in industry when 8-hour daily TWA exposures equal or exceed 85 dBA (the OSHA action level); in such cases, all affected employees must be included. But such programs should not simply be installed to maintain compliance with the law; their overarching objective should be the complete prevention of occupational noise-induced hearing loss in employees. An effective hearing conservation program stems from a strategic, programmatic effort that is initiated, organized, implemented, and maintained by the employer with cooperation and commitment from affected employees. Typically, audiologists or similarly trained professionals, occupational hygienists, noise control engineers, hearing protection vendors, plant managers and supervisors, and other support personnel are involved in administering a successful hearing conservation program.

A well-accepted hearing conservation approach is to address the noise exposure problem from a *systems* perspective, wherein empirical noise measurements provide the data that drive the implementation of countermeasures against the noise (including engineering controls, administrative strategies, and hearing protection). Subsequently, noise exposure data and audiometric data, which reflect the effectiveness of those countermeasures, serve as feedback for program adjustments. Employees' noise exposures must be monitored regularly and updated at any point when work-related changes may affect exposures. In addition, audiograms must be conducted annually and records kept. Employees must be informed of their exposures and their audiometric performance and trained in the use of hearing protection. If there are indications that employees may be suffering NIHL based on results of their annual audiograms, they must be referred to appropriate health care professionals and remedial actions must be taken with respect to their work exposures.

These are only a few of the elements of a hearing conservation program; an in-depth discussion may be found in Casali and Robinson (1999) and in Berger et al. (2003).

Conventional Hearing Protection Devices

OSHA (1983) requires that a selection of hearing protection devices (HPDs) that are suitable for the noise and work situation must be made available to all employees whose daily TWA exposures meet or exceed 85 dBA. Such HPDs are also useful outside the workplace to protect against noises produced by power tools, lawn care equipment, recreational vehicles, target shooting and hunting equipment, spectator events, and other exposures. In this section, only *conventional* HPDs are discussed; in other words, those which attenuate noise by strictly passive and static means (i.e., no electronics, moving parts, filters, or nonlinear elements). New technologies in hearing protection have emerged with the objective of providing better hearing while still affording protection again NIHL. We discuss these in a later section.

Earplugs consist of vinyl, silicone, spun fiberglass, cotton/wax combinations, and closed-cell foam products that are inserted into the ear canal to form a noise-blocking seal. Proper fit to the user's ears and training in insertion procedures are critical to the success of earplugs. A related device is the *semi-insert* or *ear canal cap,* which consists of earplug-like pods that are positioned at the rim of the ear canal and held in place by a lightweight headband. The headband is useful for storing the device around the neck when the user moves out of the noise.

Earmuffs consist of earcups, usually of a rigid plastic material with an absorptive liner, that completely enclose the outer ear and seal around it with foam- or fluid-filled cushions. A headband connects the earcups; on some models, this band is adjustable so that it can be worn over the head, behind the neck, or under the chin, depending on the presence of other headgear, such as a welder's mask.

In general terms, as a group, earplugs provide greater attenuation than do earmuffs below about 500 Hz and equivalent or greater protection above 2000 Hz. At intermediate frequencies, earmuffs typically have the advantage in attenuation. Earmuffs are generally more easily fit by the user than either earplugs or canal caps, and depending on the temperature and humidity of the environment, the earmuff can be uncomfortable (in hot or high humidity) or a welcome ear insulator (in a cold environment). Semi-inserts generally offer less attenuation and comfort than earplugs or earmuffs, but because they are readily storable around the neck, they are convenient for those workers who frequently move in and out of noise. A thorough review of HPDs and their applications may be found in Berger and Casali (1997).

Regardless of type, HPD effectiveness depends heavily on the proper fitting and use of the devices (Park & Casali, 1991). Therefore, the employer is required to provide annual training to all affected employees in the fitting, care, and use of HPDs (OSHA, 1983). Hearing protector use is *mandatory* if the worker has not had a baseline audiogram and is at or above 85 dBA TWA, has experienced a standard threshold shift (STS), or has a TWA exposure that meets or exceeds 90 dBA. For a worker with an STS, the HPD must attenuate the noise to 85 dBA TWA or below; otherwise, the HPD must reduce the noise to at least 90 dBA TWA.

The protective adequacy of an HPD for a given noise exposure must be determined by applying the attenuation data that the Environmental Protection Agency (EPA, 1979) requires be included on protector packaging in the United States. These data must be obtained from psychophysical real-ear-attenuation-at-threshold (REAT) tests performed on human listeners at nine 1/3-octave bands with centers from 125 to 8000 Hz. The difference between the thresholds with the HPD on and without it constitutes the attenuation at a given frequency. Spectral attenuation statistics (means and standard deviations) and the single-number noise reduction rating (NRR), which is computed therefrom, are provided. The ratings are the primary means by which end-users compare different HPDs on a common basis and make determinations of whether adequate protection and OSHA compliance will be attained for a given noise environment.

A more accurate method of determining HPD adequacy is to use octave band measurements of the noise and the spectral mean and standard deviation attenuation data to determine the protected exposure level under the HPD. This is called the *NIOSH long method* or the *octave band method*. Computational procedures appear in NIOSH (1975). Because this method requires octave band measurements of the noise, preferably with each noise band's data in TWA form, the data requirements are large and the method is not widely applied in industry. However, because the noise spectrum is compared against the attenuation spectrum of the HPD, a "matching" of exposure to protector can be obtained.

The NRR represents a means of collapsing the spectral attenuation data into one broadband attenuation estimate that can easily be applied against broadband dBC or dBA TWA noise exposure measurements (dBC and dBA weightings are depicted in Figure 3.2 of Earshen, 2003). In the calculation of the NRR, the mean attenuation is reduced by two standard deviations; this translates into an estimate of protection theoretically achievable by 98% of the population (EPA, 1979). The NRR is primarily intended to be subtracted from the dBC exposure TWA to estimate the protected exposure level in dBA, as per the following equation:

Workplace TWA in dBC − NRR = Protected TWA in dBA (3)

Unfortunately, because OSHA regulations require that noise exposure monitoring be performed in dBA, the dBC values may not be readily available to the hearing conservationist. When the TWA values are in dBA, the noise reduction rating can still be applied, albeit with some loss of accuracy.

With dBA data, a 7 dB "safety" correction is applied to the NRR to account for the largest typical differences between C- and A-weighted dB measurements of industrial noise. The equation is as follows:

Workplace TWA in dBA − (NRR − 7) = Protected TWA in dBA (4)

Although the aforementioned methods are promulgated by OSHA (1983) for determining HPD adequacy for a given noise situation, a word of caution is offered here. The data that appear on HPD packaging are obtained under optimal laboratory conditions with properly fitted protectors and trained human subjects. In no way do the *experimenter-fit* protocol and other aspects of the currently required (by the EPA) test procedure (ANSI S3.19-1974) represent the conditions under which HPDs are selected, fit, and used in the workplace (Park & Casali, 1991). *Therefore, the attenuation data used in the octave band or NRR formulae discussed earlier are typically inflated and cannot be assumed as representative of the protection that will be achieved in the field.*

In Figure 7.3, we display the results of a review of research studies in which manufacturers' on-package NRRs were compared against NRRs computed from actual subjects taken with their HPDs from field settings (Berger, 2003b). Clearly, the differences between these laboratory measurements and field estimates of HPD attenuation are large, more so for earplugs than for earmuffs, and the hearing conservationist must take this into account when selecting protectors. ANSI Working Group S12/WG11 developed a new testing standard, ANSI S12.6-1997(R2002), which has a "Method B" provision for subject (not experimenter) fitting of the HPD and relatively naive (untrained) subjects. Furthermore, ANSI S12.6-1997(R2002) is much improved compared with the current ANSI S3.19-1974 testing standard in experimental controls and human factors protocol, such as the new requirement that the manufacturer's instructions be the only source of instruction for the subject.

The ANSI S12.6-1997(R2002) Method B (Subject-Fit) testing protocol has been demonstrated to yield attenuation data that are more representative of those achievable under workplace conditions in which a quality HCP is operated. At the time of this writing, this new standard was under consideration by the EPA (EPA, 2003) for adoption into law as the required protocol for producing the data to be utilized in labeling HPD performance. This would replace the obsolete ANSI S3.19-1974 procedures cited in EPA (1979). More information on HPD testing, labeling, and selection for particular noise exposures may be found in Berger (2003b).

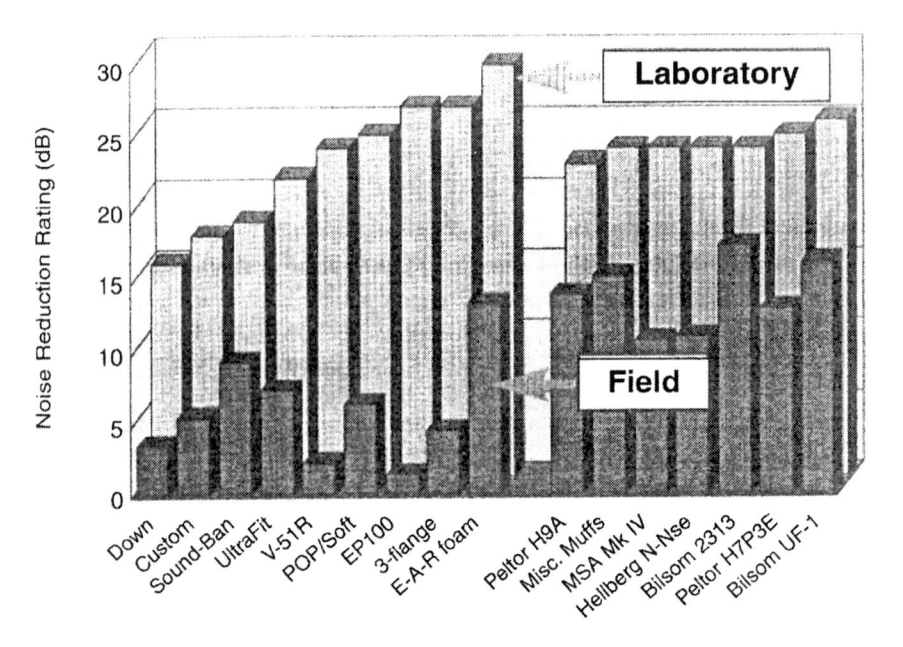

Figure 7.3. Comparison of hearing protection device NRRs by device type: Manufacturers' laboratory data versus real-world field data. Adapted with permission from Berger, E. H. (2003), Fig. 10.18, pg. 421, in E. H. Berger, L. H. Royster, J. D. Royster, D. P. Driscoll, & M. Layne (Eds.), *The Noise Manual, Revised 5th Ed.*, Fairfax, VA: American Industrial Hygiene Association.

NOISE INFLUENCE ON HEARING LOSS

As stated at the outset, noise-induced hearing loss is one of the most widespread occupational maladies in the United States, if not the world. Unfortunately, the control of noise sources in both type and number has not kept pace with the proliferation of industrial and service sector development, or with new noise sources produced by transportation and the military. And partly because of the fact that before 1971 there were no U.S. federal regulations governing noise exposure in *general industry* (rather, only for federal workers and contractors), many workers over age 50 now suffer hearing loss resulting from the effects of occupational noise.

Of course, the total noise exposure from both occupational and nonoccupational sources determines the NIHL that a victim experiences. Of the estimated 28 million Americans who exhibit significant hearing loss caused by a variety of etiologies, such as pathology of the ear and hereditary tendencies, more than 10 million have losses that are directly attributable to noise exposure (NIH, 1990). The noise-related losses are *preventable* in nearly all cases. The majority of loss is likely caused by on-the-job exposures, but leisure noise sources do contribute a significant amount of energy to the total noise exposure of some individuals. Although the effects of noise exposure are serious and must be reckoned with by human factors/ergonomics and safety professionals, one fact is encouraging: Process- or machine-produced noise, as well as most sources of leisure

Protection and Enhancement of Hearing in Noise

noise, are physical stimuli that can be avoided, reduced, or eliminated, and therefore, NIHL is preventable with effective abatement and protection strategies.

Types and Etiologies of Noise-Induced Hearing Loss

The major concern of the HF/E or safety professional is to prevent hearing loss that stems from occupational noise exposure. However, it is important to recognize that hearing loss may also emanate from a number of sources other than noise, including infections and diseases specific to the ear, most frequently originating in the middle or conductive portion; other bodily diseases, such as multiple sclerosis, which injures the neural part of the ear; ototoxic drugs, of which the mycin family is a prominent member; exposure to certain chemicals and industrial solvents; hereditary factors; head trauma; sudden hyperbaric- or altitude-induced pressure changes; and aging of the ear (presbycusis).

Furthermore, not all noise exposure occurs on the job. Many workers are exposed to hazardous levels during leisure activities, from such sources as automobile or motorcycle racing, personal stereo headsets and car stereos, firearms, and power tools. The effects of noise on hearing are readily subdivided into acoustic trauma and temporary or permanent threshold shifts (Melnick, 1991).

Acoustic trauma. Immediate and often profound organic damage to the ear from an extremely intense acoustic event such as an explosion is known as *acoustic trauma*. The damage may be to the conductive chain of the ear, including rupture of the eardrum or dislodging of the ossicles (small bones) of the middle ear. In many cases, conductive losses can be compensated for with a hearing aid and/or can be surgically corrected.

Neural damage may also occur involving a dislodging of the hair cells and/or breakdown of the neural organ (Organ of Corti) itself. Unfortunately, neural loss is irrecoverable and not typically compensable with a hearing aid.

Acoustic trauma represents a severe injury, but fortunately, its occurrence is relatively uncommon, even in industrial, construction, mining, and military settings.

Noise-induced threshold shift. A *threshold shift* is defined as an elevation of hearing level from the individual's baseline hearing level and constitutes a loss of hearing sensitivity. *Noise-induced temporary threshold shift* (NITTS) is, by definition, recoverable with time away from the noise. The elevation of threshold is temporary and is usually considered a temporary malfunction of cochlear transduction, such that caused by overstimulation of the stereocilia of hair cells. Although the individual may not notice the temporary loss of sensitivity, NITTS is a cardinal sign of overexposure to noise. It may occur over the course of a full workday in noise or even after a few minutes of exposure to intense noise.

Although the relationships are complex and individual differences are large, NITTS does depend on the level, duration, and spectrum of the noise, as well as the audiometric test frequency in question (Melnick, 1991).

With *noise-induced permanent threshold shift* (NIPTS) there is no possibility of recovery. NIPTS can manifest suddenly as a result of acoustic trauma; however, noises that cause NIPTS most typically constitute exposures that are repeated over a long period and have a cumulative effect on hearing sensitivity. In fact, the losses are often insidious, in that they occur in small steps over a number of years of overexposure and the individual may not be aware until it is too late. This type of exposure produces permanent neural damage, and although there are individual differences as to the magnitude of loss and audiometric frequencies affected, the typical pattern for NIPTS is a prominent elevation of threshold at or near the 4000 Hz audiometric frequency (sometimes called the *4 kHz notch*), followed by a spreading of loss to adjacent frequencies of 3000 and 6000 Hz.

Drawing on a classic study on workers in the jute weaving industry (Taylor, Pearson, Mair, & Burns, 1964), in Figure 7.4 the temporal profile of NIPTS is depicted as a family of audiometric threshold shift curves, with each curve representing a different number of years of exposure. As noise exposure continues over time, the hearing loss spreads over a wider frequency bandwidth inclusive of midrange and high frequencies, which encompasses the range of most auditory warning signals. In some cases, the hearing loss renders it unsafe or unproductive for the victim to work in certain occupational settings where the hearing of certain signals is requisite to the job. Unfortunately, the power of the consonants of speech sounds, which heavily influence the intelligibility of human speech, also lie in the frequency range that is typically affected by NIPTS; this compromises the victim's ability to understand speech. This is the tragedy of NIPTS: The worker's ability to communicate is hampered, often severely and always irrecoverably.

Hearing loss is a particularly troubling disability because its presence is not overt; therefore, the victim is often unintentionally excluded from conversations and may miss important auditory signals because others either are unaware of the loss or simply forget about the need to help compensate for it.

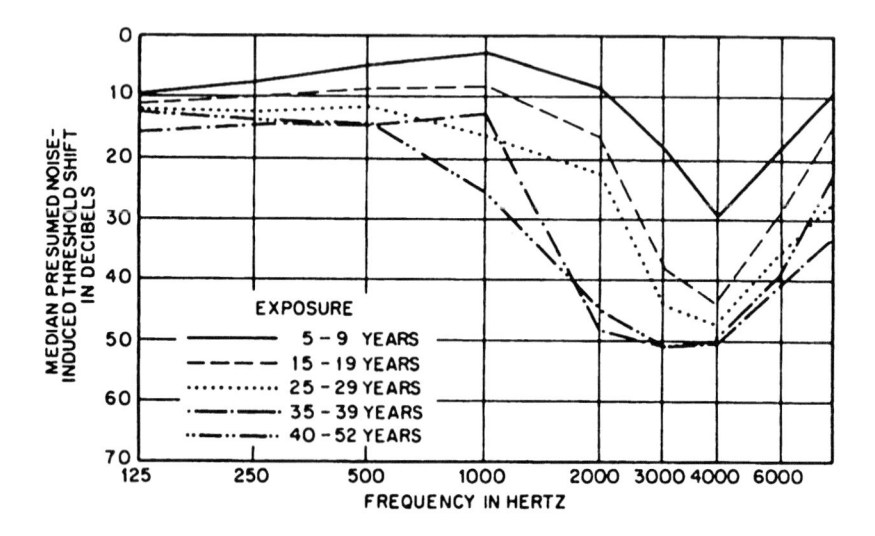

Figure 7.4. Cumulative auditory effects of years of noise exposure in a jute weaving industry. Adapted with permission from Taylor, W., Pearson, J., Mair, A., & Burns, W. (1964). Study of noise and hearing in jute weavers. *Journal of the Acoustical Society of America, 38,* 113–120.

Concomitant Auditory Injuries: Tinnitus and Hyperacusis

Following exposure to high-intensity noise, some individuals will notice that ordinary sounds are perceived as muffled, and in some cases, they may experience a ringing, whistling, or hissing sound in the ears, known as *tinnitus*. Tinnitus manifestations should be taken as serious indications that overexposure has occurred and that protective action should be taken if similar exposures are encountered in the future. Tinnitus may also occur by itself or in conjunction with NIPTS. It often manifests in the absence of external acoustic stimulation and often accompanies hearing loss. Some individuals report that tinnitus is always present, pervading their lives. It thus has the potential to be disruptive and, in severe cases, debilitating. Estimates indicate that more than 36 million Americans experience tinnitus (some with hearing loss as well). It is a malady that is compensable under workers compensation schedules in some states (Berger, 2003a).

Hyperacusis is more rare than tinnitus but is typically debilitating. It refers to hearing that is extremely sensitive to sound. Hyperacusis can manifest in many ways, but a number of victims report that their hearing became painfully sensitive to sounds of even normal levels after exposure to a particular noise event. Therefore, at least for some, hyperacusis can be directly traced to noise exposure.

Sufferers sometimes must wear HPDs when performing normal activities, such as walking on noisy city streets, visiting movie theaters, or washing dishes in a sink, because such activities produce sounds that are painfully loud to them. However, the use of HPDs is not a "cure" by any means, and hyperacusis sufferers should consider the treatment protocols that have emerged in the past few years. Note that hyperacusis sufferers often exhibit normal audiograms, even though their reaction to sound is one of hypersensitivity.

NOISE INFLUENCE ON PERFORMANCE, BODILY RESPONSE, AND ANNOYANCE

Noise Influence on Task Performance and Productivity

It is important to recognize that, among other deleterious effects, noise can degrade operator task performance. Research studies concerning the effects of noise on performance are primarily laboratory-based and task/noise-specific; therefore, extrapolation of the results to actual industrial settings is somewhat tenuous (Sanders & McCormick, 1993). Nonetheless, on the negative side, noise is known to mask task-related acoustic cues and to cause distraction and disruption of conscious thought. On the positive side, noise may at least initially heighten operator arousal and thereby improve performance on tasks that do not require substantial cognitive processing (Poulton, 1978).

To realize reliable negative effects of noise on task performance, except on tasks that rely heavily on short-term memory, the level of noise must be fairly high, usually 90 dBA or greater. Simple and repetitive tasks often show no deleterious performance effects (and sometimes improvements) in the presence of noise, whereas difficult tasks that rely on perception and information processing on the part of the operator will often exhibit performance degradation (Sanders & McCormick). It is generally accepted that unexpected or aperiodic noise causes greater degradation than does predictable, periodic, or continuous noise, and that the startle response created by sudden noise can be disruptive.

Several controlled studies point to the deleterious workplace-related effects of noise on metrics reflecting productivity, some of which we describe in this section. Bhattacharya, Roy, Tripathi, and Chatterjee (1985) studied 100 weavers in India who were found to perform better on dexterity and coordination tasks with hearing protection than without it in approximately 103 dBA noise. This is a very high level of noise, for which OSHA (1983) would only permit 1.3 hours of exposure in an 8-hour workday without mandatory hearing protection.

In a seminal study in 1935, Weston and Adams studied the effects of hearing protection on the work performance of weavers in England in 96 dBA noise. Alternating on a weekly basis, weavers either wore or did not wear hearing protection for a six-month period. When hearing protection was used, a 12% improvement in work efficiency was measured. In another study, these authors had one group of weavers not wear hearing protection for a full year, while another group did; the latter group exhibited a 7.5% improvement in efficiency.

Finally, a case study is worthy of mention. Staples (1981) reported on a bearing grinder that generated very high noise levels, from 103 to 114 dBA. After it was fitted with an enclosure, noise was reduced by about 20 dB. Obviously, the hearing hazard risk was reduced, but there were ancillary benefits, including an increase of about 20% in production, as well as reduced absenteeism, which had been very high prior to the noise control efforts.

Noise Influence on Work-Related Injury and Absenteeism

Other industrial field studies have addressed noise effects in addition to those related strictly to productivity. Berger (2003a) provided a thorough review of these, and several salient studies are noted here.

In a large study on almost 2,500 Egyptian weavers, Noweir (1984) reported that workers exposed to lower noise levels (from noise levels of 80 to 99 dBA in different departments) exhibited fewer disciplinary issues and less absenteeism than did those with higher exposures. However, the benefits of lower noise levels on production were very small (about 1%). A longitudinal study by Schmidt, Royster, and Pearson (1982) examined occupational injury data five years prior to and five years after the installation of a hearing conservation program in a yarn plant. There was a significant reduction in injuries from before to after the program.

Cohen (1976) performed a field study on boiler workers in 95 dBA exposures who were evaluated over a two-year period, before and after the implementation of a hearing conservation program that included HPD usage. Following the hearing conservation program, workers in the experimental group showed reduced job-related injuries, lower absenteeism, and fewer medical problems than prior to program implementation. A low-noise control group did not show reduced absenteeism, providing evidence that the reduction of noise exposures in the high-noise group likely resulted in their absenteeism decreases. It is also interesting to note that because job-related injuries were actually lower with the use of HPDs, there was no evidence that the presence of the protectors compromised the hearing of important work-related cues that might be of an accident-avoidance warning nature. This issue is further discussed later in this chapter.

Although data are lacking on the role of hearing protection in actual industrial accidents, at least two studies have implicated the role that *hearing loss* plays in accidents. In 1990, in studying 300 Dutch workers who experienced occupational injuries compared with 300 who did not, van Charante and Mulder reported that hearing-impaired workers who had greater than 20 dB loss at 4000 Hz had a 90% increased risk of work-related injury over normal-hearing workers. Similar results were found in a large National Health Interview Survey on approximately 450,000 workers from 18 to 65 years old, which demonstrated that self-reported hearing impairment was associated with a 55% increased risk (1.55 odds ratio) of work-related injury (Zwerling, Whitten, Davis, & Sprince, 1997). Somewhat surprising, only blindness and complete deafness—which displayed odds ratios of 3.21 and 2.19, respectively—had higher risk factors for workplace injury than did hearing impairment. These results provide relational evidence that the sense of hearing is important to workplace safety, yet another reason that hearing should be protected (Berger, 2003a).

Because of a lack of comprehensive accident analyses and records in industry, many of which do not mention noise as an attendant or contributing factor, it is difficult to draw general conclusions about how noise has affected the history of workplace accidents and the resultant injuries and property damage incurred. However, it is believed that unprotected workers who are exposed to high noise levels are involved in more lost-time accidents, demonstrate lower work productivity, and experience more problems than those with lower noise exposures (Berger, 2003a). Work (or other situations) that require the hearing of warning signals and/or understanding of speech may be compromised if the noise levels are sufficiently high to produce masking or distraction effects. Thus, later in this chapter, we discuss methods and hearing protection technologies for evaluating and improving hearing in noise.

Noise Influence on Nonauditory Health

Noise has been linked to physiological problems other than those of the hearing sense, including increased hypertension and blood pressure, especially when the prolonged daily exposures exceed 85 dBA (e.g., Smoorenburg et al., 1996). Other responses include increased fatigue and digestive problems, although there is disagreement as to the validity and proper interpretation of data from the associated body of research evidence (Berger, 2003a). Almost all physiological responses of this nature are symptomatic of stress-related disorders. In one study of note, Melamid and Bruhis (1996) reported that textile workers who were exposed to noise levels of 85–95 dBA exhibited reduced fatigue, lowered after-work irritability, and reduced corticol levels when hearing protection was used on the job, compared with unprotected conditions.

Because the presence of high noise levels often induces other stressful feelings (such as sleep disturbance and interference with conversation in the home, and fear of missing oncoming vehicles or warning signals on the job), there are second-order effects of noise on physiological functioning that are difficult to predict. The reader is referred to Kryter (1994) for a detailed discussion of other nonauditory health effects of noise.

Noise Influence on Human Annoyance

Noise has frequently given rise to vigorous complaints in many settings, including office environments, aircraft cabins, and homes. Such complaints are manifestations of what is known as noise-induced *annoyance,* which has given rise to a host of products such as white/pink noise generators for masking undesirable noise sources in office and home settings, noise-canceling headsets, and noise barriers to reduce sound propagation over distances and through walls. In the populated community, noise is a common source of disturbance, and for this reason many rural and urban communities have instituted noise ordinances and/or zoning restrictions that regulate the maximum noise levels that can result from certain sources and in certain land areas. In communities with no such regulations, residents who are disturbed by noise sources, such as industrial plants or spectator events, often have no other recourse than to bring civil lawsuits for remedy (Casali, 1999).

The principal rationale for limiting noise in communities is to reduce sleep and speech interference and to avoid annoyance (Driscoll, Stewart, & Anderson, 2003). More detailed information on the subject of noise annoyance may be found in Driscoll et al. (2003), Fidell and Pearsons (1997), and Casali (1999).

SIGNAL AUDIBILITY IN NOISE

General Concepts in Signal Audibility

Signal-to-noise ratio. One of the most noticeable effects of noise is its interference with speech communications and the hearing of nonverbal signals. Workers often complain that they must shout to be heard and that they cannot hear others trying to communicate with them. Likewise, noise interferes with the detection of signals such as alarms for general-area evacuation and warnings in buildings, enunciators, on-equipment alarms, on-person alarms (such as for firefighters), and machine-related sounds on which industrial workers rely for feedback.

In a car or truck, the hearing of external signals, such as emergency vehicle sirens or train horns or in-vehicle warning alarms or messages, may be compromised by the ambient noise levels. The ratio (actually the algebraic difference) of the total speech or signal level to the total noise level, termed the *signal (or speech)-to-noise ratio* (S/N), is a critical parameter in determining whether speech or signals will be heard in noise. An S/N of 5 dB means that the signal is 5 dB greater than the noise, and an S/N of -5 dB means that the signal is 5 dB lower than the noise.

Masking and masked threshold. Technically, masking is defined as the increase (in dB) of the threshold of a desired signal or speech *(the masked sound)* in the presence of an interfering sound (the *masking sound* or *masker*). For example, in the presence of noisy traffic alongside a busy street, the volume of an auditory pedestrian crossing signal must be sufficiently louder than the traffic to enable a pedestrian to hear it, whereas a lower volume will be audible (and probably more comfortable) when there is no traffic present.

It is also possible for one signal to mask another signal if both are active at the same time. In a controlled laboratory test scenario, a signal that is about 6 dB above the masked threshold will result in near-perfect detection performance (Sorkin, 1987). It is important to remember that the masked threshold is, in fact, a *threshold*; it is not the level at which the signal is clearly audible.

For the ensuing discussion, a functional definition of an auditory threshold is the dB level at which the stimulus is *just audible* to an individual listening intently for it in the specified conditions. If the threshold is determined in "silence," as is the case during an audiometric examination, it is referred to as an *absolute* threshold. If, on the other hand, the threshold is determined in the presence of noise, it is referred to as a *masked* threshold.

Analysis of Signal Detectability in Noise

Fundamentally, *detection* of an auditory signal is prerequisite to any other function performed on or about that signal, such as *discrimination* of it from other signals, *identification* of its source, *recognition* of its intended meaning or urgency, *localization* of its placement in space, and/or *judgment* of its speed. S/N ratio is one of the most critical parameters that determine a signal's detectability in a noise, but there are many other factors as well. These include the spectral content of the signal and noise (affecting *critical bandwidth*), temporal characteristics of signal and noise, duration of the signal's presentation, listener's hearing ability, demands on the listener's attention, criticality of the situation at hand, and the attenuation of hearing protectors, if used. These factors are discussed in detail in Robinson and Casali (2003).

In the following discussion, we concentrate primarily on the most important issue of spectral content of the signal and noise and how that content influences masking effects.

Masking: Signal levels, noise levels, and contrast. Generally speaking, the greater the dB level of the background noise relative to the signal (inclusive of speech), the more difficult it will be to hear the signal. Conversely, if the level of the background noise is reduced and/or the level of the signal is increased, the masked signal will be more readily audible. In some cases, ambient noise can be reduced through engineering controls, and in the same or other cases, it may be possible to increase the intensity of the signals.

Although most off-the-shelf auditory warning devices have a preset output level, it is possible to increase the effective level of the devices by placing multiple alarms or warning devices throughout a coverage area instead of relying on one centrally located device. This *distributed* approach can also be used for variable-output systems such as public-address loudspeakers, because simply increasing the output of such systems often results in distortion of the amplified speech signal, thereby reducing intelligibility. Increasing the signal level without adding more sound sources can have the undesirable effect of increasing the noise exposures of people in the area of the signal. If the signal levels are extremely high—for example, over 105 dB—exposed individuals could experience temporary threshold shifts if they are in the vicinity of the device when it is sounding.

Distortion within the inner ear is directly related to the signal and noise levels. At very high sound levels, the cochlea becomes overloaded and cannot accurately transduce or discriminate different forms of acoustic energy (e.g., signal and noise) reaching it. This results in the phenomenon known as *cochlear distortion*. In order for a signal, including speech, to be audible at very high noise levels, it must be presented at a higher level, relative to the background noise, than would be necessary at lower noise levels. This is one reason why it is best to make reduction of background noise a high priority in occupational or other environments.

In addition to manipulating the levels of the auditory displays, alarms, warnings, and background noise, it is also possible to increase the likelihood of detection of an auditory display or alarm by manipulating its spectrum so that it *contrasts* with the background noise and other common workplace sounds. In a series of experiments, Wilkins and Martin (1982, 1985) found that the contrast of a signal with both the background noise and irrelevant signals was an important parameter in determining the detectability of a signal. For example, in an environment characterized by high-frequency noise such as sawing and planing operations in a wood mill, it might be best to select a warning device with strong low-frequency components, perhaps in the 700–800-Hz range. On the other hand, for low-frequency noise such as might be encountered in the vicinity of large-capacity ventilation fans, an alarm with strong midfrequency components in the 1000–2500-Hz range might be a better choice. It is also prudent to provide signal and noise contrast in the temporal dimension; this is one reason that certain signals, such as backup alarms on construction equipment, are pulsed in on-off beeping fashion over time.

Masking: Upward spread. When considering the masking of a tonal signal by a tonal noise or a narrow band of noise, masking is greatest in the immediate vicinity of the masking tone or, in the case of a band-limited noise, the center frequency of the band. (This is one reason that increasing the contrast in frequency between the signal and noise can increase the audibility of a signal.) However, the masking effect does spread out above and below this frequency, being greater at the frequencies above the frequency of the masking noise than at frequencies below the frequency of the masking noise (e.g., see Figure 6 in Egan & Hake, 1950). This phenomenon is referred to as the *upward spread of masking*, and it becomes more pronounced as the level of the masking noise increases, probably because of cochlear distortion.

In practical situations, masking by pure tones would seldom be a problem, except when the noise contains strong tonal components, or if two warnings with similar frequencies were activated simultaneously. Upward spread of masking also can occur when broadband and band-limited noises are used as maskers.

Masking: Broadband noise. A common form of masking that is characteristic of typical industrial workplaces or building spaces such as conference rooms or auditoria occurs when a signal or speech is masked by a broadband noise. In examining the masking of pure-tone stimuli by white noise (white noise, flat by Hz, sounds very much like static on a radio or TV tuned to a frequency or channel with no signal), Hawkins and Stevens (1950) found that masking was directly proportional to the level of the masking noise, irrespective of the frequency of the masked tone. In other words, if a given background white noise level increased the threshold of a 2500-Hz tone by 35 dB, then the threshold of a 1000-Hz tone would also be increased by 35 dB. Furthermore, they found that for the noise levels investigated, masking increased linearly with the level of the white noise, meaning that if the level of the masking noise were increased by 10 dB, the masked thresholds of the tones also increased by 10 dB. The bottom line is that broadband noise such as white noise, because of its inclusion of all frequencies, serves as a very effective masker of tonal signals and speech. Thus, its abatement often needs to be of high priority.

Signal audibility analysis method based on critical band masking. Fletcher (1940) developed what would become *critical band theory*, which has formed the fundamental basis for explaining how signals are masked by narrow-band noise. According to this theory, the ear behaves as if it contains a series of overlapping auditory filters, with the bandwidth of each filter being proportional to its center frequency. When masking of pure tones by broadband noise is considered, only a narrow "critical band" of the noise centered at the frequency of the tone is effective as a masker, and the width of the band is dependent only on the frequency of the tone being masked. In other words, the masked threshold of a pure tone could be predicted simply by knowing the frequency of the tone and the spectrum level (dB per Hz) of the masking noise, assuming the noise spectrum is reasonably flat in the region around the tone. Thus, the masked threshold of a tone in white noise would simply be

$$L_{mt} = L_{ps} + 10 \log_{10}(BW) \tag{5}$$

where L_{mt} is the masked threshold, L_{ps} is the spectrum level of the masking noise, and BW is the width of the auditory filter (critical band) centered on the signal tone (Glasberg & Moore, 1990, Figure 7):

$$BW \ (Hz) = 24.7 \ (4.37 \ [\text{signal frequency (Hz)}] + 1)$$

Strictly speaking, this relationship applies only when the masking noise is flat (equal energy per Hz) and when the masked signal has a duration greater than 0.1 second. However, an acceptable approximation may be obtained for other noise conditions as long as the spectrum level in the critical band does not vary by more than 6 dB (Sorkin, 1987). In many environments, the background noise is likely to be sufficiently constant and can often be presumed to be flat in the critical band for a given signal. The exception to this assumption is a situation in which the noise has prominent tonal components and/or fluctuates a great deal.

The spectrum level of the noise in each of the 1/3-octave bands containing the signal components is *not* the same as the band level measured using an octave-band or 1/3-octave-band analyzer. Spectrum level refers to the level per Hz, or the level that would be measured if the noise were measured using a filter 1-Hz wide. If it is assumed that the noise is flat within the bandwidth of the 1/3-octave-band filter, then the spectrum level can be estimated using the following equation:

$$L_{ps} = 10 \log_{10} \frac{10^{L_{pb}/10}}{BW_{1/3}} \tag{6}$$

where L_{ps} is the spectrum level of the noise within the 1/3-octave band; $BW_{1/3}$ is the bandwidth of the 1/3-octave band, calculated by multiplying the center frequency (f_c) of the band by 0.232; and L_{pb} is the sound pressure level measured in the 1/3-octave band in question.

Finally, if the signal levels measured in one or more of the 1/3-octave bands considered exceed these masked threshold levels, then the signal should be audible.

Signal audibility analysis method based on ISO 7731-2003(E). In designing specific auditory alarms for specific applications, the human factors engineer should first consult any prevailing regulations or standards, of which there are many in acoustics. For example, the Department of Defense (DoD MIL standards), National Fire Protection Association (NFPA), Society of Automotive Engineers (SAE), Underwriters Laboratories (UL), American National Standards Institute (ANSI), and International Organization for Standardization (ISO) are examples of organizations that have promulgated standards to guide the design of auditory warning signals for specific applications, such as on-vehicle warning displays, sirens, evacuation alarms, fire alarms, and bodily-worn PPE safety alarms.

In the absence of a prevailing regulation or standard for a specific application, the HF/E professional should refer to a more general standard for warning and danger signals. In this regard, for performing an *analysis* of nearly any acoustic alarm to determine its predicted audibility in a specific noise, perhaps the most comprehensive standard is ISO 7731-2003(E), *Danger Signals for Public and Work Areas—Auditory Danger Signals* (ISO, 2003). This standard provides guidelines for calculating the masked threshold of audibility, specifies the spectral content and minimum signal-to-masked threshold ratios required for clear audibility, and invokes special considerations (i.e., requiring prominent signal content below 1500 Hz) for individuals suffering from hearing loss or those wearing HPDs. In general, the standard specifies that a danger signal shall include frequency components in the 500–2500-Hz range; it is further recommended that there be two dominant frequency components within the subset range of 500–1500 Hz. In addition, the standard states that the signal shall not exceed 118 dBA in the signal reception region. All measurements for both the ambient noise and the signal are to be obtained with the sound-level meter or spectrum analyzer on its slow response setting, and the maximum reading obtained on each trial (assuming a representative number of trials) is to be used in the calculations. Whenever possible, the ambient noise and signal should be measured separately.

Section 4.2.2 of ISO 7731, which provides the performance criteria for what constitutes a "clearly audible" signal, is perhaps the most comprehensive and important part of the standard. First, in a broadband sense, the dBA level of the danger signal is not to be less than 65 dBA at any point in the signal reception region. Second, the signal must also satisfy at least one of the following: (a) In a broadband sense, the dBA level of the danger signal must exceed the dBA level of the ambient noise by greater than 15 dB. (b) If spectral measurements are obtained in octave bands, the dB level of the signal in at

Protection and Enhancement of Hearing in Noise

least one octave band shall exceed the effective masked threshold by at least 10 dB in the octave band under consideration. (c) If spectral measurements are obtained in 1/3-octave bands, the dB level of the signal in at least one 1/3-octave band shall exceed the effective masked threshold by at least 13 dB in the 1/3-octave band under consideration.

Because of the fact that more narrow-frequency resolution is provided, the prediction of masked threshold and thus the subsequent calculation for audibility of a danger signal is most precisely and accurately performed with the 1/3-octave spectrum analysis method, followed by the octave spectrum analysis method, and then by the broadband dBA method. It is also important to realize that to impart features of distinctiveness and recognition into the danger signal (as required by ISO 7731), narrow-band or tonal characteristics in the signal will be advantageous. Such characteristics are best measured with the 1/3-octave technique, which yields the most precise data for analyzing the dominant spectral features of the signal against similar frequency content in the noise.

Furthermore, we advocate that the broadband dBA method be used only as a last resort in the absence of spectral measurements. This is because it does not discern dominant spectral content in the signal or noise, and thus critical band masking effects are not subject to analysis. In addition, the broadband dBA method does not consider the upward spread of masking.

Application of ISO 7731-2003(E) is best illustrated by an example. A common warning signal is a standard backup alarm typically found on commercial trucks and construction or industrial equipment. It has strong tonal components at 1000 and 1250 Hz and strong harmonic components at 2000 and 2500 Hz. The alarm has a 1-s period and a 50% duty cycle (that is, it is "on" for 50% of its period). The levels in all other 1/3-octave bands are sufficiently below those in the bands mentioned as to be inconsequential. The levels needed for audibility of this signal will be determined for application in a hypothetical noise spectrum represented by its octave band and 1/3-octave-band levels, shown in Table 7.1 in columns 2 and 4, respectively.

Step 1: Starting at the lowest octave band or 1/3-octave-band level available, the masked threshold (L_{mt1}) for a signal in that band is

$$L_{mt1} = L_{pb1}, \tag{7}$$

where L_{pb1} is the noise level measured in the octave or 1/3-octave band in question.

Step 2: For each successive octave band or 1/3-octave-band filter n, the masked threshold (L_{mtn}) is the noise level in that band or the masked threshold in the preceding band, less a constant; whichever is *greater*

$$L_{mtn} = \max(L_{pbn}; L_{mtn-1} - C), \tag{8}$$

where C = 7.5 dB for octave-band data or 2.5 dB for 1/3-octave-band data.

This procedure takes upward spread of masking into account by comparing the level in the band in question with the masked threshold level in the preceding band. For example, for the 1250-Hz row of column 3 in Table 7.1, it can be seen that the masked threshold of the previous band (1000 Hz) determines, via Equation 8, the masked threshold (77.1 dB) of the 1250-Hz band because of upward masking effects.

Table 7.1. Masked Threshold Calculations According to ISO 7731-2003(E) for Octave-Band and 1/3-Octave-Band Methods

1 Center Freq. (Hz)*	2 1/3-Octave-Band level (dB)	3 Masked Threshold (dB)**	4 Octave-Band level (dB)	5 Masked Threshold (dB)**
25	52.0	52.0		
31.5	50.7	50.7	54.7	54.7
40	42.9	48.2		
50	56.4	56.4		
63	86.8	86.8	88.5	88.5
80	83.7	84.3		
100	79.7	81.8		
125	83.7	83.7	87.1	87.1
160	82.8	82.8		
200	76.5	80.3		
250	81.4	81.4	85.1	85.1
315	81.6	81.6		
400	76.3	79.1		
500	77.3	77.3	80.7	80.7
630	73.1	74.8		
800	74.4	74.4		
1000	79.6	**79.6**	81.5	**81.5**
1250	73.4	**77.1**		
1600	82.6	82.6		
2000	80.1	**80.1**	87.9	**87.9**
2500	85.3	**85.3**		
3150	83.7	83.7		
4000	85.7	85.7	90.9	90.9
5000	88.0	88.0		
6300	74.2	85.5		
8000	77.3	83.0	79.1	83.4
10000	58.7	80.5		
12500	67.4	78.0		
16000	48.7	75.5	67.6	75.9
20000	53.3	73.0		

Note. *Frequencies in **boldface** type are octave-band center frequencies. **Thresholds in **boldface** type are the masked thresholds for the dominant signal components of the hypothetical backup alarm described in the text.

All the masked thresholds for octave bands and 1/3-octave bands of noise for the example are shown in columns 3 and 5, respectively, of Table 7.1. For the purposes of the example signal (a backup alarm), only the thresholds for the 1/3-octave bands centered at 1000, 1250, 2000, and 2500 Hz and the threshold for the octave bands centered at 1000 and 2000 Hz are relevant because these are the signal's dominant component bands. (But if the signal had possessed significant energy below 1000 Hz, the 1/3-octave bands centered at 500, 630, and 800 Hz would require attention, as would the octave band centered at 500 Hz.)

The conclusion is that if the signal levels measured in one or more of these bands exceed the calculated masked threshold levels (as indicated by boldface type), then the backup alarm is predicted to be "barely audible." More important, to next determine the necessary sound level output of the alarm to render it "clearly audible" per ISO 7731, to simplify, we will assume that the backup alarm's dominant frequencies (1000, 1250, 2000, and 2500 Hz) themselves cannot change but that their dB output can be raised. Thus, based on the 1/3-octave analysis, in order for the alarm to be reliably audible, *the signal level would have to be at least the following, in at least one of these four 1/3-octave bands centered at* 1000 Hz: 79.6 + 13 = 92.6 dB; 1250 Hz: 77.1 + 13 = 90.1 dB; 2000 Hz: 80.1 + 13 = 93.1 dB; 2500 Hz: 85.3 + 13 = 98.3 dB. Or, based on the octave analysis, in order for the alarm to be reliably audible, *the signal level would have to be at least the following, in at least one of these two octave bands centered at* 1000 Hz: 81.5 + 10 = 91.5 dB; 2000 Hz: 87.9 + 10 = 97.9 dB.

Both the critical band and ISO 7731 methods may be used to calculate masked thresholds with and without hearing protection devices. For ISO 7731, calculating a protected masked threshold for a particular signal requires the following: (a) subtracting the attenuation of the HPD from the noise spectrum to obtain the noise spectrum effective when the HPD is worn; (b) calculating a masked threshold for each signal component using the procedures outlined in the preceding discussion, which results in the signal-component levels that would be *just audible* to the listener when the HPD is worn, and (c) adding the attenuation of the HPD to the signal-component thresholds to provide an estimate of the environmental (exterior to the HPD) signal-component levels that would be required to produce the under-HPD threshold levels calculated in Step 2. These levels would then need to be inflated by 13 dB for 1/3-octave measurements or 10 dB for octave measurements, as the case may be.

Although not difficult, this procedure does require a reasonably reliable estimate of the actual attenuation provided by the HPD. The manufacturer's data supplied with the HPD are unsuitable for this purpose because they overestimate the real-world performance of the HPD, as explained previously. Furthermore, if a 1/3-octave-band masking computation is desired, the manufacturer's attenuation data, which are available for only nine selected 1/3-octave bands, are insufficient for the computation.

It is important to keep in mind that beyond the rather extensive masked threshold and audibility calculations discussed earlier, ISO 7731-2003(E) also includes recommendations for signal temporal characteristics, unambiguous meaning, discriminability, inclusion of signal energy below 1500 Hz to accommodate high-frequency hearing loss and hearing protectors, and addition of redundant visual signals if the ambient noise exceeds 100 dBA.

The broadband S/N recommendation of 15 dB of ISO 7731 is generally in keeping with those of auditory researchers. For example, Sorkin (1987) suggested that signal levels 6–10 dB above masked threshold are adequate to ensure 100% detectability, whereas signals that are approximately 15 dB above their masked threshold will elicit rapid operator response. Sorkin also suggested that signals more than 30 dB above the masked threshold could result in an unwanted startle response and that no signal should exceed 115 dB. (This suggested upper limit on signal level is consistent with OSHA hearing conservation requirements [OSHA, 1983], which prohibit exposure to continuous noise levels greater than 115 dBA.) These recommendations are in line with those of other authors (Deatherage, 1972; Wilkins & Martin, 1982).

Masked thresholds estimated using ISO 7731-2003(E) are not necessarily exact, nor are they intended to be. They do provide conservative masked threshold estimates for a large segment of the population representing a wide range of hearing levels for nonspecific noise environments and signals.

Analysis/Evaluation of Speech Intelligibility in Noise

Many of the concepts we have presented that relate to the masking of nonspeech signals by noise apply equally well to the masking of speech, so they will not be repeated here. However, for the spoken message, the concern is not simply audibility or detection but, rather, intelligibility. The listener must understand *what* was said, not simply know that something was said. Imagine the dilemma of a pilot who hears his or her aircraft call sign from air traffic control but then does not understand the change in routing directive that follows.

Speech is a very complex broadband signal whose components are not only differentially susceptible to noise but are also highly dependent on the gender of the speaker and the content and context of the message. Many influential design factors impinge on speech quality and intelligibility, including speech-to-noise ratio, speech bandwidth and fidelity, acoustical environment (such as room directionality, reverberation time, and shadow zones), and communications system fidelity and transmission quality. For example, as reverberation time increases, speech intelligibility will decrease, and amplification will not necessarily compensate. In addition, other human factors must be considered, such as the effects of hearing protection use by the speaker and/or listener, hearing loss of the speaker and/or listener, stress, competing attentional demands, and the vocal output effort of the speaker.

The evaluation of an acoustical environment for speech intelligibility performance is very complex and beyond the scope of this chapter. The reader is referred to the following references on the subject: Casali (2006), Robinson and Casali (2003), and Kryter (1994). A number of useful standards exist that apply to speech intelligibility testing and analysis. For *experimental* testing of communications systems using human subjects, the reader is referred to ANSI S3.2-1989, *American National Standard Methods for Measuring the Intelligibility of Speech over Communications Systems*. A standard that provides a comprehensive *analytical* method for evaluating speech communications systems and environments is ANSI S3.5-1997 (R2002), *Methods for the Calculation of the Speech Intelligibility Index*.

General Guidance for Design of Signals for Use in Noise

The following principles regarding masking effects on nonverbal signals and speech are offered as a summary for general guidance.

1. Because of *direct masking,* the greatest increase in masked threshold occurs for nonverbal signal frequencies that are equal or near to the predominant frequencies of the masking noise. Therefore, warning signals should not utilize tonal frequencies equivalent to those

of the masker. Preferably, the signal should be in the most sensitive range of human hearing, approximately 1000–4000 Hz, unless the noise energy is intense at these frequencies.
2. If the signal and masker are tonal in nature, the primary masking effect is at the fundamental frequency of the masker and at its harmonics. For instance, if a masking noise has primary frequency content at 1000 Hz, this frequency and its harmonics (2000, 3000, 4000, etc.) should be avoided as signal frequencies.
3. The greater the noise level of the masker, the greater the increase in masked threshold of the signal. A general rule of thumb is that the S/N ratio at the listener's ear should (at a minimum) be about 15 dB above masked threshold for reliable signal detection. However, in noise levels above about 80 dBA, the signal levels required to maintain a signal-to-noise ratio of 15 dB above masked threshold may increase the risk to hearing, especially if signal presentation occurs frequently. Therefore, if lower S/Ns become necessary, it is best to design contrasting signals that are unlike the masker in frequency and temporal characteristics, and which have modulated or alternating frequencies to grab attention.
4. To avoid verbal communications interference and operator annoyance, warning signals should not exceed the masked threshold by more than 30 dB (Sorkin, 1987).
5. As the noise level of the masker increases, the primary change in the masking effect is that it spreads upward in frequency, often causing signal frequencies that are higher than the masker to be missed (i.e., *upward masking*). Given that most warning signal guidelines recommend that primarily midrange and high-frequency signals (about 1000–4000 Hz) be used for detectability, it is important to consider that the masking effects of noise of lower frequencies can spread upward and cause interference in this range. Therefore, if the noise has its most significant energy in this range, a lower-frequency signal—say, 500 Hz—may be necessary. However, it must be kept in mind that the ear is not as sensitive to low frequencies, so the signal level must be set carefully to ensure reliable audibility.
6. Masking effects can also spread downward in frequency, causing signal frequencies below those of the masker to be raised in threshold (i.e., remote masking). The effect is greatest at signal frequencies that are subharmonics of the masker. With typical industrial noise sources, remote masking is generally less of a problem than is direct or upward masking.
7. If hearing loss or hearing protection (or both) is involved, the signal should have prominent frequency components below about 1500 Hz.
8. In extremely loud environments of about 110 dB and above, nonauditory signal channels such as visual and vibrotactile should be considered as alternatives to auditory displays. They should also be used for redundancy in some lower-level noises in which the auditory signal may be overlooked or it blends in as the background noise varies, and also when individuals who have hearing loss must attend to the signal.
9. When signals must be localized in space, it is particularly important to include prominent energy content above 3000 Hz, where intensity differences between the ears caused by head shadowing can provide localization cues. If possible, energy content below 1500 Hz is also desirable; in this frequency range, the phase (timing) differences of the sound wave are differentially distinguishable between the two ears, which provides a localization cue.

AUGMENTED HEARING PROTECTION TECHNOLOGY FOR IMPROVED HEARING IN NOISE

Up to this point, we have dealt with the dangers to human hearing imposed by noise, the need for protecting one's hearing in noise, and the difficulties in designing signals and communications systems to provide audibility and intelligibility in noise. Now we turn to the issue of actually selecting and providing the needed hearing protection devices (HPDs), but with the added objective of improving the user's hearing in certain noise situations through technology advancements in hearing protection, commonly called *augmented HPDs*. In 1996, Casali and Berger published an overview of the technologies used in HPD design as of that date. Since that time, new technologies have emerged and existing ones have improved, and in the remainder of this chapter, we discuss those technologies along with potential applications.

Conventional Passive HPD Design and Testing

It is important to recognize that so-called conventional HPDs reduce noise at the ear solely by passive means, and the attenuation provided is the same regardless of incident sound level; that is, the devices are *level independent* or *amplitude insensitive*. In the United States, HPDs are tested at the threshold of hearing using the previously discussed and EPA-required REAT standard (ANSI S3.19-1974; Experimenter-Fit), and the attenuation achieved at threshold remains the same (or linear) throughout most of the dynamic range of noises normally encountered in industry and other settings (EPA, 1979). Exceptions include extremely high impulses of noise that may excite the HPD on the human head, an example being the separation of an earmuff cushion from the side of the head as an explosion's or high-caliber weapon's blast wave passes it. It is also noteworthy that most conventional HPDs—including earplugs, earmuffs, and supra-aural devices—have spectral attenuation curves that increase nonlinearly (providing more attenuation) as a function of sound frequency.

A major impetus for the development of *augmented HPDs* has been the sometimes negative influence that conventional HPDs have on the hearing ability of users. They have often been implicated in compromised auditory perception, degraded signal detection, reduced speech communication abilities, and reduced situational awareness. Depending on circumstances, these effects can create hazards for the wearer or, at the very least, resistance to the use of hearing protection.

Nonetheless, to combat the damaging effects of high-intensity noise on hearing, the promulgation in 1983 of the OSHA Hearing Conservation Amendment (OSHA, 1983) has caused the use of HPDs to proliferate in U.S. industrial workplaces. Similar occupational requirements have been legislated in many other countries, also bringing increased reliance on HPDs. Likewise, HPDs have been a staple piece of personal protection equipment in the U.S. military for at least five decades. Recently, there is indication that HPDs are becoming more popular among the general public; for example, for reducing noise annoyance and improving audio signal fidelity on airplanes and for engaging in loud recreational activities such as target shooting, power tool operation, and noisy spectator events.

Conventional HPD effects on hearing ability for speech and signals. In some situations, too much attenuation may be provided by an HPD for a particular noise situation, with the concomitant effect that the user's hearing is unnecessarily degraded. In lay terms, this is commonly referred to as *overprotection*. In noise exposures that are coupled with hearing critical job demands, there is often a fine balance between inadequate attenuation, or underprotection, and overprotection.

Overall, the research evidence on normal hearers generally suggests that conventional passive HPDs have little or no degrading effect on the wearer's understanding of external speech and signals in ambient noise levels above about 80 dBA. Studies also indicate that conventional passive HPDs may even yield some improvements, with a crossover between disadvantage to advantage between 80 and 90 dBA (Casali & Horylev, 1987; Hormann, Lazarus-Mainka, Schubeius, & Lazarus, 1984; Howell & Martin, 1975; Suter, 1989). However, HPDs often cause increased misunderstanding and poorer detection (compared with unprotected conditions) in regions with lower sound levels, where HPDs are not typically needed for hearing defense anyway but may be applied to reduce annoyance.

In intermittent noise, HPDs may be worn during quiet periods so that when a loud noise occurs, the wearer will be protected. However, during those quiet periods, conventional passive HPDs typically reduce hearing acuity. In certain of these cases, the family of *level-dependent augmented HPDs* can be beneficial, those that provide minimal or moderate attenuation (or, alternatively, more amplification of external sounds) during quiet but increased attenuation (or less amplification) as noise increases.

Noise- and age-induced hearing losses generally occur in the high-frequency regions first. For those so impaired, the effects of HPDs on speech perception are not clear-cut. Because the protector further raises their already elevated thresholds for mid- to high-frequency speech sounds, hearing-impaired individuals are usually disadvantaged in their hearing by conventional HPDs. Though there is no consensus across studies, certain reviews have concluded that sufficiently hearing-impaired individuals will usually experience additional reductions in communications abilities with conventional HPDs worn in noise (Suter, 1989). In some instances, HPDs with electronic hearing-assistive circuits— sometimes called *electronic sound-transmission HPDs* or *sound restoration HPDs*—can be offered to hearing-impaired individuals to determine if their hearing is improved, especially in quiet to moderate noise levels below about 85 dBA, with such devices while they still receive needed protection.

Conventional passive HPDs cannot differentiate and selectively pass speech or nonverbal signal energy (as opposed to noise energy) at a given frequency. Therefore, conventional HPDs do not improve the signal-to-noise ratio in a given frequency band, which is the most important factor for achieving reliable detection or intelligibility. Conventional HPDs attenuate high-frequency sound more than low-frequency sound, thereby attenuating the power of high-frequency consonant sounds that are important for word discrimination while allowing low-frequency noise through. This enables an associated upward spread of masking to occur if the penetrating noise levels are high enough.

Active noise reduction (ANR) devices, which afford electronic phase-derived cancellation of noises that are primarily below about 1000 Hz, can improve the low-frequency attenuation of passive HPDs. In doing so, they may reduce the upward spread of masking of low-frequency noise into the speech and warning signal bandwidths. They can also reduce noise annoyance in certain environments that are dominated by low frequencies, such as jet aircraft cockpits and passenger cabins (Gower & Casali, 1994; Nixon, McKinley, & Steuver, 1992).

Conventional HPDs exhibit attenuation profiles that increase in value as frequency increases, so an imbalance from the listener's perspective is created because the relative amplitudes of different frequencies are heard differently than they would be without the HPD. In other words, they sound more "bassy" (Casali & Berger, 1996). Thus, the spectral quality of a sound is altered, and sound interpretation, which is important in certain jobs (e.g., machining, mining, and performing music) that rely on aural inspection, may suffer as a result. This is one of the reasons that *uniform* (or *flat*) *attenuation HPDs* have been developed as an augmentation technology.

Because some of the high-frequency binaural cues (especially above about 4000 Hz) that depend on the cartilaginous outer ear (pinnae) are altered by HPDs, judgments of sound direction and distance may be compromised with their use. Earmuffs, which completely obscure the pinnae, radically interfere with localization in the vertical plane and also tend to cause horizontal plane errors in both contralateral (left-right) and ipsilateral (front-back) judgments (Suter, 1989). Earplugs may result in some ipsilateral judgment errors, but generally they cause fewer localization problems than do muffs. Exceptions exist, however, in that at least one high-attenuation earplug disrupts localization in magnitude similar to muffs (Mershon & Lin, 1987). In an effort to compensate for the lost pinnae-derived cues for sound localization that are typically destroyed with application of an earmuff, *dichotic sound transmission HPDs*, with a sound pickup microphone on each earmuff cup, can be utilized to maintain some of the binaural cues.

To reduce worker complaints and to mitigate to some extent the aforementioned dilemma of underprotection versus overprotection, certain features have been developed and integrated into HPDs. These advancements can be collectively termed *augmentations*, hence the term *augmented HPDs*. For example, through the use of adjustable vent leakage paths, the attenuation provided by adjustable-attenuation HPDs can be tailored to prevailing ambient noise levels, job demands, and/or user hearing abilities. Frequency-selective filters can be incorporated to aid speech communication and/or signal detection. Flat-attenuation devices allow more natural hearing, which is an important consideration for some users, such as musicians. By reducing excessive low-frequency noise, ANR devices can reduce noise annoyance and sometimes alleviate the masking of speech.

The goal of these and other features just discussed, though not always realized in practice, is to foster the use of hearing protection by producing devices that are more acceptable to the user population and amenable to the working environment, as well as to afford better hearing under a protected state.

In Table 7.2, we provide a simple framework for classifying augmented HPDs into a dichotomy of passive (nonelectronic) and active (electronic) devices, with subgroups under each.

Uniform Attenuation HPDs

As mentioned earlier, the attenuation of conventional passive HPDs is nonlinear and generally increases as frequency increases, as in the three earplug functions labeled "fiberglass," "premolded," and "foam" in Figure 7.5. With these HPDs, although the sounds are reduced in level, they are also changed in a nonuniform manner across the spectrum so that the wearer's hearing of the sound spectrum is distorted. Because many auditory cues

Table 7.2. Augmented HPD Technologies Classified by Type

Passive (Nonelectronic) HPDs
  Uniform (spectrally flat) attenuation devices
  Level-dependent (amplitude-sensitive) devices
  Adjustable-aattenuation devices
  Verifiable-attenuation devices

Active (Electronic) HPDs
  Level-dependent (amplitude-sensitive) sound transmission devices
    (also called sound restoration devices)
  Active noise reduction (ANR) devices
  Adjustable-attenuation (or filtering, hearing-assistive, etc.) devices

Note. Some technologies may fit under more than one category.

depend on spectral shape for informational content (e.g., pitch perception by musicians), conventional HPDs may compromise these cues.

In an attempt to counter these effects, spectrally flat- or uniform-attenuation HPDs such as the ER-15 and ER-20 Musician's™ earplugs (Figure 7.5) or the ER-20 Hi-Fi™ earplug have been developed (Casali & Berger, 1996). These devices utilize acoustical networks to provide essentially flat attenuation over the range of frequencies from 125 to 8000 Hz, as shown in the top two functions of Figure 7.5. To date, there is no published research regarding the effectiveness of these products for hearing acuity. However, anecdotal evidence indicates that in certain situations when the hearing of spectral information is needed, uniform attenuation earplugs are preferred over conventional earplugs. One caution, however, is that because of the overall "moderate" levels of attenuation provided, they should

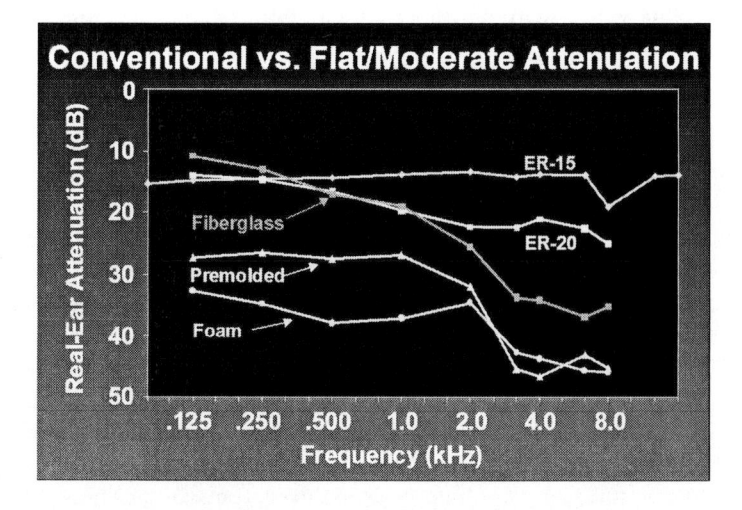

Figure 7.5. Spectral attenuation obtained with REAT procedures for two uniform attenuation, custom-molded earplugs (ER-15, ER-20) and three conventional earplugs: premolded, user-molded foam, and spun fiberglass. (Courtesy of E. H. Berger, E-A-R/AEARO Corporation.)

not be relied on for regular hearing protection purposes in high noise levels (that is, above about 100 dBA).

Level-Dependent HPDs

Level-dependent (also sometimes called *amplitude-sensitive*) HPDs are designed to alter their attenuation characteristics as the ambient noise level changes (increasing attenuation as noise level increases). Such devices may be passive (relying on acoustical networks or mechanical valves for their unique attenuation characteristics) or electronic, and they are embodied in both earplug and earmuff designs. Typically, these devices offer little or no attenuation at low to moderate noise levels, but as ambient noise levels increase, their attenuation increases to some maximum level.

Passive level-dependent HPDs. In passive level-dependent HPDs, a dynamically functional valve or, alternatively, a sharp-edged orifice provides a controlled leakage path into the HPD. At low noise levels, the passive attenuation of most of these devices behaves the same as that of a *leaky protector*, offering minimal attenuation below about 1000 Hz. This minimal attenuation is all that is available to protect the wearer's hearing at sound levels below about 110 dB, and as such, it may be insufficient. But given that such devices are intended to be used primarily in *intermittent, impulsive* noise, this should not be a problem as long as the off periods are relatively quiet (i.e., below about 85 dBA). At elevated sound pressure levels (above about 110–120 dB, such as might occur during a gunshot), in the valve-type devices, the valve is designed to close quickly. However, because of inertial and frictional effects (depending on its design), full valve closure may lag behind the steep pressure rise time of an explosive detonation (Casali & Berger, 1996).

In the orifice-type devices, there are no moving parts and, therefore, theoretically no time lags in response to a sharp pressure transient. But as shown in Figure 7.6, a critical performance parameter is the transition sound level at which the attenuation increases significantly (Allen & Berger, 1990). Essentially, as the incident sound level moves into the transition range, the flow through the orifice changes from laminar to more and more turbulent, effectively "closing" the orifice and thus sharply increasing the attenuation of the device. As illustrated in Figure 7.6, at sound levels beyond the transition sound level, insertion loss increases at a rate of up to about half the increase in sound level. The increase in attenuation continues to a point at which the measured insertion loss approaches that of the equivalent HPD with its nonlinear element sealed shut. An additional advantage is that some orifice-based devices offer roughly flat attenuation (Allen & Berger, 1990).

Passive level-dependent earplugs have existed for many years (e.g., the Gunfender™; Mosko & Fletcher, 1971), and more recently, a new orifice-based design has been embodied as a double-ended earplug for military use: the Combat Arms™ device (Figure 7.7, top; Babeu, Binseel, Mermagen, & Letowski, 2004). This product is intended to provide, in a single earplug unit, situation-dependent hearing protection to accommodate soldiers who are exposed to continuous noise at times, such as during transport in a heavy vehicle, and who are exposed to intermittent, impulsive noise at other times (typically gunfire), while at the same time enabling them to auditorially discern signals and speech in their environment. Thus, one side of the earplug (the green end) is inserted (leaving the bright yellow

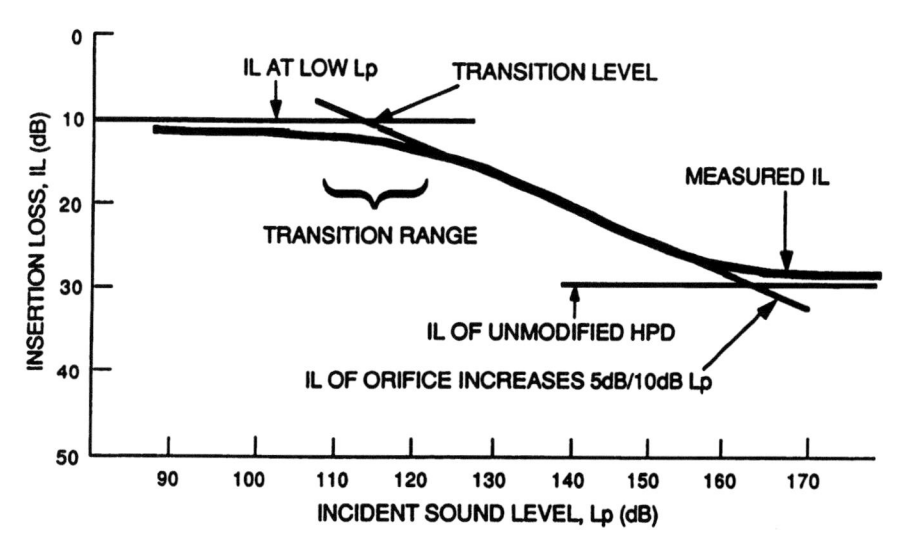

Protection and Enhancement of Hearing in Noise 227

Figure 7.6. Representative insertion loss (IL) and illustration of the transition level at a single frequency, as a function of incident sound level (Lp), for an amplitude-sensitive hearing protector with a nonlinear orifice. Adapted with permission from Allen, C. H., & Berger, E. H. (1990). Development of a unique passive hearing protector with level-dependent and flat attenuation characteristics. *Noise Control Engineering Journal, 34(3):* 97–105.

end exposed) during continuous noise exposure, providing attenuation that is achieved with a standard, passive, flanged earplug (Figure 7.7, bottom; E. H. Berger, EARCAL attenuation measurements on the Combat Arms earplug, personal communication, January 12, 2004)—that is, more protection than is available with a level-dependent earplug in its nontransition state. Conversely, the yellow end is inserted (with the camouflage green end exposed) during reconnaissance and other missions when the sense of hearing is acutely needed but when exposure to gunfire may be imminent and sudden. As afforded by its orifice, which penetrates an occlusion within a duct that runs through the stem of the yellow earplug, this configuration provides moderate attenuation during quiet conditions (level-dependent function of Figure 7.7) but increased attenuation during a gunshot, akin to (but not the same value as) the conventional attenuation function depicted in Figure 7.7.

Research on soldiers' hearing ability with the Combat Arms earplug has been limited, but its yellow, level-dependent end has been reported in one study to provide better sound (azimuth) localization in noise and in quiet than an E-A-R Classic™ foam earplug, but not better than the Combat Arms earplug's conventional triple-flanged end, which is essentially an E-A-R Ultrafit™ design (Babeu et al., 2004). Another study aimed at determining the protective effectiveness of the Combat Arms earplug as worn under an infantry soldier's helmet (Model PASGT) against gunfire produced from an M4 carbine (Babeu, 2005). This study demonstrated that the yellow level-dependent end yielded a protected peak pressure level of 133 dB and insertion loss of 31 dB, compared with 120 dB and 34 dB, respectively, for the green conventional earplug end. By comparison, the unprotected condition (but still wearing the helmet) yielded peak pressure of 164 dB and insertion loss of 0 dB.

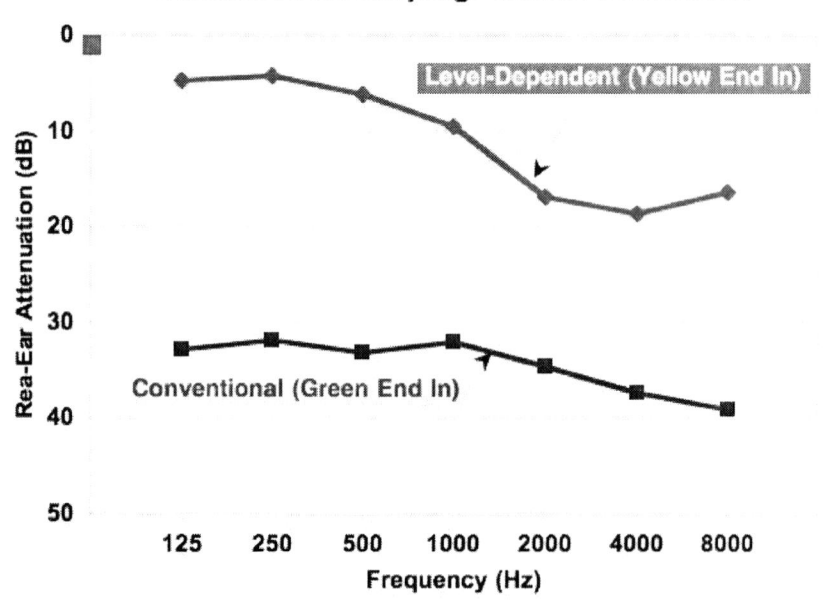

Figure 7.7. **Photograph:** *Combat Arms™ earplug. Left: double-ended earplug consisting of green (upper) end (non-level-dependent) for insertion during continuous noise exposures and yellow (lower) end for insertion during missions when hearing is needed but gunfire may suddenly occur. Right: green (upper) flange removed, revealing ported stem, which contains an orifice that penetrates the duct's occluding member, for effecting level-dependent attenuation when yellow (lower) end is inserted.* **Graph:** *Combat Arms earplug attenuation. Lower function: green end (non-level-dependent); upper function: yellow end (level-dependent), with attenuation measured in resting state. Photo credit: Babeu, L. A., Binseel, M. S., Mermagen T. J., & Letowski, T. R. (2004). Sound localization with the Combat Arms™ earplug. Proceedings of the 29th Annual National Hearing Conservation Association Conference. Graph credit: E. H. Berger, EARCAL attenuation measurements on the Combat Arms earplug, personal communication. January 12, 2004.*

Protection and Enhancement of Hearing in Noise

Active/electronic level-dependent HPDs. Typically earmuff based, these electronic devices incorporate a microphone and output-limiting amplifier to transmit external sounds to earphones mounted within the earcups. The electronics can be designed to pass and boost only certain sounds, such as the critical speech band or critical warning signal frequencies. The limiting amplifier maintains a predetermined (in some cases, user-adjustable) gain, often limiting the earphone output to about 82–85 dBA, unless the ambient noise reaches a cutoff level of 115–120 dBA, at which point the electronics cease to function. At this point, the device essentially becomes a passive HPD. These functional states are illustrated in Figure 7.8.

Ideally, a level-dependent sound transmission HPD should exhibit flat frequency response and distortion-free amplification (without spurious electronic noise) across its passband, as well as high signal-to-noise ratios at levels below its predetermined cutoff level. The cutoff level itself should be relatively low, typically about 80–85 dBA (to ensure that transmitted sound does not overexpose the wearer), and be achieved quickly (i.e., the "attack" or "reaction" time to full shutoff should be on the order of 10 ms or less). The transition from amplification to shutoff should occur without transients or oscillations. The passband of the electronics should be adequate to accommodate desired signals, but not so wide as to pass unnecessary and undesirable noise to the listener. Typically, this passband is in the region of 300–5000 Hz or, in some cases, narrower. The external microphones (preferably two, one for each ear to permit dichotic listening, which assists in sound localization) should be minimally affected by wind or normal movement of the head.

Such devices have the potential for improving the hearing of hearing-impaired listeners in quiet or moderate noise levels, acting much like a hearing aid. However, normal-hearing listeners may not realize similar benefits because of the potential for the residual electronic noise to mask desired signals; the electronics are always on and amplifying unless the incident noise level is high. Like their passive counterparts, some of these devices are well suited for impulsive noise if they have sufficiently sharp (rapid) shut-off profiles,

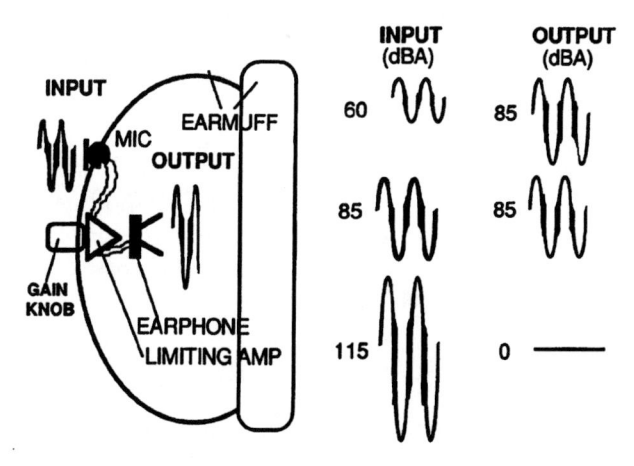

Figure 7.8. General operating characteristics of an electronic level-dependent sound-transmission (or sound restoration) earmuff.

but they are less suited for sounds with long on-durations, which can produce distortion artifacts that become objectionable as they are heard over time.

To address this issue, rather than having a design in which the electronic circuitry is on continuously, one very recent approach (as embodied in the Peltor PTL™ earmuff) has been to incorporate a push-to-listen pushbutton on the outside of the earmuff. This turns on the sound transmission circuitry, latching it on for a short period (e.g., 30 s), and then it shuts itself off. The advantage of this design is that the wearer can decide, on demand, to utilize the passband amplification for a short period while receiving no sound pass-through at other times. The disadvantage is obvious: The wearer may miss desired sounds if the button is not pressed. From purely a human factors standpoint, designs that automatically swing as needed between amplification and shut-off states are more desirable because the operator cannot always determine when each state is needed.

Research on level-dependent HPDs. Murphy and Franks (2002) reported on an experiment in which they evaluated the attenuation of six electroacoustic level-dependent earmuffs (Bilsom 707 Impact II™, Bilsom Targo Electronic™, Howard Leight Thunder™, Howard Leight Leightning™, Peltor Tactical 6S™, Peltor Tactical 7™) and one assistive listening device (hearing aid) against gunfire (ten weapons, eight handguns of various calibers and two 12-gauge shotguns) using an acoustic head and torso simulator (HATS). The devices were tested at three gain (volume) settings (off, unity, and maximum gain). They found that peak attenuation ranged from about 22 to 34 dB across devices, peak attenuation was only slightly dependent on the gain setting of the electronics, and peak attenuation differences between maximum gain versus gain off settings were within about 3 dB for most devices.

As part of the Combat Arms earplug study mentioned earlier, Babeu (2005) also tested the attenuation performance of an electronic level-dependent earmuff (Peltor Com Tac™) in response to military rifle fire. It afforded a protected peak pressure level of 136 dB and an insertion loss attenuation of 28 dB, which was comparable to the performance of the passive level-dependent Combat Arms earplug, which can be considered a competitive alternative. However, the level-dependent earmuff was found to be 16 dB higher in peak pressure exposure than the conventional passive earplug, and it offered 6 dB less insertion loss.

Casali and Wright (1995) compared a Peltor T7-SR™ level-dependent, electronic sound restoration earmuff and an E-A-R Ultra 9000™ passive level-dependent earmuff in a signal detection study using audiometrically normal subjects who listened for a vehicle backup alarm. In addition to the two level-dependent muffs, the subjects underwent the same signal detection task with two conventional earmuffs. These conventional muffs were the basic muff designs that had been modified to produce the level-dependent designs.

The masked threshold results demonstrated that there was no advantage to wearing either the electronic muff or the level-dependent passive muff when attempting to detect a backup alarm in continuously presented pink noise (flat by octaves) levels of 75, 85, and 95 dBA. In other words, there were no differences in masked thresholds between the four earmuffs tested. These results were somewhat surprising, especially for the Ultra 9000 earmuff, which has considerably lower attenuation than its conventional counterpart muff at these sound levels, given that its attenuation does not become nonlinear (i.e., increase) until about 110 dB (Figure 7.6). In the case of the Peltor T7-SR electronic sound-restoration earmuff, even though it *did* amplify the backup alarm signal in 75 and 85 dBA noise, there was no measurable detection advantage over listening through a conventional muff. It is possible that this is attributable to the fact that the noise was also amplified by the electronic muff, and thus there was no S/N ratio advantage when the electronics were operational.

Finally, because the electronic muff did amplify sound, and given that its gain setting was under the control of the subjects, there was an additional need to determine if the sound restoration circuit had potential for increasing the subjects' noise exposure. After each masked threshold test, the sound levels produced at each subject-selected gain control setting were determined by removing the earmuff and placing it on the artificial ears of an acoustic manikin that housed measurement microphones. Under the earmuff, dBC, dBA, and 1000 Hz-centered 1/3-octave-band noise measurements were then made, with gain status (*on* at the subject's setting or *off*) as the independent variable. The gain on versus gain off differences were largest (5 dB higher with gain on) at 1000 Hz, but overall, the contribution of the gain to the noise exposure, as measured by OSHA noise dose, was negligible at all three pink noise levels. Thus, the presence of the sound restoration gain/filter circuit did not increase noise exposures when this particular electronic earmuff was used (Casali & Wright, 1995). Similar results have since been reported by others.

Active Noise Reduction HPDs

Active noise reduction (ANR) relies on the principle of destructive interference of equal amplitude but 180° out-of-phase sound waves at a given point in space. In the case of hearing protectors, the cancellation is established at or within the outer ear. Although the principal technique of ANR was patented in the 1930s, it is only in the past decade that major advances in miniature semiconductor technology and high-speed signal processing have enabled ANR-based HPDs and communications headsets to become viable products.

ANR has been incorporated into two types of personal systems, those designed solely for hearing protection and those designed for one- or two-way communications. Both types are further dichotomized into open-back (or supra-aural) and closed-back (or circumaural earmuff) variations. In the former, a lightweight headband connects ANR microphone/earphone assemblies that are surrounded by foam pads that rest on the pinnae. There are no earmuff cups to afford passive protection, so the open-back devices provide only active noise reduction. If there is an electronic failure, little or no protection is provided by the open-back device.

Closed-back devices, which represent most ANR-based HPDs, are typically based on a passive noise-attenuating earmuff that houses the ANR transducers and, in some cases, the ANR signal-processing electronics. If backup attenuation is needed from an ANR HPD in the event of an electronic failure of the ANR circuit, the closed-back HPD is advantageous because of the passive attenuation afforded by its earcup.

Analog ANR HPDs. A generic block diagram depicting the typical components of an analog electronics, feedback-type, muff-based ANR HPD appears in Figure 7.9. The

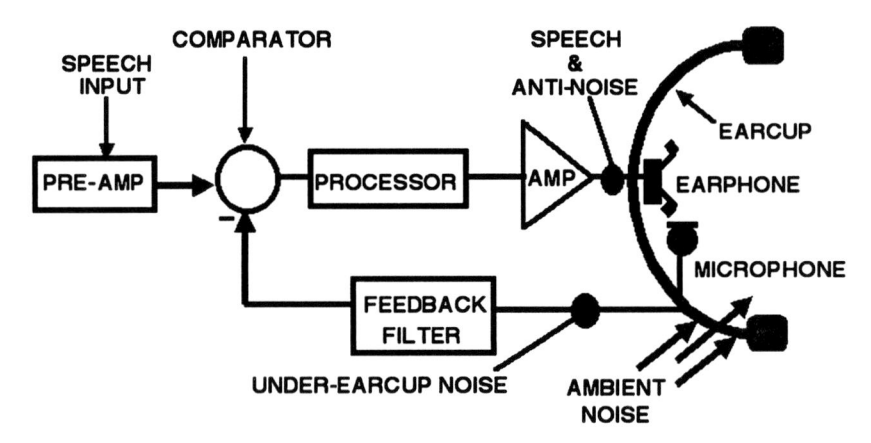

Figure 7.9. Generic block diagram of an analog, feedback-type active noise reduction earmuff with speech input.

example shown is a closed-loop feedback system that receives input from a sensing microphone that detects the noise penetrating the passive barrier posed by the earmuff. The signal is then fed back through a phase compensation filter, which reverses the phase, to an amplifier that provides the necessary gain. Finally, the signal is output as an *antinoise* signal through an earphone loudspeaker to effect cancellation inside the earcup. Although most all ANR devices have been built in earmuff or supra-aural headset configurations, earplug examples have been prototyped recently. In contrast to the common ANR closed-loop feedback configuration, open-loop, feed-forward systems are also available; these are typically of the lightweight headset (i.e., open-back) variety.

Because of the phase shifts that can be attributed to transducer (i.e., microphone, earphone) location differences, and/or throughput delays in signal processing, establishing the correct phase relationship between the cancellation signal and the offending noise becomes more difficult as the bandwidth of the noise increases (i.e., as wavelengths get shorter). For this reason, ANR has been most effective against low-frequency noise. For example, with analog ANR devices, maximal attenuation values of about 22 dB are typically found to be in a range from about 100 to 250 Hz. However, with some very recent digital or hybrid (analog/digital) devices, attenuation of higher than 25 dB in this frequency range has been noted by the first author in laboratory tests at Virginia Tech. These values drop to essentially no attenuation above about 1000 Hz (Nixon et al., 1992), although the bandwidth is increasing as the technology develops (Casali & Robinson, 1994). Across many ANR headset tests by the first author since the early 1990s, it is typical to find some noise enhancement (i.e., amplification attributable to the ANR) of 2–6 dB, but in some cases more or less with ANR devices, occurring in the midrange frequencies from about 1000 Hz up to as high as 4000 Hz. An example of the passive and total (passive + active) attenuation of a closed-back ANR earmuff appears in Figure 7.10.

Digital ANR HPDs. With advances in the speed, power, reliability, and miniaturization of digital signal-processing components, digital technology has demonstrated its

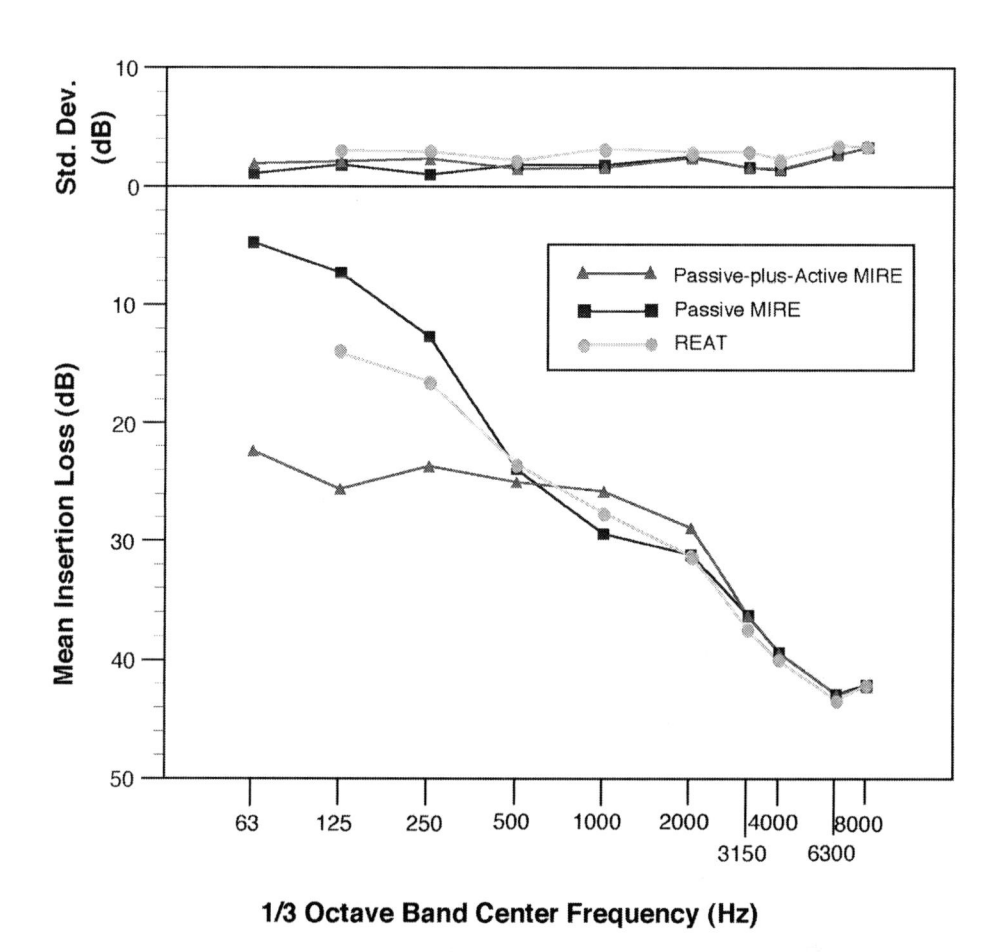

Protection and Enhancement of Hearing in Noise

Figure 7.10. Spectral attenuation of a typical closed-back ANR headset (NCT PA-3000) as obtained with techniques indicated in legend and explained in text. Graph credit: Casali, 2005.

effectiveness for improving the capabilities of ANR-based HPDs, particularly with regard to precise tuning of the control system using software for optimizing the cancellation of specific sound frequencies. Advantages with the use of digital technology, as generally compared with analog, reside largely in its ability to perform complex computations with high precision. In addition, electronic components are less affected by temperature variations and remain more stable, performance tolerances can be held very tight, and it is possible to use continuously adaptive filters based on statistical algorithms (Denenberg & Claybaugh, 1993). Some ANR HPDs incorporate hybrid analog/digital designs.

Research on ANR HPDs. In certain types of noises, especially those characterized by high overall levels and/or frequency content that is heavily skewed toward frequencies of less than 1000 Hz, ANR-based hearing protectors (vs. conventional passive HPDs) may offer some advantages in the detection of certain auditory warning signals.

In this vein, Casali, Robinson, Dabney, and Gauger (2004) demonstrated, using normal-hearing subjects, that in 100-dBA broadband noise, a Bose ANR headset produced lower masked thresholds (i.e., better detection) for a vehicle backup alarm than a conventional earmuff with much weaker low-frequency attenuation. Furthermore, in "red" noise, which has considerable low-frequency bias as evidenced by a dBC-dBA value of 9, the ANR headset again demonstrated a signal detection advantage over a conventional earmuff. However, in the same noise conditions, the ANR device was not associated with lower masked thresholds than was a high-attenuation conventional foam earplug.

The bottom line is that for very low frequency noises of high levels, ANR may offer a signal detection advantage over some conventional protectors that do not have sufficient low-frequency attenuation to prevent the upward spread of noise masking.

Issues in attenuation performance testing for ANR HPDs. At present, standardized and labeled attenuation data and noise reduction ratings are not available for ANR HPDs because the EPA-required test for HPD product labeling—namely, ANSI S3.19-1974—does not accommodate the electronically provided (i.e., active) attenuation component of ANR-based products (Casali, 2005). As discussed earlier in this chapter, such attenuation data are necessary for ensuring that hearing protectors are adequate for use in given occupational noise exposures and that they are also useful for matching HPDs to military, leisure, or noise exposure situations. Real-ear-attenuation-at-threshold (REAT) techniques, as specified in ANSI S3.19-1974, use human subjects as listeners and can be applied to determine the passive component of ANR HPD attenuation (i.e., with the electronics shut off). Microphone-in-real-ear (MIRE) techniques, which use human subjects as fixtures for in-ear measurement microphones (e.g., by applying ANSI S12.42-1995), can be used to measure both the passive (ANR off) and total (ANR on) attenuation of the device, as described in Casali (2005).

However, there are disadvantages to this approach. For example, the MIRE approach does not account for the bone conduction of sound that is included in the measurement only when human listening protocols (i.e., REAT tests) are applied. Regardless of whether MIRE or REAT protocols are used to measure the passive attenuation, the active or the ANR component of the attenuation can then be computed using the following relationship, with each term measured in dB attenuation:

$$[\text{Active component} = \text{MIRE total} - \{\text{MIRE passive (or REAT passive)}\}] \quad (9)$$

One could conclude that ANR manufacturers are penalized by the lack of standardized test procedures and EPA labeling guidelines because this precludes them from selling their products as industrial or consumer hearing protection devices. However, some manufacturers, based on the fact that they target their sales to applications other than industrial markets (e.g., military, general aviation, consumer noise annoyance), may not necessarily want an NRR.

In Figure 7.10, REAT and MIRE attenuation for a typical closed-back circumaural ANR earmuff is shown (Casali, 2005). Readily apparent in Figure 7.10 is the difference between the MIRE and REAT attenuation at 125 and 250 Hz. This difference is likely attributable to physiological noise masking the test stimulus during the occluded REAT test, which renders those occluded thresholds as artificially higher and thus produces a higher attenuation. Also evident in the figure is the slight reduction in total attenuation at 1000 and 2000 Hz when the ANR device is turned on. Although Figure 7.10 illustrates the total (MIRE) and passive (both REAT and MIRE) attenuation, as shown in Equation 9, the active component of the device can be separated from the total attenuation simply by subtracting the MIRE passive attenuation from the MIRE total attenuation.

It is important to recognize, as exemplified in Figure 7.10, that ANR is effective as a noise reduction technique primarily for low-frequency noise that is below about 1000 Hz. This is somewhat fortuitous, given that the passive materials required to attenuate such frequencies are generally heavy and bulky and, as such, do not bode well for a device designed to be worn on the head. Because of their biased performance, ANR devices are most amenable to sources that are dominated by low frequencies, examples of which are large diesel engines, aircraft turbofans and cockpit wind effects, and certain heavy machinery.

Although ANR devices cannot currently be sold in the United States under EPA labeling regulations as hearing protectors (except as to their passive attenuation only) because of the aforementioned lack of appropriate ANR testing standards and labeling regulations, they are now being applied for various purposes in both the public and the private sector. ANR headsets are available for noise annoyance mitigation (as airline passenger and personal stereo headsets) and as communications headsets for commercial, military, and civilian aviation, and they are used to combat severe noise environments in the military (particularly in armored vehicles and rotorcraft). Special-purpose ANR devices are also (or have been) commercially available for telephone operators and telemarketers to reduce patient noise exposure in MRI machines, and even to reduce siren noise for emergency vehicle crews. There is some debate as to the potential for the application of ANR products to industrial noise markets and, therefore, whether an NRR-like rating is really necessary, given that relatively few industries are characterized by noise with dBC-dBA values greater than 6, which are evidence of low-frequency spectral dominance.

Variable-Attenuation and Verifiable-Attenuation HPDs

In addition to the specialized hearing protectors discussed earlier, there are several new types of HPDs that allow users to "adjust" or "tune" their attenuation for the specific noise threat at hand. Some of these products also include microphone measurement systems to enable determination, or "verification in situ," of the amount of attenuation provided to the user.

Consider that the amount of attenuation needed by noise-exposed individuals may vary depending on the severity of their exposures, the importance of hearing auditory signals and/or speech in their jobs, and, in some cases, their own hearing abilities. However, conventional HPDs provide a relatively constant amount of attenuation without regard to these factors. To help tailor the amount of protection for an individual, earplugs have been developed which afford the HPD fitting professional and/or the actual end-user some level of control over the amount of attenuation achieved. These devices incorporate a leakage path that can be adjusted by setting a valve that obstructs a tunnel or vent cut through the body of the plug. It can also be adjusted by selecting from among a choice of available filters or dampers that are inserted into the vent. A complete discussion appears in Casali (2005). We briefly introduce three examples in this section.

A Dutch earplug, the Ergotec Variphone™, has an adjustable-needle valve design in an acrylic custom-molded impression of the user's ear canal. This product affords a continuously variable attenuation setting, allowing the HPD to be optimized as noise exposures change. Another example is the Canadian-manufactured Custom Protect Ear dB Blocker Vented™, which is an earplug made from custom impressions; the final product is molded in silicone with a special vent passing through the canal portion (Lancaster & Casali, 2005). Into this vent can be inserted several different filter dampers, singly or in combination, which are intended to provide a selection of different spectra of attenuation depending on the wearer's needs.

Finally, another example of a selectable-damper design is the Sonomax SonoCustom™, manufactured in Canada (Voix, Zeidan, & Cloutier, 2005). This unique earplug is semi-custom-molded to the user's ears in situ by selecting one of several blank sizes that best fits the ear and then molding the blank to custom-fit the ear by injecting a silicone filler into a bladder or skin that expands from within to fill the convolutions of the concha and outer ear canal. This process obviates the need for obtaining deep ear impressions, which must be sent back to a laboratory for manufacture of the final earplug. On the other hand, it does not typically provide the same deep ear canal fit of a true custom-molded earplug for which the impression is obtained with an ear dam. The Sonomax earplug has a pass-through vent into which one of several attenuation damper cartridges can be inserted, ranging in resistance from 330 to 4700 ohms, plus a full block, thus providing discretely adjustable attenuation.

Both the Custom Protect Ear dB Blocker Vented and the Sonomax SonoCustom earplug have the added advantage that their attenuation can be verified using MIRE measurement techniques that can be implemented in the field. This is a significant development, given that the actual protection level afforded to the wearer can be verified and then adjusted to accommodate specific noise exposure and work-related hearing needs.

CONCLUSIONS

As we have discussed in this chapter, noise causes many problems for humans, including but not limited to annoyance, hearing loss, physiological effects, task performance degradation, and interference with the hearing and understanding of auditory signals and speech. For people to reliably hear signals and speech communications in intense noise is obviously a challenge, but in many circumstances, that hearing is critical to safety, job performance, and/or mission success.

Furthermore, many noise exposures require the use of hearing protection devices to guard against the insidious occurrence of noise-induced hearing loss. In this chapter, we have provided guidance to aid human factors and safety practitioners in addressing the hearing-related problems associated with noise exposures, with particular emphasis on signal and speech communication system design considerations to yield the best opportunity for audibility and intelligibility. Several acoustical standards to assist in this endeavor have been recommended and explained. In addition, when the need for hearing protection is indicated, the practitioner should consider the guidance provided herein for selecting augmented passive and electronic hearing protection devices that can yield improved hearing in certain circumstances.

In all cases, careful and accurate measurement of the noise exposure situation, particularly in terms of spectral content, as well as of the signals that need to be heard, are prerequisite to the selection of augmented hearing protection devices and to the design of auditory signals to be heard in noise.

ACKNOWLEDGMENTS

Portions of this chapter are adapted in part from Casali (2005), Casali (2006), and Robinson and Casali (2003).

REFERENCES

Allen, C. H., & Berger, E. H. (1990). Development of a unique passive hearing protector with level-dependent and flat attenuation characteristics. *Noise Control Engineering Journal, 34*(3), 97–105.

ANSI S3.19-1974. *Method for the measurement of real-ear protection of hearing protectors and physical attenuation of earmuffs.* New York: American National Standards Institute.

ANSI S1.4-1983 (R2001). *Specification for sound level meters.* New York: American National Standards Institute.

ANSI S3.2-1989. *American National Standard methods for measuring the intelligibility of speech over communications systems.* New York: American National Standards Institute.

ANSI S1.1-1994 (R2004). *Acoustical terminology.* New York: American National Standards Institute.

ANSI S12.42-1995. *Microphone-in-real-ear and acoustic test fixture methods for the measurement of insertion loss of circumaural hearing protection devices.* New York: American National Standards Institute.

ANSI S3.5-1997 (R2002). *Methods for the calculation of the speech intelligibility index.* New York: American National Standards Institute.

ANSI S12.6-1997 (R2002). *Methods for measuring the real-ear attenuation of hearing protectors.* New York: American National Standards Institute.

Babeu, L. A. (2005). Level dependent hearing protection and distortion product otoacoustic emissions (DPOAE) after small arms fire. *Proceedings of the 30th Annual National Hearing Conservation Association Conference* [CD]. Denver, CO: National Hearing Conservation Association.

Babeu, L. A., Binseel, M. S., Mermagen T. J., & Letowski, T. R. (2004). Sound localization with the Combat Arm™ earplug. *Proceedings of the 29th Annual National Hearing Conservation Association Conference.* Denver, CO: National Hearing Conservation Association.

Berger, E. H. (2003a). Noise control and hearing conservation: Why do it? In E. H. Berger, L. H. Royster, J. D. Royster, D. P. Driscoll, & M. Layne (Eds.), *The noise manual* (rev. 5th ed., pp. 2–17). Fairfax, VA: American Industrial Hygiene Association.

Berger, E. H. (2003b). Hearing protection devices. In E. H. Berger, L. H. Royster, J. D. Royster, D. P. Driscoll, & M. Layne (Eds.), *The noise manual* (rev. 5th ed., pp. 379–454). Fairfax, VA: American Industrial Hygiene Association.

Berger, E. H., & Casali, J. G. (1997). Hearing protection devices. In M. J. Crocker (Ed.), *Encyclopedia of acoustics* (pp. 967–981). New York: Wiley.

Berger, E. H., Royster, L. H., Royster, J. D., Driscoll, D. P., & Layne. M. (Eds.) (2003). *The noise manual* (rev. 5th ed.). Fairfax, VA: American Industrial Hygiene Association.

Bhattacharya, S. K., Roy, A., Tripathi, S. R., & Chatterjee, S. K. (1985, July). Behavioral measurements in textile weavers wearing hearing protectors. *Indian Journal of Medical Research, 82,* 55–64.

Casali, J. G. (1999). Litigating community noise annoyance: A human factors perspective. In *Proceedings of the Human Factors and Ergonomics Society 42nd Annual Meeting* (pp. 612–616). Santa Monica, CA: Human Factors and Ergonomics Society.

Casali, J. G. (2005). Advancements in hearing protection: Technology, applications, and challenges for performance testing and product labeling. In *Proceedings of the 2005 International Congress and Exhibition on Noise Control Engineering* (on CD; pp. 2097–2118). Rio de Janeiro, Brazil: Brazilian Acoustical Society.

Casali, J. G. (2006). Sound and noise. In G. Salvendy (Ed.), *Handbook of human factors* (3rd ed.; pp. 612–642). New York: Wiley.

Casali, J. G., & Berger, E. H. (1996). Technology advancements in hearing protection: Active noise reduction, frequency/amplitude-sensitivity, and uniform attenuation. *American Industrial Hygiene Association Journal, 57,* 175–185.

Casali, J. G., & Horylev, M. J. (1987). Speech discrimination in noise: The influence of hearing protection. In *Proceedings of the Human Factors Society 31st Annual Meeting* (pp. 1246–1250). Santa Monica, CA: Human Factors and Ergonomics Society.

Casali, J. G., & Robinson, G. S. (1994). Narrow-band digital active noise reduction in a siren-canceling headset: Real-ear and acoustic manikin insertion loss. *Noise Control Engineering Journal, 42*(3), 101–115.

Casali, J. G., & Robinson, G. S. (1999). Noise in industry: Auditory effects, measurement, regulations, and management. In W. Karwowski & W. Marras (Eds.), *Handbook of occupational ergonomics* (pp. 1661–1692). Boca Raton, FL: CRC Press.

Casali, J. G., Robinson, G. S., Dabney, E. C., & Gauger, D. (2004). Effect of electronic ANR and conventional hearing protectors on vehicle backup alarm detection in noise. *Human Factors, 46,* 1–10.

Casali, J. G., & Wright, W. H. (1995). Do amplitude-sensitive hearing protectors improve detectability of vehicle backup alarms in noise? In *Proceedings of the Human Factors and Ergonomics Society 39th Annual Meeting* (pp. 994–998). Santa Monica, CA: Human Factors and Ergonomics Society.

Cohen, A. (1976). The influence of a company hearing protection program on extra-auditory problems in workers. *Journal of Safety Research, 8*(4), 146–162.

Crocker, M. (Ed.). (1998). *Handbook of acoustics.* New York: Wiley.

Deatherage, B. H. (1972). Auditory and other sensory forms of information presentation. In H. P. Van Cott & R. G. Kincade (Eds.), *Human engineering guide to equipment design* (pp. 123–160). New York: Wiley.

Denenberg, J. N., & Claybaugh, D. J. (1993). A selective canceling headset for use in emergency vehicles. *Institute of Noise Control Engineering Journal/Japan, 17*(2), 1993.

Driscoll, D. P., Stewart, N. D., & Anderson, R. R. (2003). Community noise. In E. H. Berger, L. H. Royster, J. D. Royster, D. P. Driscoll, & M. Layne (Eds.), *The noise manual* (rev. 5th ed., pp. 601–637). Fairfax, VA: American Industrial Hygiene Association.

Dummer, W. (1997). Occupational health and workman's compensation in Chile. *Applied Occupational and Environmental Hygiene, 12*(12), 805–812.

Earshen, J. J. (2003). Sound measurement: Instrumentation and noise descriptors. In E. H. Berger, L. H. Royster, J. D. Royster, D. P. Driscoll, & M. Layne (Eds.), *The noise manual* (rev. 5th ed., pp. 41–100). Fairfax, VA: American Industrial Hygiene Association.

Egan, J. P., & Hake, H. W. (1950). On the masking pattern of a simple auditory stimulus. *Journal of the Acoustical Society of America, 22,* 622–630.

EPA. (1979). Noise labeling requirements for hearing protectors. Environmental Protection Agency, 40CFR211, *Federal Register, 44*(190), 56130–56147.

EPA. (2003). *EPA Docket OAR-2003-0024.* U.S. Environmental Protection Agency Workshop on Hearing Protection Devices, Washington, D.C., March 27–28.

Fidell, S. M., & Pearsons, K. S. (1997). Community response to environmental noise. In M. Crocker (Ed.), *Encyclopedia of acoustics* (pp. 1083–1091). New York: Wiley.

Fletcher, H. (1940). Auditory patterns. *Reviews of Modern Physics, 12,* 47–65.

Glasberg, B. R., & Moore, B. C. J. (1990). Derivation of auditory filter shapes from notched-noise data. *Hearing Research, 47,* 103–138.

Gower, D., & Casali, J. G. (1994). Speech intelligibility and protective effectiveness of selected active noise reduction and conventional communications headsets. *Human Factors 36,* 350–367.

Harris, C. M. (1991). *Handbook of acoustical measurements and noise control.* New York: McGraw-Hill.

Hawkins, J. E., & Stevens, S. S. (1950). The masking of pure tones and of speech by white noise. *Journal of the Acoustical Society of America, 22*(1), 6–13.

Hormann, H., Lazarus-Mainka, G., Schubeius, M., & Lazarus, H. (1984). The effect of noise and the wearing of ear protectors on verbal communication. *Noise Control Engineering Journal 23*(2), 69–77.

Howell, K., & Martin, A. M. (1975). An investigation of the effects of hearing protectors on vocal communication in noise. *Journal of Sound and Vibration 41*(2), 181–196.

ISO 7731–2003(E). *Danger signals for public and work areas—Auditory danger signals.* Geneva: International Organization for Standardization.

Kryter, K. D. (1994). *The handbook of hearing and the effects of noise.* New York: Academic Press.

Lancaster, J. A., & Casali, J. G. (2005). An integrated MIRE field measurement technique for predicting real-ear attenuation of a custom-molded earplug: Instrumentation and validation. In *Proceedings of the 30th Annual National Hearing Conservation Association Conference* [CD]. Denver, CO: National Hearing Conservation Association.

Laughery, K. R., & Wogalter, M. S. (2006). Warning design and effectiveness: Current status and new directions. In R. C. Williges (Ed.), *Reviews of human factors and ergonomics, Volume 2* (pp. 241–271). Santa Monica, CA: Human Factors and Ergonomics Society.

Melamid, S., & Bruhis, S. (1996). The effects of chronic industrial noise exposure on urinary corticol, fatigue, and irritability. *Journal of Occupational and Environmental Medicine, 38,* 252–256.

Melnick, W. (1991). Hearing loss from noise exposure. In C. M. Harris (Ed.), *Handbook of acoustical measurements and noise control* (pp. 18.1–18.19) New York: McGraw-Hill.

Mershon, D. H., & Lin, L. J. (1987). Directional localization in high ambient noise with and without the use of hearing protectors. *Ergonomics, 30,* 1161–1173.

Mosko, J. D., & Fletcher, J. L. (1971). Evaluation of the Gunfender earplug: Temporary threshold shift and speech intelligibility. *Journal of the Acoustical Society of America, 49:* 1732–1733.

MSHA. (1999). Health standards for occupational noise exposure; Final Rule. Mine Safety and Health Administration, 30CFR Part 62, 64. *Federal Register,* 49548–49634, 49636–49637.

Murphy, W. J., & Franks, J. R. (2002). Do sound restoration earmuffs provide adequate protection for gunshot noise? *Journal of the Acoustical Society of America, 112*(5, pt 2), 2294.

National Institute for Occupational Safety and Health (NIOSH). (1975). *List of personal hearing protectors and attenuation data* (National Institute for Occupational Safety and Health-HEW Publication No. 76-120; pp. 21–37). Washington, DC: Author.

National Institute for Occupational Safety and Health (NIOSH). (1996). *National occupational research agenda* (National Institute for Occupational Safety and Health DHHA [NIOSH] Publication No. 96-115). Cincinnati, OH: Author.

National Institute for Occupational Safety and Health (NIOSH). (1998). A proposed national strategy for the prevention of noise-induced hearing loss. In *Proposed national strategies for the prevention of leading work-related diseases, Part 2* (pp. 51–63). Cincinnati, OH: Author.

National Institutes of Health (NIH). (1990). Noise and hearing loss. National Institutes of Health Consensus Development Panel. *Journal of the American Medical Association, 263,* 3185–3190.

Nixon, C. W., McKinley, R. L., & Steuver, J. W. (1992). Performance of active noise reduction headsets. In A. L. Dancer, D. Henderson, R. J. Salvi, & R. P. Hamernik (Eds.), *Noise-induced hearing loss* (pp. 389–400). St. Louis: MosbyYear Book.

Noweir, M. H. (1984). Noise exposure as related to productivity, disciplinary actions, absenteeism, and accidents among textile workers. *Journal of Safety Research, 15*(4), 163–174.

Occupational Safety and Health Administration (OSHA). (1971a). *Occupational noise exposure (general industry).* Occupational Safety and Health Administration, 29CFR1910.95. *Federal Register, 36*(10), 10518.

Occupational Safety and Health Administration (OSHA). (1971b). *Occupational noise exposure (construction industry).* Occupational Safety and Health Administration, 29CFR1926.52. *Federal Register.*

Occupational Safety and Health Administration (OSHA). (1983). *Occupational noise exposure; hearing conservation amendment; final rule.* Occupational Safety and Health Administration, 29CFR1910.95. *Federal Register, 48*(46), 9738–9785.

Park, M. Y., & Casali, J. G. (1991). A controlled investigation of in-field attenuation performance of selected insert, earmuff, and canal cap hearing protectors. *Human Factors, 33,* 693–714.

Poulton, E. (1978). A new look at the effects of noise: A rejoinder. *Psychological Bulletin, 85,* 1068–1079.

Robinson, G. S., & Casali, J. G. (2003). Speech communications and signal detection in noise. In E. H. Berger, L. H. Royster, J. D. Royster, D. P. Driscoll, & M. Layne (Eds.), *The noise manual* (rev. 5th ed., pp. 567–600). Fairfax, VA: American Industrial Hygiene Association.

Sanders, M. S., & McCormick, E. J. (1993). *Human factors in engineering and design* (7th ed.). New York: McGraw-Hill.

Schmidt, J. W., Royster, L. H., & Pearson, R. G. (1982). Impact of an industrial hearing conservation program on occupational injuries. *Sound and Vibration, 16*(5), 16–20.

Smoorenburg, G. F., Axelsson, A., Babisch, W., Diamond, I. G., Ising, H., Marth, E., Miedema, H.M.E., Ohrstrom, E., Rice, C.G., Abbing, E.W.R., van de Wiel, J.A.G., & Passchier-Vermeer, W. (1996). Effects of noise on health. *Noise News International 4*(3), 137–150.

Sorkin, R. D. (1987). Design of auditory and tactile displays. In G. Salvendy (Ed.), *Handbook of human factors* (pp. 549–576). New York: McGraw-Hill.

Staples, N. (1981). Hearing conservation—Is management shortchanging those at risk? *Noise and Vibration Control Worldwide, 12*(6), 236–238.

Suter, A. H. (1989). *The effects of hearing protectors on speech communication and the perception of warning signals* (AMCMS Code 611102.74A0011). Aberdeen Proving Ground, MD: U.S. Army Human Engineering Laboratory.

Taylor, W., Pearson, J., Mair, A., & Burns, W. (1964). Study of noise and hearing in jute weavers. *Journal of the Acoustical Society of America, 38*, 113–120.

van Charante, A. W. M., & Mulder, P. G. H. (1990). Perceptual acuity and the risk of industrial accidents. *American Journal of Epidemiology, 121*, 652–663.

Voix, J., Zeidan, J., & Cloutier, A. (2005). A new approach to the objective assessment of HPD performance in the field. In *Proceedings of the 30th Annual National Hearing Conservation Association Conference* [CD]. Denver, CO: National Hearing Conservation Association.

Workers' Compensation Board of British Columbia (WCB). (1998). *Occupational diseases in British Columbia, 1979–1996.* Vancouver, Canada: Author.

Weston, H. C., & Adams, S. (1935). *The performance of weavers under varying conditions of noise* (Report No. 70). London: Medical Research Council Industrial Health Research Board.

World Health Organization (WHO). (1995). Community noise. In B. Berglund & T. Lindvall (Eds.), *Archives of the Center for Sensory Research, 2*(1). Stockholm: University of Stockholm.

Wilkins, P. A., & Martin, A. M. (1982). The effects of hearing protection on the perception of warning sounds. In P. W. Alberti (Ed.), *Personal hearing protection in industry* (pp. 339–369). New York: Raven Press.

Wilkins, P. A., & Martin, A. M. (1985). The role of acoustical characteristics in the perception of warning sounds and the effects of wearing hearing protection. *Journal of Sound and Vibration, 100*(2), 181–190.

Zwerling, C., Whitten, P. S., Davis, C. S., & Sprince, N. L. (1997). Occupational injuries among workers with disabilities, the National Health Interview Survey, 1985–1994. *Journal of the American Medical Association, 278*, 2163–2166.

CHAPTER 8

Designing Effective Warnings

By Kenneth R. Laughery & Michael S. Wogalter

Since the early 1980s there has been an increased interest in research on warnings. This chapter has several objectives. First, we describe the purpose of warnings and where warnings fit with other safety considerations, such as design and guarding. Next, we present a model that incorporates both communication and information-processing concepts, which is characteristic of theoretical orientations that have guided much of the warning research. The research and application issues have generally focused on two themes: design factors and non-design factors that influence warning effectiveness. Third, we review the progress and status of research and application, with an emphasis on identifying those factors that appear to be most important in determining warning effectiveness. Finally, we conclude with a discussion of some of the challenges and opportunities facing warning designers and researchers in the future.

Concern for public safety has increased in the United States since the 1960s. This concern has been manifested in various ways. Local, state, and federal laws have been introduced to address safety issues. U.S. government agencies such as the Consumer Product Safety Commission (CPSC), the Occupational Safety and Health Administration (OSHA), the Food and Drug Administration (FDA), and the Environmental Protection Agency (EPA) have been assigned responsibilities for public safety in various domains. Regulations, standards, and guidelines concerning product and environmental safety have been promulgated by these agencies and by private organizations such as the American National Standards Institute (ANSI) and Underwriters Laboratories (UL).

Another outcome of the increased concern for safety is the greater attention to and use of warnings. A substantial body of published scientific research on topics related to warning design and effectiveness has accompanied the growing use of warnings as a tool for achieving environmental and product safety. A significant portion of this research has been carried out by human factors/ergonomics (HF/E) specialists and published in HF/E literature.

Several noteworthy reviews and collections of the research literature on warnings have been published. Lehto and Miller (1986) provided a review of early literature on warnings. DeJoy (1989) reported an analysis of the implications of the early warning effectiveness research. A text by Edworthy and Adams (1996) contains a general review of visual and auditory warnings. Other reviews published in the mid- and late-1990s include Laughery and Wogalter (1997); Parsons, Seminara, and Wogalter (1999); Wogalter, DeJoy, and Laughery (1999); and Wogalter and Laughery (1996).

More recent reviews of the warnings literature are Rogers, Lamson, and Rousseau (2000) and Wogalter and Laughery (in press, 2006). Two collections of papers published in the *Proceedings of the Human Factors and Ergonomics Society* have also been assembled and published (Laughery, Wogalter, & Young, 1994; Wogalter, Young, & Laughery, 2001).

Finally, a substantial collection of papers reviewing the warnings literature was prepared specifically for a handbook edited by Wogalter (2006).

Warning research questions have tended to focus on factors that influence whether or not a warning will be effective. At the same time, however, a generally accepted underlying theoretical context for the research drawing on communication theory and human information-processing theory has served as a means for organizing the research and as a tool for explaining and predicting warning failures. This theoretical orientation will be described in a separate section.

In order for a warning to be effective, it must accomplish certain things. Generally, a warning must capture attention; that is, it must be noticed and encoded. With some exceptions, people do not typically search for or seek out warnings. Thus, warnings must be sufficiently conspicuous, and they must have characteristics that encourage encoding the content. Warnings must also provide the information needed for recipients to make informed decisions regarding compliance. Compliance decisions can be viewed as based on cost-benefit trade-offs. The costs involved can take the form of effort, time, money, and so on. The benefits of compliance can include avoiding negative health effects, injuries, or property damage. One reason that people do not comply with a warning is that the perceived costs of compliance are judged to outweigh the benefits. Thus, a focus in warning design is to provide the information needed for compliance decisions to be made rationally and wisely. Efforts have been made to apply signal detection and decision theory to warnings (Lehto, 2006).

Whether or not (and how) a person complies with a warning depends not merely on the warning's characteristics but also on many additional factors, such as the user's experience, familiarity with the product or situation, competence or ability to carry out the action, and the perceived costs (effort, time, money) of complying. Others have addressed warning compliance from a similar perspective. Edworthy (1998) developed a decision model based on utility theory for evaluating decisions in the warning process. Similarly, Riley (2006) and Cameron and DeJoy (2006) addressed motivational processes in compliance decisions.

In the language of communication theory, the concept of the medium (or channel) directly relates to warnings. Warnings can be conveyed directly in many media, such as on labels, in product manuals, on signs, in videos, and on computer screens. Alternatively, warnings may be conveyed indirectly, such as from another person who had been exposed to a warning earlier. Regardless of the methods of conveyance, warnings are usually delivered to individuals through the visual or auditory modality, although other sensory modalities are occasionally employed. For example, the propane and natural gas delivered to consumers can be neither seen nor smelled; therefore, an odorant (ethyl mercaptan) is added to allow the use of olfaction to detect leaks. The tactual (including the kinesthetic and haptic) senses are used for built-in vibratory feedback in airplanes when path and speed might result in a dangerous stall.

In this chapter, we review research addressing the design and effectiveness of warnings for products and environments. With some exceptions, these are warnings presented through the visual modality. Chapter 6 by Morrow, North, and Wickens (2006) in *Reviews of Human Factors and Ergonomics, Volume 1* and Chapter 7 by Casali and Gerges (2006) in the current volume describe research and application of auditory warnings in the contexts of hospital procedures and hearing protection. Because of the existence of these substantial reviews and because research on warnings in the visual and auditory modalities has not greatly overlapped (in part because they are different modalities and have substantially different properties), in this chapter, we focus on visual warnings.

It is important to note the concept of a warning system. The notion of a warning being a sign, a label, a paragraph or picture in a manual, or an auditory alarm is too narrow a view of how such safety information does, or should, get transmitted. A warning system for a particular setting or product may consist of a number of components. The system may include a printed statement on a box, a package insert, a sign on a barrier, a verbal message at the point of purchase, a siren, flashing lights, and so forth. How the components of the warning system interact and complement each other is one of the significant aspects of warning design.

The different components may play different roles in the communication process. Some components may be intended to capture attention and direct the person to another component where more information is presented. A prominent statement on the front label of a toxic solvent container may direct the consumer to read the warning statement on the back label for more detailed information. An auditory alarm may alert (capture the attention of) the control room operator to access a visual display panel for emergency safety information.

Similarly, different components may be intended for different target audiences. Prescription drug warnings in the *Physician's Desk Reference* (an industry compendium of FDA-approved medicine labeling) under the headings of "contraindications" and "side effects" could (and should) employ terminology appropriate to prescribing physicians, whereas warnings on the label of a drug container intended for consumers should use less technical language.

Research on the design and effectiveness of warnings is neither simple nor easy. Ethical constraints and measurement issues abound. It is unethical to expose research participants to actual hazards while manipulating warning systems to assess effects on compliance. Dependent measures are often indirect, which may include assessments of comprehension, beliefs, behavioral intentions, and simulated performance. Although such methodologies play an important role in warnings research, they also leave one with concerns, such as the fidelity of simulations and the extent to which beliefs and intentions are valid predictors of behavior. These issues and related considerations have been addressed by researchers including Young and Lovvoll (1999), Wogalter and Dingus (1999), and Smith-Jackson and Wogalter (2006).

PURPOSE OF WARNINGS

Warnings can be thought of as safety communications. There are four levels of analysis at which the purpose of warnings may be addressed.

Safer World

At the most general level of analysis, a purpose of warnings is to make the world a safer place. At this level, warnings play a societal role. Improved health and reduced accidents and injuries are ways of measuring and talking about the intent of warnings. Government requirements such as warnings on cigarette packages are examples of efforts to reduce long-term health effects associated with the product. The warning requirements regarding air bags in vehicles can be thought of as an effort to make vehicles safer for the public.

Provide Information

Another purpose of warnings is to provide information. They are, after all, communications. Among the issues here is what should be communicated. Much research and analysis has been reported in recent years addressing such matters. A fair amount of agreement seems to have emerged regarding some of the kinds of information that a warning should provide; included are information about the hazard, information about the potential consequences, and instructions regarding safe and unsafe behavior. In other words, the warning should provide the information people need to make judgments regarding the level of risk involved in a particular environment or use of a product so they can then use that information in making judgments about the level of risk they are willing to accept or not accept (a cost-benefit analysis).

Influence Behavior

A third purpose of warnings can be viewed as an effort to influence or control the behavior of the persons to whom it is directed; that is, to promote safe behavior. Consider the implementation of a "fasten seat belt" warning. If the vehicle occupant does not fasten the seat belt, the warning with respect to its behavioral purpose could be viewed as a failure. If a homeowner using a drain cleaner containing sulfuric acid to clear a clogged drain does not wear rubber gloves and goggles as instructed by the warning, then the attainment of the behavioral goal of a warning could be viewed as a failure. In short, this purpose of the warning focuses on behavior and whether it achieves that intent. It is closely tied to the instructions component of the warning (i.e., what the warning tells people to do or not do).

Reminder

A fourth purpose of warnings is to serve as a reminder. This point can be thought of in terms of a distinction between knowledge and awareness. A person may *know* about a hazard, its consequences, and the appropriate safety behavior, but the critical issue is whether he or she is aware of it at the proper time. Thus, warnings may be intended as reminders; that is, to call into awareness the hazard information that may otherwise be latent in long-term memory or unavailable because of other demands on attention. An example is the auditory signal and visual symbol in automobiles intended to remind occupants to fasten their seat belts.

BRIEF HISTORY OF WARNINGS RESEARCH AND APPLICATIONS

Egilman and Bohme (2006) provided a brief but interesting history of warnings during the period 1900–1980. They noted that the precursors to warnings during this period were auditory and visual signals designed to prevent rail accidents. They pointed out that major legislation, trends in tort law, and corporate strategies concerning warnings were factors in the development of warnings during the first half of the 20th century in the United States. Examples of legislation are the Pure Food and Drug Act in 1906, which was significant in establishing the federal government's regulatory role, and the Federal Caustic Poison Act (FCPA) in 1927, which addressed the effects of chemicals used in households.

During the period 1900–1910, warnings also started to gain prominence when employees began to be successful in suing employers for injuries at work. The introduction of warnings offered employers a defense based on the assumption of risk; more specifically, after being warned, the worker knows the dangers and accepts the risk.

Some of the early efforts to formulate guidelines for warning signs occurred during the first half of the century. Hansen (1914) published a book that included guidelines for the use of warning signs in the industrial workplace. In 1928, the National Safety Council (NSC) published a pamphlet that provided guidelines for the design and use of warning signs (NSC, 1928). The *Manual on Uniform Traffic Control Devices for Streets and Highways* was published in 1935 and contained new guidelines for the construction of warning signs (American Association of State Highway Officials and National Conference on Street and Highway Safety, 1935). Another example a decade later was the first *Manual L-1: A Guide for the Preparation of Warning Labels for Hazardous Chemicals* published by the Manufacturing Chemists Association (MCA, 1945).

The period 1950 through 1980 witnessed a number of developments on the warnings front, including an increased role of government regulation. The Federal Insecticide and Rodenticide Act dealt with government regulation of pesticides and provided warning language for toxic pesticides. The 1960 Federal Hazardous Substances Labeling Act covered a broad array of flammable, toxic, irritating, or corrosive substances. The role of the FDA in warning regulations was increased after public concern in 1962 over thalidomide, a tranquilizer that caused birth defects in children whose mothers took the drug during pregnancy (Pina & Pines, 2002).

One of the outcomes of this increased FDA role was the greater use of the patient package insert as a means of warning consumers. Soon thereafter came the introduction of warning labels on cigarettes, which followed the 1964 Surgeon General's report on the dangers of smoking. Despite intense opposition from the tobacco industry, warnings on cigarette packages and on cigarette advertisements were ultimately required, including rotating warnings (Kluger, 1997).

Two additional government developments in the 1950 to 1980 time frame were the 1970 Act that established OSHA and the 1972 legislation creating CPSC. OSHA has been instrumental in requiring warnings for substances used in industrial work settings, and CPSC has set warning requirements for consumer products.

In addition to the role U.S. government agencies have played in the history of warning guidelines and requirements, the American National Standards Institute (ANSI) has been influential in establishing voluntary standards for warning signs and labels. Its standards for accident prevention signs (ANSI, 1972), labeling for industrial chemicals (ANSI, 1988), and product safety signs and labels (ANSI, 1991) have provided guidelines for warning design.

Although the foregoing sample of historical events reflects some of the developments of warnings design requirements, guidelines, and use from 1900 to 1980, relatively little formal research was carried out to serve as the basis for these efforts. Then, as mentioned earlier, in the mid-1980s there was a noteworthy upsurge in warnings research. The years since then have produced a substantial body of knowledge regarding warning design and effectiveness. During this period, the types of issues and questions addressed have broadened. Initially, the research questions were straightforward—"Do warnings work?" The research quickly began to focus on design issues that influence when they work. Issues such as how big, what colors, which signal word, and what reading level were typical of the questions addressed. A few years later, the research issues broadened to encompass other questions of effectiveness. What are the factors that influence whether or not warnings make a difference? Dependent measures included behavioral intentions as well as actual behavior. Also, theoretical contexts were introduced such as communication theory and human information-processing theory. It is this research since the 1980s that is the focus of the current chapter.

WHERE DO WARNINGS FIT IN? A SYSTEMS APPROACH

There is a concept in safety (and in human factors) called the *safety hierarchy* or, alternatively, the *hazard control hierarchy*. This concept concerns a sequence or priority of approaches for dealing with hazards. The basic sequence is first to design it out, second to guard, and third to warn. If a hazard exists with a product or in an environment, the first approach is to try to eliminate it through alternative design. If a nonflammable propellant in a can of hair spray can be substituted for a flammable carrier and still adequately function, such an alternative design would be preferred. Eliminating sharp edges on product parts or pinch points on industrial equipment are examples of eliminating hazards. But safe alternative designs are not always technologically or economically feasible.

The second approach is guarding, and the purpose is to prevent contact between people and the hazard. Guards may take several forms. Personal protective equipment such as rubber gloves and goggles, barricades on the highway, and a fence around an electrical station are examples of physical guards. Designing a task so as to prevent people from contacting the hazard is a procedural guard. An example would be the controls on a punch press that require the operator to simultaneously make two control inputs, one with each hand, thus ensuring that fingers will not be under the piston when it strokes. However, guarding—like hazard elimination through design—is not always a feasible solution.

The third line of hazard defense is to warn. Warnings are third in the priority sequence because they are generally less reliable than design or guarding solutions. Even the best warnings are not likely to be 100% effective. People at risk may not see or hear a warning, they may not understand it, or they may not be motivated to comply. Influencing human behavior is often difficult and seldom foolproof. But these concerns about reliability

Designing Effective Warnings

should not be regarded as a basis for not warning. Rather, warnings are one tool available to manufacturers and designers for dealing with environment and product safety. If they are used, their design should involve characteristics that maximize their effectiveness in reducing or preventing personal injury and property damage.

There are other approaches to dealing with hazards, such as training and personnel selection. These approaches are viewed as similar to warnings in that they mostly involve efforts intended to inform and influence behavior.

THEORETICAL APPROACHES

Two theoretical approaches or models that have been employed for theorizing about and organizing warning research and applications are communication theory and human information-processing theory. It is not our intent to provide a detailed discussion of these theoretical approaches in this chapter. Rather, we briefly describe how these approaches have been employed in organizing how warnings are viewed and researched.

The typical, basic communications model can be represented as consisting of four components: the source, the medium, the message, and the receiver. In the warnings context, these components can be viewed as follows:

- *Source*—the designer, originator, or sender of the warning message
- *Medium*—how the message is presented or displayed (visual, auditory, etc.)
- *Message*—the content of the warning
- *Receiver*—the target audience of the warning

The human information-processing framework is essentially a stages model. It consists of a sequence of stages through which warning information flows. At each stage, the information is processed and, if successful at that stage, "flows" to the next stage. If processing at any stage is not successful, it can block the flow and result in failure of the warning.

Wogalter, DeJoy, and Laughery (1999) combined the communications and human information-processing models into a single framework for warnings. A representation of their Communications-Human Information Processing (C-HIP) model is displayed in Figure 8.1. Similar models have been presented by others (Lehto & Miller, 1986; Rogers et al., 2000).

From Figure 8.1, five receiver stages are defined: attention (notice and encode), comprehension, attitudes/beliefs, motivation, and behavior. As noted, one implication of the model is that if information flowing through the processing stages is blocked or fails at any stage, the warning may fail. However, the process may not be so simple as this linear process might suggest. The feedback loops shown on the right of the diagram are intended to indicate that what happens at one stage may influence the processing at other stages. For example, if a warning is noticed and encoded (the attention stage) but the person realizes that he or she did not understand it (the comprehension stage), or if there is uncertainty about the potential consequences in making a compliance decision (the attitudes/beliefs stage), that person may go back and read it again.

The C-HIP model has been useful in organizing and describing factors that influence

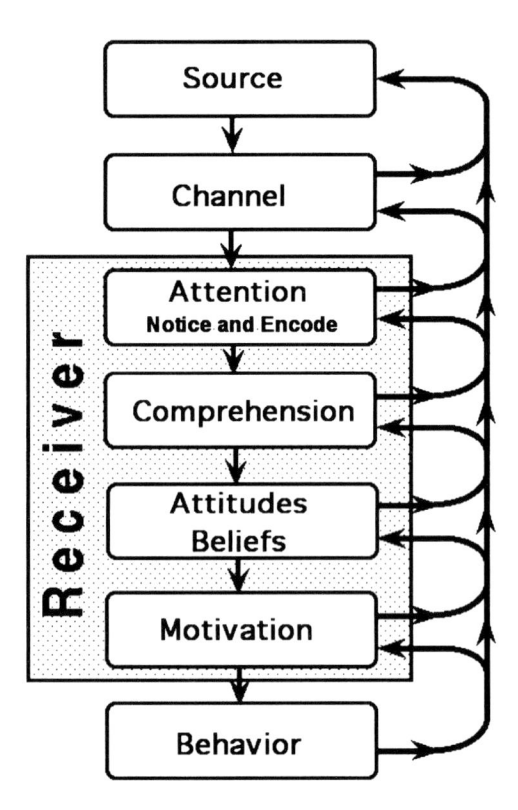

Figure 8.1. C-HIP model.

warning effectiveness. Wogalter and Laughery (2006) employed a model similar to the one shown in Figure 8.1 as a basis for organizing their review of the warnings research literature. For example, considerations at the attention stage include warning design factors such as location, size, color, and pictorials that influence whether or not a warning gets noticed and encoded. Similarly, Rogers, Lamson, and Rousseau (2000) organized their review of the literature on the basis of four components of the warning process: notice, encode, comprehend, and comply. They identified 19 person and 38 warning variable effects on the four stages of warning processing and compliance.

Although such analyses have proven useful in organizing and describing factors that influence warning effectiveness, they are also useful in diagnosing warning inadequacies. For example, a warning may be noticed, read, and understood but still fail to elicit the appropriate safety behavior because of discrepant beliefs and attitudes held by the receiver. According to the C-HIP model, a warning will be processed successfully at this stage if it agrees with the receiver's existing beliefs and attitudes. If the warning information does not concur with existing attitudes and beliefs, however, it will have to alter the receiver's attitudes and beliefs in order to be effective.

It is not our intent in this chapter to attempt another review of all the warning design variables and all the target audience variables that influence warning effectiveness. As

Designing Effective Warnings

already noted, several reviews—some very recent—have been published. Rather, our goal is to identify relevant factors and emphasize those that appear to be most important. Although one can always state "more work needs to be done," it is our conclusion that much has been accomplished in the last 20–25 years in understanding the warning process. It is what can be gleaned from this understanding that this chapter attempts to present.

In the introduction, we noted that in order for a warning to be effective, it must capture attention and must provide the information needed for the receiver to make an informed decision regarding compliance. This latter point can be regarded as a cost-benefit trade-off decision. We also noted that nondesign factors (situation and target audience characteristics) influence whether a warning is attended as well as the outcome of the cost-benefit decision. The following review and analysis is organized around design factors and nondesign factors and their influence on warning effectiveness.

DESIGN FACTORS THAT INFLUENCE WARNING EFFECTIVENESS

In this section, we review research that has addressed the characteristics of warning systems that have a role in whether or not they are effective. The focus is on those design factors that influence the attention goal of warnings and the costs-benefit trade-off decisions regarding compliance.

Attention (Noticing and Encoding)

A number of design factors influence whether or not a warning will be noticed and encoded. *Attention* in this context includes not only whether the warning is seen, heard, smelled, and so on but also whether the information in the warning is encoded (read, listened to, stored in memory, etc.). Several of the influential design factors are what one would expect: size, location/placement, color/contrast, signal word, and the presence of a pictorial. Other factors have also been studied, including length and interactivity.

Size. Bigger is generally better, although what usually matters is the size of the warning relative to other displayed information. Boldness—a form of size—can also be a factor. Barlow and Wogalter (1991, 1993) showed that bigger print size benefited subsequent recall (encoding), and Young and Wogalter (1990) found that print warnings with bigger, bolder print led to better memory for owner's manual warnings. It is likely that such effects are partly attributable to the print's bigger size making it more conspicuous to the reader.

Location/placement. A general principle is that warnings located close to the hazard both physically and in time are more likely to be noticed and encoded. Frantz and Rhoades (1993) found that a warning label placed on a product (file drawer) was noticed more often than when the label was on the shipping carton. Alcohol warnings located on the front of a beverage container are more likely to be noticed than warnings on secondary (back or side) labels (Laughery, Young, Vaubel, & Brelsford, 1993).

There may be times when space for a warning is limited, as with small product containers such as pharmaceuticals. Methods available to increase the surface area for print warnings include adding tags or extended labels (Barlow & Wogalter, 1991; Wogalter, Magurno, Dietrich, & Scott, 1999). Another method is to put some minimum critical information on a primary label and direct the user to additional warning information in a secondary source, such as an owner's manual or package insert. Wogalter, Barlow, and Murphy (1995) showed that such procedures can be effective.

Other location/placement factors that influence attention to warnings have been studied. For a more complete review of these factors and their effects, see Rogers et al. (2000) and Wogalter and Laughery (2006). Overall, the principle to be kept in mind in deciding location/placement of a warning is to place it physically and temporally where and when it is most likely to be encountered.

Color/contrast. Generally, color or other forms of contrast are associated with greater noticeability of warnings (Braun & Silver, 1995; Young, 1991). Also, color seems to have influences beyond attracting attention. The color red has been consistently found to have the highest hazard connotation (e.g., Klein, Braun, Peterson, & Silver, 1993).

It is not surprising that the ANSI Z535 (2002) standard relies on color in the signal word panel of warnings to attract attention, given that color or other forms of contrast result in a greater likelihood that a warning will be noticed and encoded. Besides the color red, the ANSI Z535 (2002) standard notes other colors that should be used in warnings, most notably orange and yellow.

Signal word. Signal words in warnings are used to attract attention and provide a general indication of hazard level. In the United States, standards such as ANSI Z535 (2002) and guidelines such as those by FMC Corporation (1985) recommend that warnings include one of the signal words "CAUTION," "WARNING," or "DANGER." These signal words are widely employed in product and environmental warnings. The word CAUTION is intended for hazards in which minor injury or damage to property *might* occur; WARNING is intended for hazards that *might* cause serious injury; and, DANGER is intended for hazards that *will* cause serious injury.

Research indicates that the presence of the word DANGER is more likely to attract attention than CAUTION and WARNING or no signal word (e.g., Adams, Bochner, & Bilik, 1998). People do not readily differentiate between CAUTION and WARNING with regard to hazard level, but both terms are interpreted as connoting lower hazard levels than DANGER (Wogalter & Silver, 1995).

Pictorials. Pictorials (also known as *symbols, graphics,* and other names) in a warning may take several forms, among them actual photographs, directly representative drawings, and abstract symbols. Generally, research shows that pictorials can serve two primary functions in warnings: They can help to attract attention to the warning, and they can convey content information. Guidelines such as ANSI Z535 (2002) and FMC (1985) place considerable emphasis on the use of pictorials to communicate hazard information.

A number of studies have shown that pictorials in warnings can be effective in capturing attention (e.g., Davies, Haines, Norris, & Wilson, 1998; Jaynes & Boles, 1990; Kalsher,

Designing Effective Warnings

Wogalter, & Racicot, 1996; Laughery & Young, 1991; Laughery, Young, Vaubel, & Brelsford, 1993; Young, 1991). Perhaps related to the effects of pictorials on attention is the finding that people prefer warnings that contain pictorials compared with warnings without them (Kalsher et al., 1996). Generally, pictorials enhance the conspicuousness of a warning. For example, Laughery and Young (1991) reported that pictorials combined with color and borders were more effective in attracting attention than the individual features separately.

Pictorials are also useful in enhancing encoding and helping to increase comprehension (e.g., Boersema & Zwaga, 1989; Collins, 1983; Laux, Mayer, & Thompson, 1989; Wolff & Wogalter, 1998; Zwaga & Easterby, 1984). Pictorials may be especially helpful when the target audience includes those who are illiterate and/or non-English readers. Also, pictorials can potentially be useful in circumstances when ideas need to be conveyed quickly, such as during highway travel.

As in text, pictorials may communicate the hazard, consequences, and instructional information. Two pictorials intended to communicate slip and fall and vapor inhalation hazards are shown in Figure 8.2. Figure 8.3 presents two pictorials intended to communicate consequences: electrocution and a hand injury resulting from a pinch point hazard. Figure 8.4 shows two pictorials that communicate instructional information: "Do not drink water" and "Wear a face shield."

The pictorials in Figures 8.2, 8.3, and 8.4 are examples of direct representation; that is, the information represented by the pictorial is expected to be recognized and understood by the target audience on the basis of general experience and knowledge. Figure 8.5, on the other hand, presents a pictorial for biohazard that must be learned in order to be understood. Some pictorials, such as a skull and crossbones, may fall in between, in the sense many people may relate the image to safety or health hazard information, but the connection to a poison hazard may require considerable inference or learning. For this reason, when the skull-and-crossbones hazard is used, it is often accompanied by the signal word POISON, which makes it adequate to readers of English.

As a general principle, pictorials that directly represent the information are preferred, particularly for general target audiences. Pictorials that require inference or learning are less likely to be recognized or understood.

Pictorials are an exceptionally valuable tool for communicating warning information, and they may be particularly useful in warning those who are illiterate or those who do not read English, but it is not always simple to develop pictorials that can be understood.

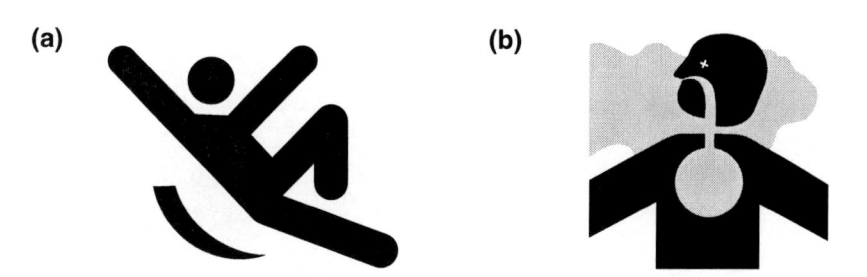

Figure 8.2. Symbols displaying hazards: (a) slip-and-fall hazard, (b) inhalation hazard.

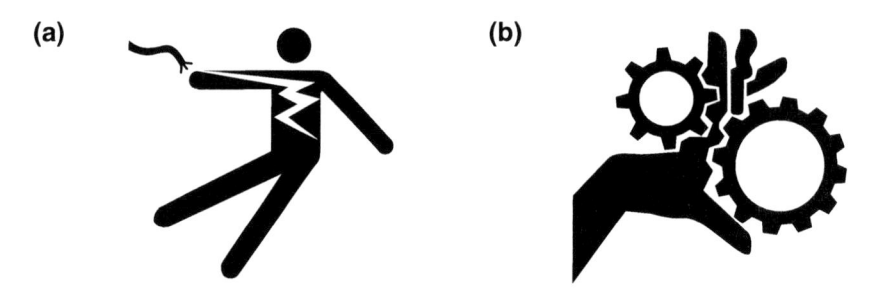

Figure 8.3. Symbols displaying consequences: (a) electrocution, (b) hand injury.

The question of an acceptable level of comprehension has been addressed in the ANSI Z535 (2002) standard. This standard suggests an acceptability criterion of 85% correct comprehension. However, this value is a goal, and if comprehension is less than 85%, the pictorial may still be helpful for attracting attention.

An important consideration is that the pictorial should not communicate incorrect information; that is, the probability of misinterpretation should be at a minimum. ANSI Z535 (2002) recommends having no more than 5% critical confusion errors (opposite or potentially dangerous answers). Indeed, the error rate is more important than the simple correct comprehension rate.

Message length. Brevity has been a generally accepted criterion for warnings; that is, the warnings should be no longer than necessary to communicate the needed information (Laughery & Wogalter, 1997). One frequent assumption is that the longer a warning message, the less likely it is to be read and encoded. However, research addressing this issue is inconclusive. Silver, Leonard, Ponsi, and Wogalter (1991) reported a positive correlation between warning message length and willingness to read. One explanation offered was that a longer warning suggests a greater hazard level, thus resulting in a greater willingness to read. On the other hand, Chen, Gilson, and Mouloua (1997) manipulated the number of warning messages on consumer products and found that the perceived risk declined as the number of low-risk messages increased beyond five.

A concept related to message length is *overwarning*. Overwarning typically refers to a large number of warnings associated with a product or with an environment. The assumption is that people may not attend to them or may become highly selective, attending

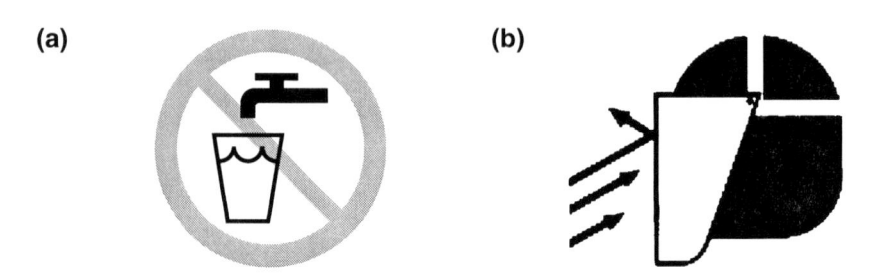

Figure 8.4. Symbols displaying instructions: (a) Do not drink the water, (b) Wear face shield.

Designing Effective Warnings

Figure 8.5. Symbol for biohazard.

only to some. Sometimes the term is applied more broadly, referring to the notion that the world is filled with warnings. Although both notions of overwarning have some face validity, the former interpretation has some empirical support (Chen et al., 1997). There are few empirical data on the issue of "the world is filled with warnings." Nevertheless, overwarning in the latter respect may be a valid concern, and unnecessary warnings should be avoided.

Physical interactivity. A technique that has been somewhat successful in increasing the likelihood that people will notice and encode a warning is to require them to interact with the warning physically in some way. Frantz and Rhoades (1993) reported that a warning that had to be removed before the file drawers could be used increased its noticeability compared with other locations without the required physical interaction. Gill, Barbera, and Precht (1987) and Duffy, Kalsher, and Wogalter (1995) also reported studies showing that a warning label that had to be moved before a product could be used resulted in the label being noticed and read more than when the label was simply on the product.

Compliance Decisions

Research on warning compliance has generally employed one or both of two dependent measures: behavioral intentions and behavior. Two reviews have specifically addressed this research (Kalsher & Williams, 2006; Silver & Braun, 1999).

In the preceding section, we focused on a number of factors that affect noticeability and encoding. Obviously, factors that influence noticeability and encoding are important in determining whether compliance behavior occurs. If a given warning label or sign is not noticed or encoded, it cannot have a direct effect on behavior. Thus, one would correctly expect that the factors reviewed earlier would also be positively correlated with likelihood to comply. Research results have generally supported this expectation (Kalsher & Williams, 2006; Rogers et al., 2000; Silver & Braun, 1999).

In this section, we address the issue of compliance likelihood in terms of people's cost-benefit trade-off decisions. This decision making occurs as a consequence of warning information that interacts with people's attitudes and beliefs and that then feeds into the cost-benefit analysis determining the compliance outcome. In the following subsections, we focus on the information that makes up the content of warnings and how compliance is influenced by this information.

Pictorials. Although most studies on pictorials have been concerned with attention and comprehension, some research has also explored the effects of pictorials on warning compliance. For example, Jaynes and Boles (1990); Otsubo, 1988; and Wogalter, Begley, Scancorelli, and Brelsford (1997) reported studies showing that the presence of pictorials increased compliance, compared with warnings without pictorials.

Explicitness. A design factor that has emerged as especially important to warning effectiveness is explicitness of content information. A recent review by Laughery and Smith (2006) summarized the findings of a number of studies addressing this topic. *Explicitness* in this context is defined as information that is specific, detailed, clearly stated, and leaves nothing implied. One of the first investigations to refer to this concept was by Sherer and Rogers (1984). In their study, concreteness (explicitness) appeared to increase the perceived severity of possible injury and concrete information was better remembered than abstract information. Lehto and Miller (1988) also suggested using explicit over abstract or general formulation of warnings.

As an example, suppose a person working in a job environment uses a chemical product that emits toxic vapors, the inhalation of which can lead to severe and permanent lung damage. Also, suppose it is important to wear a particular type of respirator when working with or around the chemical. The following warning text contains hazard, consequence, and instruction information:

> Dangerous Environment
> May Cause Health Problems
> Take Precautionary Measures

Certainly this warning will be of little or no use to the person exposed to the hazard. It could be considered a classic example of a vague or nonexplicit warning. The hazard statement "Dangerous Environment" reveals little about what the safety problem is; the consequences statement, "May Cause Health Problems," only notes a potential problem having to do with health; and the instruction "Take Precautionary Measures" is virtually useless in telling the user what to do or what not to do.

Consider the following as a possible alternative warning:

> Toxic Vapors
> Can Lead to Severe Lung Damage
> Always Wear Type 1234 Respirator in Area

The point of contrasting these two examples is to emphasize the importance of providing information at a level of specificity or explicitness that will enable people to make informed judgments and decisions.

In their review of the research, Laughery and Smith (2006) addressed the importance of explicit information for all three warning content categories: hazards, consequences, and instructions. They concluded that explicitness in all three categories plays an important

Designing Effective Warnings

role in compliance, and they offered the following principles regarding explicitness that may be useful for the warnings designer:

- Do not assume "everybody knows."
- Do not rely on inference.
- Be careful about assuming that hazards and consequences are open and obvious.
- People do not always remember the appropriate safety information at the appropriate time. Reminders may be needed.
- *Explicit* is not necessarily synonymous with *quantitative*.
- Technical jargon is usually not a good way to achieve explicitness, especially for a general target audience.

From a motivational perspective, it is not surprising that more explicit information influences compliance. More specific information about hazards and consequences enables people to make better-informed cost-benefit trade-off decisions regarding the need to comply. Thus, one would expect explicit information to be especially important when the consequences are more severe, an expectation the research shows to be valid. Further, more explicit instructions enable people to better understand and carry out appropriate actions, which is also a common research finding.

NONDESIGN FACTORS THAT INFLUENCE WARNING EFFECTIVENESS

In this section, we review research that has addressed the characteristics of target audiences and situations that have a role in whether or not warnings are effective. The focus is on those factors that influence attention to warnings and the cost-benefit trade-off decisions regarding compliance.

Attention (Noticing and Encoding)

A number of target audience factors influence whether or not a warning will be noticed and encoded. Sensory capabilities or limitations as well as cognitive competencies are relevant factors. Two other factors that have been found to be important are perceived hazard and familiarity.

Perceived hazard. An important variable in whether or not people will look for and read warnings is their a priori hazard perceptions associated with a product or environment. In much of the research, perceived hazard or hazardousness has been defined as a composite variable that takes into account the likelihood of the hazard and the severity of the potential consequences. The greater the level of perceived hazard, the more likely people will look for, notice, and process a warning (Wogalter, Brelsford, Desaulniers, & Laughery, 1991; Wogalter, Brems, & Martin, 1993). Otsubo (1988) found that individuals were more likely to report having noticed a warning on a product (circular saw) that was perceived as more dangerous than another product (jigsaw). Other studies have reported a similar relationship between perceived hazard and attention to warnings (e.g., Wogalter, Jarrard, & Simpson, 1994).

Familiarity. Familiarity refers to experience with a particular or similar product or environment from which relevant information has been acquired. The concept of familiarity in the context of warnings is related to perceived hazard. Godfrey and Laughery (1984) found that the more familiar women were with tampons, the less likely they were to notice warnings regarding toxic shock syndrome. Similarly, Godfrey, Allender, Laughery, and Smith (1983) reported that people who had greater familiarity with products rated them to be less hazardous and were less likely to look at warning labels. Others have reported a negative relationship between familiarity and warning-related processes (e.g., Johnson, 1992).

A related and converse point should be kept in mind; namely, people more familiar with a product or environment may be more likely to notice a warning because they are more frequently exposed to it. Goldhaber and deTurck (1988) reported that middle school students who dove into the shallow ends of pools were more likely to have noticed (but ignored) a no-diving warning at the shallow end. Similarly, Greenfield and Kaskutas (1993) found that heavy drinkers were more likely to report having noticed the warning labels on alcoholic beverages than were moderate drinkers or nondrinkers. These two studies have implications for the effects of familiarity and perceived hazard on compliance decisions.

A likely explanation for the effects of perceived hazard and familiarity on attention to warnings is simply that people are more likely to seek such information when a threat is perceived to be greater. Greater familiarity, assuming no negative experiences in the past, may result in lower levels of perceived hazard and, in turn, less motivation to seek warning information. More simply, if people are looking for a warning, they are more likely to notice and encode one that is present.

Other nondesign or target audience variables such as gender and age have been shown to influence attention to warnings, but these effects have not been nearly as consistent or robust as the effects of perceived hazard and familiarity. For a review of the gender and age variables, see Rogers et al. (2000) and Smith-Jackson (2006a).

Compliance Decisions

We address the issue of compliance with warnings in terms of cost-benefit trade-off decisions. Several target audience or situational variables have been found to influence compliance. A general review of this research can be found in Rogers et al. (2000) and Wogalter and Laughery (in press, 2006). We focus on three factors: familiarity, modeling, and cost of compliance.

Familiarity. In warning research, familiarity has been found to have somewhat equivocal results. Assuming there have been no negative experiences with a product or environment, research mostly seems to indicate that people who have greater familiarity with a product or environment are less likely to comply with a warning. For example, Wogalter, Barlow, and Murphy (1995) found that experienced computer users were less likely to comply with antistatic warnings associated with the installation of a disk drive than were less experienced users. Harrell (2003) found that mothers who had reported previously allowing their children to stand in grocery carts were less likely to comply with a warning not to allow this behavior than mothers who were not so experienced. Other studies reporting similar findings include Burnett, Purswell, Purswell, and Krenek (1988); Goldhaber and DeTurck (1988); Lehto and Foley (1991); and Wogalter, Brelsford, Desaulniers, and Laughery (1991).

However, some research has also shown that familiarity with a product increases compliance with warning information. For example, Ortiz, Resnick, and Kengskiil (2000) found that when participants were asked to apply pesticides to plants, those who were familiar with the product were more likely to comply with a warning to use personal protective equipment than were those who were less familiar. As noted in our earlier discussion of familiarity, the effects of this variable are probably mediated by the level of perceived hazard. The state of affairs described by the notion "familiarity breeds contempt" may be at work because greater familiarity leads to lower levels of perceived threat, which, in turn, results in less compliance. Stated differently, in the cost-benefit trade-off decision, familiarity results in lower perceived costs associated with noncompliance.

Modeling. People's behavior, including warning compliance, is influenced by their social context and the behavior of others. People tend to model the safe or unsafe behaviors of others they observe. A number of studies have been reported showing a robust effect of modeling as a factor in warning compliance. The use of protective equipment has been a context for several studies addressing the modeling issue. Wogalter, Allison, and McKenna (1989); deTurck, Chich, and Hsu (1999); and Edworthy and Dale (2000) all reported results showing greater compliance in using protective equipment when others were observed using the equipment.

How does the modeling concept and effect fit with the cost-benefit trade-off decision addressed here? One possible explanation is that observing the behavior of others is a form of instruction regarding what is appropriate or inappropriate behavior in a particular context. If a passenger gets into the vehicle and observes the driver fasten the seat belt, the passenger is more likely to do the same. The effect may also be a form of social influence; that is, one may be motivated to behave the same as others.

Cost of compliance. The cost of complying with a warning may take many forms, including money, time, effort, and convenience. A substantial amount of research has explored the effects of such costs on compliance and generally has found that the effects are robust. If one views the decision to comply or not comply with a warning as a cost-benefit analysis, then the compliance costs represent half the equation, and such costs have an important influence on the decision outcome.

Research on the cost of compliance was reviewed by Silver and Braun (1999) as well as by Rogers et al. (2000). We will not review details of specific studies here, except to note that researchers have explored this variable and its effects on warning compliance in a variety of settings. These settings include the use of goggles on racquetball courts (Dingus, Wreggit, & Hathaway, 1993; Hathaway & Dingus, 1992), using cleaning solvents (Dingus, Hathaway, & Hunn, 1991), avoiding broken doors (Godfrey, Rothstein, &

Laughery, 1985), using office equipment (Wogalter et al., 1987), working with power tools (Zeitlin, 1994), working in a chemistry lab (Magurno & Wogalter, 1994), and wearing helmets with all-terrain vehicles (Lehto & Foley, 1991). Almost without exception, the research shows that lower costs of compliance lead to significantly greater compliance with warnings.

DISCUSSION

At the outset of this chapter, we pointed out that two primary goals a warning should accomplish are to capture attention and to provide the information needed for people to make informed decisions regarding compliance. Attention is viewed here as concerned with both noticing and encoding the warning. The compliance decisions are viewed as cost-benefit trade-offs in which the pluses and minuses of complying are taken into account. Following brief background coverage of purpose, history, system context, and theoretical perspective, we focused on the design variables and the situational/target audience variables that have emerged as most significant to accomplishing the attention and informed decision goals.

Generally, the design variables that influence attention to warnings are those one would expect—size, location, color/contrast, and the use of signal words. Others that have been researched and found to be important are pictorials, length, and interactivity. Target audience variables also influence a warning's success in being noticed and encoded. Two that seem to matter most are the level of perceived hazard and familiarity. We suggest that if the warning system designer appropriately takes into account these various factors, the probability that the intended audience will notice and encode the warning will be relatively high.

Regarding the decision to comply, for a warning to be effective, clearly, it must be noticed, encoded, and understood. The various attention variables and the use of pictorials have been shown to have a positive effect on compliance. Regarding content, the explicitness of the hazard, consequences, and instructional variables appears to be important. This conclusion seems to be especially true with regard to the explicitness of consequences information when the negative outcome may be severe. Three situational or target audience factors seem to merit special emphasis regarding their effects on the compliance decision outcome: familiarity, modeling, and cost of compliance. Indeed, the consistent and robust cost of compliance effect marks it as one of the most important considerations for the warning designer to keep in mind.

The fact that a person does not follow the actions recommended in the instructions of a warning does not necessarily mean the warning has not been successful. Although, as noted, one of the purposes of a warning is to describe and motivate safe behavior, there may be occasions when it is rational—or at least understandable—for someone not to comply even though the warning information has been encoded and understood. In short, the person may decide to take the risk.

For example, assume the drain of your sink is clogged. You go to the store and buy drain cleaner (sulfuric acid), take it home, and then read the label. You read that the material is toxic and that you could suffer chemical burns if you get it on your skin. You learn from the warning that you should wear rubber gloves and goggles while using it. But you do not have rubber gloves or goggles, and the store where you might be able to get them is some distance away. So there are costs of complying: some money, some time, and some inconvenience. Instead of complying by going and purchasing the protective items, you simply decide to be very careful in handling the drain cleaner while using it. In doing so, you would have made the decision that the compliance costs are greater than the noncompliance costs. Although the desired compliance goals have not been fulfilled (rubber gloves and goggles were not worn), in terms of the informed decision goal for warnings, this warning has been successful.

The principles of warnings are not product- or environment-specific. There is no separate set of guidelines or criteria for warnings that address hazards, consequences, and instructions for an environment containing toxic gas or a slippery walking surface. Similarly, the same principles apply to products, whether one is dealing with lawnmowers, toxic solvents, vehicle air bags, prescriptive medications, or punch presses. The specifics for how a warning is presented to capture attention and the content of the hazard, consequence, and instructional information will vary with the product, the environment, and the target audience, but the general guidelines apply.

THE FUTURE

Although warning research to date has resulted in good progress in understanding design and effectiveness issues, its focus has been somewhat traditional. The issues and variables explored have merited the time and attention they have received, but there are several challenges and opportunities that can and should move closer to the center of the research stage. Among the challenges are target audience diversity, and the opportunities include the greater use of technology in warnings.

Target Audience Diversity

One of the challenges in warnings design concerns the need to communicate to larger and more diverse target audiences; this is a consequence of factors such as growing international trade. Language barriers, illiteracy, and cultural considerations represent a part of this more global challenge. One approach to dealing with language barriers is to present warnings in more than one language. It is common to find warnings accompanying products marketed in Canada printed in both English and French. In some areas of the United States with substantial Hispanic populations, warnings are often presented in both English and Spanish. But the warning designer must be mindful of the potential that too much information will make it difficult to access the warning in the language appropriate for a particular recipient. In other words, the structure and organization of the warning system may become more important as a result of increasing the number of languages. How to organize and present such information merits research attention to determine effective multilanguage warnings.

Pictorials are an obvious approach to addressing language barrier and literacy issues. As noted earlier, research has been reported indicating that pictorials can enhance the likelihood of noticing warnings. An ongoing challenge for using pictorials in communicating warning information is comprehension. At the international level, research is needed to better establish how various pictorials are understood across cultures. Are there universals? Is it widely understood that the red circle with a slash (or a slash by itself) means prohibition? Certainly a common method across the world is the use of pictorials. Figure 8.6 shows several signs photographed in various countries that make use of pictorial communication.

Cultural differences may represent a variety of challenges to the warning designer. For example, there may be a range of views or customs regarding the amount of responsibility an individual is expected to take for his or her own safety. Differences in belief systems may influence how warning systems should be designed and what may be expected in terms of their effectiveness (Smith-Jackson, 2006b).

Another domain that warrants research effort is in the area of children and their caregivers. Although very young children cannot be expected to understand warnings (and their health and safety must be the responsibility of caregivers), older children with developing cognitive capacities might be able to assist in the goal of injury reduction if warnings are developed within those capacities. A warning about airbag dangers to smaller passengers sitting in the front seat that is understandable by 8- to 12-year-olds may enhance consequent compliance likelihood. Considerations regarding children's warnings are provided in Kalsher and Wogalter (in press).

Technology and Warnings

In this chapter, we have considered what has been learned about warnings from the upsurge in interest and research that began in the early 1980s. There are exceptions, but most visual warnings for products and environments are signs, labels, manuals, and so on. Further, such warnings are generally static and passive. However, in this context it is noteworthy that people's perceptual and cognitive systems are less attuned to stimuli that do not change. Current and new technology provides opportunities for more dynamic ways of warning people that would be potentially more effective. Several recent papers have explored this topic (Smith-Jackson & Wogalter, 2004; Wogalter & Conzola, 2002; Wogalter & Mayhorn, 2005, 2006). Here we focus on a few of the ways that technology may enhance warnings.

We are beginning to see warnings in different contexts, such as on television and the Internet. TV commercials about prescription drug products in the United States now include warnings. On TV, warnings can combine both visual and auditory modalities. Research by Shaver (2004) and Barlow and Wogalter (1993) showed that the dual-modality presentation of both print and voice warnings enhances the communication of warning information compared with either method alone. The Internet has taken hold as a source for information acquisition and communication; research by Hicks, Wogalter, and Vigilante (2005) revealed that the Internet is one of the main sources people report they would use when seeking information about risks associated with prescription medicines (ranked third behind physicians and pharmacists).

Some potential approaches for applying technology to warnings are discussed in the following sections. The topics include dynamic warnings, use of new technology displays,

Designing Effective Warnings 261

Figure 8.6. Pictorial signs from around the world.

hazard detection using sensors, and tailoring warnings to fit individual users. In the last section, we address possible barriers to implementing technology in warnings.

Dynamic warnings. Dynamic warnings are generally more noticeable than static warnings. Human sensory and perceptual systems are better able to detect change than constancy. When something does not change over time or is no longer novel, it is less likely to attract attention because of habituation. Adding dynamic qualities to warnings will enhance their ability to attract and maintain attention. The urgency of a relatively simple fire alarm can be enhanced by adding more dynamic qualities, such as varying the frequency and temporal aspects of the auditory signal (Edworthy & Hellier, 2006; Haas & Edworthy, 2006).

Dynamic aspects of warnings should be conspicuous to attract and sustain attention. Consider the image of the school signs shown in Figure 8.7. The purpose of the static (unchanging) school zone sign in Figure 8.7a is to warn drivers to decrease their speed during the time students are in the area. The reduced speed limit is not applicable most of the time, so drivers may inadvertently violate the speed limit because in past experiences, the sign is irrelevant when they are in the area. Figure 8.7b illustrates two additional traffic lights mounted on either side of the school zone sign; these are programmed to flash only during relevant hours. The dynamic sign in Figure 8.7b will likely result in fewer violations and provide greater safety to children than the static school zone sign because the flashing light attracts attention at the appropriate times.

Displays. Unconventional methods of displaying warnings have been made possible by new technology. One relatively recent technological innovation is the availability of flat-panel displays. High-resolution plasma and liquid crystal displays are now as common as computer monitors and high-definition televisions. Large versions of flat-panel displays

Figure 8.7. School zone signs: (a) static, (b) dynamic.

Designing Effective Warnings

are being used in sports stadiums and as advertisement billboards in cities. These new electronic display technologies can be considered for warning applications. One such use is changeable message signs on highways. An example of such a sign in Rio de Janeiro is shown in Figure 8.8. Eventually, warning signs on highways and smaller signs in other applications will use high-resolution flat panel technologies. Backlit and with high contrast, electronic signs are more likely to attract attention in most ambient environmental conditions compared with static conventional signs.

In addition to attracting attention, an important benefit of such displays is that the information content displayed can be changed as needed. Roadway sign displays can be changed to give timely, pertinent information about specific traffic and road conditions ahead and what to do to reduce delays. Displays could be mounted in or outside buildings or facilities (e.g., on walls or posts) or in work environments to display warning information as appropriate. For example, electronic signs could alert factory workers to noise, hazardous airborne particles, and respiratory hazards when such hazardous conditions exist.

Flat-panel displays can also be used to present video warnings. Are warning videos useful? Some initial research in this area was conducted by Racicot and Wogalter (1995). In their study, participants were asked to mix several chemicals. Before starting, they were assigned to one of three conditions in which they either (a) watched a video of a model demonstrating the proper safe behavior of putting on protective equipment (e.g., face mask and gloves), (b) viewed a static warning sign displaying the same warning instructions as on the video monitor, or (c) saw nothing relevant to safety on the video monitor. More people wore the protective equipment in the video model condition than in the other two conditions.

Dynamic warnings have been used in vehicles for many years. Most vehicles contain simple warnings, such as a flashing light or an intermittent tone, as a reminder to wear

Figure 8.8. Changeable-message highway sign in Rio de Janeiro, Brazil.

seat belts. Although more noticeable than static stimuli, these warnings often become habituated over time. A better reminder for wearing seat belts would be a talking car or a sound that changes to maintain the conspicuity of the signal. Many newer vehicles have navigation systems with touch screen LCD displays (for example, see Figure 8.9). Although most navigation systems store and display information such as points of interest and restaurants, they could also communicate safety information and warnings. Examples are directions on how to properly install child safety seats and whether and how much one can recline the passenger seat when the vehicle is in motion.

Detectors/sensing devices. Warning effectiveness may be benefited by detection and sensing methods that are available now or likely to be available in the future. A fundamental principle is that warnings should be presented when and where the information is needed. If the warning is presented too distant from the hazard in terms of location and time, people may not recognize the connection or may not remember the hazard.

Earlier we noted that humans have sensory, perceptual, and cognitive limitations. Warning systems that include detector (sensing) devices can take on the burden of noticing when a warning is needed (Wogalter & Mayhorn, 2005, 2006). Numerous kinds of sensor systems are available to detect heat, cold, wet, gas vapors, motion, weight, and so forth. One example of a warning sign that could use a sensing system is the caution sign used in the United States that states "Bridge Ices Before Road." Some of these signs are permanently placed and displayed. The photograph in Figure 8.10 was taken in the middle of summer during a heat wave in Raleigh, North Carolina. A better method would be to use a temperature detector that presents the message when conditions are conducive for the hazard; in other words, the message is displayed only when the temperature is near or below freezing. Another example concerns inexpensive motion detectors that are sold in stores for outdoor security lighting. Such detectors could also be used to initiate warnings when individuals enter a specific hazardous area, such as might exist in work environments.

Figure 8.9. In-vehicle navigation system displaying a warning.

Designing Effective Warnings

Figure 8.10. Static sign for a bridge in Raleigh, North Carolina.

A significant benefit of using detectors and sensing systems is that they can supplement peoples' sensory, perceptual, and cognitive abilities. Humans do not have a natural capability to detect radiation and carbon monoxide (CO), so there are devices to do that job (Geiger counters and CO detectors). These and other kinds of detection equipment can play an important role in safety by compensating for peoples' limitations. When hazardous conditions are detected, warning systems can then be activated.

Tailoring warnings to the user. In this section, we describe examples of tailoring warnings to users. The idea is that different people have different needs, and as a result, different warnings should be presented. Sometimes these differences are attributable to varying individual characteristics or capabilities, but the differences may also be based on varying situations.

Warnings could also be personalized. Research by Wogalter, Racicot, Kalsher, and Simpson (1994) suggests that relevance is associated with warning compliance. Relevance is a belief that the warning is applicable. In the Wogalter et al. (1994) study, when a participant's name was included within an electronically presented warning, compliance was greater than when the warning was generic and nonpersonalized. Information from smart cards can provide personal information by, for example, embedding the name of the targeted individual into a presented warning. Automated check-in terminals at airports that note the passenger's name after inserting a credit card are examples of such a technique. This approach shortcuts the decision-making process on whether the message is intended for, or applicable to, the individual personally.

A sophisticated extension of tailoring is to modify the warning based on the person's experience and skill level. An expert may not need a warning, or if a warning is to be given, it can be more technical and contain abbreviated information to inform and remind. For the novice, the information may need to be simple and limited in scope to avoid overloading attention resources and memory capacity. Use of a prioritization strategy would limit the presentation of certain information so that only the most critical is given. However, warning systems could also make linked information available if a more detailed description is desired.

Potential barriers. Although the potential for future technology-based warning systems is substantial, there are a number of barriers that could delay or prevent implementation. Some of the systems described in this chapter are simple, but some are more complex. Further, some may be expensive. Undoubtedly, however, the cost will go down and the sophistication will go up. As a consequence of display and other technological advancements, along with reduced costs, technology's involvement with warnings will eventually be more widespread.

However, the methods of implementation and their appropriateness must also be considered. Some of the issues of concern are warning intrusiveness and annoyance as well as maintenance. Inappropriate or false warnings must be avoided. Likewise, failure to present necessary warnings could be disastrous, and so backup systems may be needed. As the sophistication of electronic warning systems improves, the control of presentations that are in error, such as false alarms and misses, should also decrease.

Some sophisticated systems involve the collection of personal information that could generate privacy concerns. Such issues are complex, and a balance will be needed between maintaining privacy and promoting safety.

CONCLUSIONS

Future technology-enriched warning systems will have properties different from and better than those of traditional static warnings. These improved capabilities will include dynamic modification of message content, compensation for human limitations, interactivity, and personalization through tailoring warnings to meet the needs of particular users. The end result will be an increased capacity to warn users of potential or existing hazards.

We have presented a number of ways in which technology can enhance warning effectiveness. The use of flat-panel displays, video technology, and in-vehicle systems were described as examples of technology that might be implemented to improve warning delivery and presentation. Moreover, the inclusion of detectors and sensor technology in future warning systems should facilitate identification and earlier detection of potential hazards. Future warning systems can provide assistive support for sensory, perceptual, and cognitive limitations that is tailored to meet the needs of specific users. The goal is to deliver accurate, appropriate warning information in a timely fashion where and when it is needed to prevent injury, illness, and damage to property.

Although the promise of next-generation, technology-enhanced warning systems for improving safety is exciting, there are potential barriers to implementation. Besides financial costs, the largest barrier is the balance between privacy concerns stemming from acquisition of user information and the need to effectively warn users about hazards. Although such ethical considerations are beyond the scope of this chapter, warning designers should be aware of this issue when implementing new warning designs.

In this final section, we have addressed how technology-based warning systems might be developed and enhance hazard communications. We started by describing why dynamic systems would likely be more effective than common, static warnings. Future warnings can benefit users by supplementing and compensating for various limitations in the detection, identification, and comprehension of hazards. Clearly, this aspect of warnings and the future pursuit of advanced, technology-based warning systems promise to be an interesting and challenging area for research. It will also benefit safety.

REFERENCES

Adams, A., Bochner, S., & Bilik, L. (1998). The effectiveness of warning signs in hazardous work places: Cognitive and social determinants. *Applied Ergonomics, 29,* 247–254.

American Association of State Highway Officials and National Conference on Street and Highway Safety. (1935). *Manual on uniform traffic control devices for streets and highways.* Washington, DC: Author.

American National Standards Institute. (1972). *American national standard specifications for accident prevention signs, Z35.1-1772.* New York: Author.

American National Standards Institute. (1988). *American national standard for hazardous industrial chemicals—Precautionary labeling, Z129.1-1788.* New York: Author.

American National Standards Institute. (1991). *American national standard product safety signs and labels, Z535.4-1991.* New York: Author.

American National Standards Institute. (2002). *Accredited Standards Committee on Safety Signs and Colors, Z535.1-5.* New York: Author.

Barlow, T., & Wogalter, M. S. (1991). Increasing the surface area on small product containers to facilitate communication of label information and warnings. In *Proceedings of Interface '91* (pp. 88–93). Santa Monica, CA: Human Factors and Ergonomics Society

Barlow, T., & Wogalter, M. S. (1993). Alcoholic beverage warnings in magazine and television advertisements. *Journal of Consumer Research, 20,* 147–155.

Boersema, T., & Zwaga, H. J. G. (1989). Selecting comprehensible warning symbols for swimming pool slides. In *Proceedings of the Human Factors Society 33rd Annual Meeting* (pp. 994–998). Santa Monica, CA: Human Factors and Ergonomics Society.

Braun, C. C., & Silver, N. C. (1995). Interaction of signal word and color on warning labels: Differences in perceived hazard and behavioral compliance. *Ergonomics, 38,* 2207–2220.

Burnett, T. J., Purswell, J. L., Purswell, J. P., & Krenek, R. F. (1988). Hot water burn hazards: Warning label influence on user temperature adjustment. *International Journal of Cognitive Ergonomics, 2,* 145–157.

Cameron, K. A., & DeJoy, D. M. (2006). The persuasive functions of warnings: Theory and models. In M. S. Wogalter (Ed.), *Handbook of warnings* (pp. 301–312). Mahwah, NJ: Erlbaum.

Casali. J. G., & Gerges, S. (2006). Protection and enhancement of hearing in noise. In R. C. Williges (Ed.), *Reviews of human factors and ergonomics, Volume 2* (pp. 195–240). Santa Monica, CA: Human Factors and Ergonomics Society.

Chen, J. Y. C., Gilson, R. D., & Mouloua, M. (1997). Perceived risk dilution with multiple warnings. In *Proceedings of the Human Factors Society 42nd Annual Meeting* (pp. 831–835). Santa Monica, CA: Human Factors and Ergonomics Society.

Collins, B. L. (1983). Evaluation of mine-safety symbols. In *Proceedings of the Human Factors Society 27th Annual Meeting* (pp. 947–949). Santa Monica, CA: Human Factors and Ergonomics Society.

Davies, D., Haines, H., Norris, B., & Wilson, J. R. (1998). Safety pictograms: Are they getting the message across? *Applied Ergonomics, 29,* 15–23.

DeJoy, D. M. (1989). Consumer product warnings: Review and analysis of effectiveness research. In *Proceedings of the Human Factors Society 33rd Annual Meeting* (pp. 936–940). Santa Monica, CA: Human Factors and Ergonomics Society.

deTurck, M. A., Chich, I. H., & Hsu, Y. P. (1999). Three studies testing the effects of role models on product user's safety behaviors. *Human Factors, 41,* 397–412.

Dingus, T. A., Hathaway, J. A., & Hunn, B. P. (1991). A most critical warning variable: Two demonstrations of the powerful effects of cost on warning compliance. In *Proceedings of the Human Factors Society 35th Annual Meeting* (pp. 1034–1038). Santa Monica, CA: Human Factors and Ergonomics Society.

Dingus, T. A., Wreggit, S. S., & Hathaway, J. A. (1993). Warning variables affecting personal protective equipment use. *Safety Science, 16*, 655–673.

Duffy, R. R., Kalsher, M. J., & Wogalter, M. S. (1995). Increased effectiveness of an interactive warning in a realistic incidental product-use situation. *International Journal of Industrial Ergonomics, 15*, 159–166.

Edworthy, J. (1998). Warning and hazards: An integrative approach to warnings research. *International Journal of Cognitive Ergonomics, 21*, 3–18.

Edworthy, J., & Adams, A. (1996). *Warning design: A research prospective.* London: Taylor & Francis.

Edworthy, J., & Dale, S. (2000). Extending knowledge of the effects of social influence on warning compliance. In *Proceedings of the XIVth Triennial Congress of the International Ergonomics Association and 44th Annual Meeting of the Human Factors and Ergonomics Society* (pp. 4.107–4.110). Santa Monica, CA: Human Factors and Ergonomics Society.

Edworthy, J., & Hellier, E. J. (2006). Complex nonverbal auditory signals and speech warnings. In M. S. Wogalter (Ed.), *Handbook of warnings* (pp. 199–220). Mahwah, NJ: Erlbaum.

Egilman, D., & Bohme, S. R. (2006). A brief history of warnings. In M. S. Wogalter (Ed.), *Handbook of warnings* (pp. 11–20). Mahwah, NJ: Erlbaum.

FMC Corporation. (1985). *Product safety sign and label system.* Santa Clara, CA: Author.

Frantz, J. P., & Rhoades, T. P. (1993). A task-analytic approach to the temporal and spatial placement of product warnings. *Human Factors, 35*, 719–730.

Gill, R. T., Barbera, C., & Precht, T. (1987). A comparative evaluation of warning label designs. In *Proceedings of the Human Factors Society 31st Annual Meeting* (pp. 476–478). Santa Monica, CA: Human Factors and Ergonomics Society.

Godfrey, S. S., Allender, L., Laughery, K. R., & Smith, V. I. (1983). Warning messages: Will the consumer bother to look? In *Proceedings of the Human Factors Society 27th Annual Meeting* (pp. 930–934). Santa Monica, CA: Human Factors and Ergonomics Society.

Godfrey, S. S., & Laughery, K. R. (1984). The biasing effects of product familiarity on consumer awareness of hazard. In *Proceedings of the Human Factors Society 28th Annual Meeting* (pp. 483–486). Santa Monica, CA: Human Factors and Ergonomics Society.

Godfrey, S. S., Rothstein, P. R., & Laughery, K. R. (1985). Warnings: Do they make a difference? In *Proceedings of the Human Factors Society 29th Annual Meeting* (pp. 669–673). Santa Monica, CA: Human Factors and Ergonomics Society.

Goldhaber, G. M., & deTurck, M. A. (1988). Effectiveness of warning signs: "Familiarity effects." *Forensic Reports, 1*, 281–301.

Greenfield, T. K., & Kaskutas, I. A. (1993). Early impacts of alcoholic beverage warning labels: National study findings relevant to drinking and driving behavior. *Safety Science, 16*, 689–707.

Haas, E., & Edworthy, J. (2006). An introduction to auditory warnings and alarms. In M. S. Wogalter (Ed.), *Handbook of warnings* (pp. 189–198). Mahwah, NJ: Erlbaum.

Hansen, C. M. (1914). *Universal safety standards.* (2nd ed.). New York: Universal Safety Standards Publishing.

Harrell, W. A. (2003). Effect of two warning signs on adult supervision and risky activities by children in grocery shopping carts. *Psychological Reports, 92*, 889–898.

Hathaway, J. A., & Dingus, T. A. (1992). The effects of compliance cost and specific consequence information on the use of safety equipment. *Accident Analysis & Prevention, 24*, 577–584.

Hicks, K. E., Wogalter, M. S., & Vigilante, W. J., Jr. (2005). Placement of benefits and risks in prescription drug manufacturers' web sites and information source expectations. *Drug Information Journal, 39*, 267–278.

Jaynes, L. S., & Boles, D. B. (1990). The effects of symbols on warning compliance. In *Proceedings of the Human Factors Society 34th Annual Meeting* (pp. 984–987). Santa Monica, CA: Human Factors and Ergonomics Society.

Johnson, D. (1992). A warning label for scaffold users. In *Proceedings of the Human Factors Society 36th Annual Meeting* (pp. 611–615). Santa Monica, CA: Human Factors and Ergonomics Society.

Kalsher, M. J., & Williams, K. J. (2006). Behavioral compliance: Theory, methodology and results. In M. S. Wogalter (Ed.), *Handbook of warnings* (pp. 313–331). Mahwah, NJ: Erlbaum.

Kalsher, M. J., & Wogalter, M. S. (in press). Hazard control methods and warnings for caregivers and children (Chapter 14). In R. Lueder & V. Rice (Eds.), *Ergonomics for children.* Boca Raton, FL: CRC Press.

Kalsher, M. J., Wogalter, M. S., & Racicot, B. M. (1996). Pharmaceutical container labels and warnings: Preference and perceived readability of alternative designs and pictorials. *International Journal of Industrial Ergonomics, 18,* 83–90.

Klein, P. B., Braun, C. C., Peterson N., & Silver, N. C. (1993). The impact of color on warnings research. In *Proceedings of the Human Factors Society 37th Annual Meeting* (pp. 940–944). Santa Monica, CA: Human Factors and Ergonomics Society.

Kluger, R. (1997). *Ashes to ashes: America's hundred year cigarette war, the public heath, and the unabashed triumph of Philip Morris.* New York: Alfred A. Knopf.

Laughery, K. R., & Smith, D. P. (2006). Explicit information in warnings. In M. S. Wogalter (Ed.), *Handbook of warnings* (pp. 419–428). Mahwah, NJ: Erlbaum.

Laughery, K. R., & Wogalter, M. S. (1997). Warnings and risk perception. In G. Salvendy (Ed.), *Handbook of human factors and ergonomics* (2nd ed., pp. 1174–1197). New York: Wiley.

Laughery, K. R., Wogalter, M. S., & Young, S. L. (Eds.). (1994). *Human factors perspectives on warnings: Selections from Human Factors and Ergonomics Society Annual Meeting Proceedings, 1980–1993.* Santa Monica, CA: Human Factors and Ergonomics Society.

Laughery, K. R., & Young, S. L. (1991). Consumer product warnings: Design factors that influence noticeability. In *Proceedings of the 11th Congress of the International Ergonomics Association* (pp. 1104–1106). London: Taylor & Francis.

Laughery, K. R., Young, S. L., Vaubel, K. P., & Brelsford, J. W. (1993). The noticeability of warnings on alcoholic beverage containers. *Journal of Public Policy & Marketing, 12,* 38–56.

Laux, L. F., Mayer, D. L., & Thompson, N. B. (1989). Usefulness of symbols and pictorials to communicate hazard information. In *Proceedings of Interface 89—The Sixth Symposium on Human Factors and Industrial Design in Consumer Products.* (pp. 121–124). Santa Monica, CA: Human Factors and Ergonomics Society.

Lehto, M. R. (2006). Optimal warnings: An information and decision theoretic perspective. In M. S. Wogalter (Ed.), *Handbook of warnings* (pp. 89–108). Mahwah, NJ: Erlbaum.

Lehto, M. R., & Foley, J. P. (1991). Risk-taking, warning labels, training, and regulation: Are they associated with the use of helmets by all-terrain vehicle riders? *Journal of Safety Research, 22,* 191–200.

Lehto, M. R., & Miller, J. M. (1986). *Warnings, Volume 1: Fundamentals, design and evaluation methodologies.* Ann Arbor, MI: Fuller Technical.

Lehto, J. R., & Miller, J. M. (1988). The effectiveness of warning labels. *Journal of Product Liability,* 225–270.

Magurno, A. B., & Wogalter, M. S. (1994). Behavioral compliance with warnings: Effects of stress and placement. In *Proceedings of the Human Factors and Ergonomics Society 38th Annual Meeting* (pp. 826–830). Santa Monica, CA: Human Factors and Ergonomics Society.

Manufacturing Chemists Association. (1945). *Manual L-1: Guide to precautionary labeling of hazardous chemicals.* Washington, DC: Author.

Morrow, D., North, R., & Wickens, C. D. (2006). Reducing and mitigating human error in medicine. In R. S. Nickerson (Ed.), *Reviews of human factors and ergonomics, Volume 1* (pp. 254–296). Santa Monica, CA: Human Factors and Ergonomics Society.

National Safety Council. (NSC). (1928). Warning signs—Their use and maintenance. *Safe practices pamphlet series* (No. 81). Itasca, IL: Author

Ortiz, J., Resnick, M. L., & Kengskiil, K. (2000). The effects of familiarity and risk perception on workplace warning compliance. In *Proceedings of the XIVth Triennial Congress of the International Ergonomics Association and 44th Annual Meeting of the Human Factors and Ergonomics Society* (pp. 4.826–4.829). Santa Monica, CA: Human Factors and Ergonomics Society.

Otsubo, S. M. (1988). A behavioral study of warning labels for consumer products: Perceived danger and use of pictographs. In *Proceedings of the Human Factors Society 32nd Annual Meeting* (pp. 536–540). Santa Monica, CA: Human Factors and Ergonomics Society.

Parsons, S. O., Seminara, J. L., & Wogalter, M. S. (1999). A summary of warnings research. *Ergonomics in Design, 7*(1), 21–31.

Pina, K. R., & Pines, W. L. (2002). *A practical guide to food and drug law regulation* (2nd ed.). Washington DC: Food and Drug Law Institute.

Racicot, B. M., & Wogalter, M. S. (1995). Effects of a video warning sign and social modeling on behavioral compliance. *Accident Analysis and Prevention, 27,* 57–64.

Riley, D. M. (2006). Beliefs, attitudes, and motivation. In M. S. Wogalter (Ed.), *Handbook of warnings* (pp. 289–300). Mahwah, NJ: Erlbaum.

Rogers, W. A., Lamson, N., & Rousseau, G. K. (2000). Warning research: An integrative perspective. *Human Factors, 42*, 102–139.

Shaver, E. F. (2004). *Presentation modality in prescription drug direct-to-consumer television commercials.* Doctoral dissertation, North Carolina State University.

Sherer, M., & Rogers, R. W. (1984). The role of vivid information in fear appeals and attitude change. *Journal of Research in Personality, 18*, 321–334.

Silver, N. C., & Braun, C. C. (1999). Behavior. In M. S. Wogalter, D. M. DeJoy, & K. R. Laughery (Eds.), *Warnings and risk communication* (pp. 245–262). London: Taylor & Francis.

Silver, N. C., Leonard, D. C., Ponsi, K. A., & Wogalter, M. S. (1991). Warnings and purchase intentions for pest-control products. *Forensic Reports, 4*, 17–33.

Smith-Jackson, T. L. (2006a). Receiver characteristics. In M. S. Wogalter (Ed.), *Handbook of warnings* (pp. 335–344). Mahwah, NJ: Erlbaum.

Smith-Jackson, T. L. (2006b). Culture and warnings. In M. S. Wogalter (Ed.), *Handbook of warnings* (pp. 363–382). Mahwah, NJ: Erlbaum.

Smith-Jackson, T. L., & Wogalter, M. S. (2004). Potential uses of technology to communicate risk in manufacturing. *Human Factors and Ergonomics in Manufacturing, 14*, 1–14.

Smith-Jackson, T. L., & Wogalter, M. S. (2006). Methods and procedures in warning research. In M. S. Wogalter (Ed.), *Handbook of warnings* (pp. 23–33). Mahwah, NJ: Erlbaum.

Wogalter, M. S. (Ed.). (2006). *Handbook of warnings.* Mahwah, NJ: Erlbaum

Wogalter, M. S., Allison, S. T., & McKenna, N. A. (1989). Effects of cost and social influence on warning compliance. *Human Factors, 31*, 133–140.

Wogalter, M. S., Barlow, T., & Murphy, S. A. (1995). Compliance to owner's manual warnings: Influence of familiarity and the placement of a supplemental directive. *Ergonomics, 38*, 1081–1091.

Wogalter, M. S., Begley, P. B., Scancorelli, L. R., & Brelsford, J. W. (1997). Effectiveness of elevator service signs: Measurement of perceived understandability, willingness to comply, and behavior. *Applied Ergonomics, 28*, 181–187.

Wogalter, M. S., Brelsford, J. W. Desaulniers, D. R., & Laughery, K. R. (1991). Consumer product warnings: The role of hazard perception. *Journal of Safety Research, 22*, 71–82.

Wogalter, M. S., Brems, D. J., & Martin, E. G. (1993). Risk perception of common consumer products: Judgments of accident frequency and precautionary intent. *Journal of Safety Research, 24*, 97–106.

Wogalter, M. S., & Conzola, V. C. (2002). Using technology to facilitate the design and delivery of warnings. *International Journal of Systems Science, 33*, 461–466.

Wogalter, M. S., DeJoy, D. M., & Laughery, K. R. (Eds.). (1999). *Warnings and risk communication.* London: Taylor & Francis.

Wogalter, M. S., & Dingus, T. A. (1999). Methodological techniques for evaluating behavioral intentions and compliance. In M. Wogalter, D. DeJoy, & K. Laughery (Eds.), *Warnings and risk communication* (pp. 53–81). London: Taylor and Francis.

Wogalter, M. S., Godfrey, S. S., Fontenelle, G. A., Desaulniers, D. R., Rothstein, P. R., & Laughery, K. R. (1987). Effectiveness of warnings. *Human Factors, 29*, 599–612.

Wogalter, M. S., Jarrard, S. W., & Simpson, S. N. (1994). Influence of warning label signal words on perceived hazard level. *Human Factors, 36*, 547–556.

Wogalter, M. S., & Laughery, K. R. (1996). WARNING! Sign and label effectiveness. *Current Directions in Psychological Science, 5*, 33–37.

Wogalter, M. S., & Laughery, K. R. (2006). Warnings. In G. Salvendy (Ed.), *Handbook of human factors and ergonomics* (3rd ed., pp. 889–911). New York: Wiley.

Wogalter, M. S., & Laughery, K. R. (in press). Warnings. In W. Karwowski (Ed.), *Encyclopedia of human factors and ergonomics* (2nd ed.). Mahwah, NJ: Erlbaum

Wogalter, M. S., Magurno, A. B., Dietrich D., & Scott, K. (1999). Enhancing information acquisition for over-the-counter medications by making better use of container surface space. *Experimental Aging Research, 25*, 27–48.

Wogalter, M. S., & Mayhorn, C. B. (2005). Providing cognitive support with technology-based warning systems. *Ergonomics, 48*, 522–533.

Wogalter, M. S., & Mayhorn, C. B. (2006). The future of risk communication: Technology-based warning systems. In M. S. Wogalter (Ed.), *Handbook of warnings* (pp. 783–793). Mahwah, NJ: Erlbaum.

Designing Effective Warnings

Wogalter, M. S., Racicot, B. M., Kalsher, M. J., & Simpson, S. N. (1994). The role of perceived relevance in behavioral compliance in personalized warning signs. *International Journal of Industrial Ergonomics, 14*, 233–242.

Wogalter, M. S., & Silver, N. C. (1995). Warning signal words: Connoted strength and understandability by children, elders, and non-native English speakers. *Ergonomics, 38*, 2188–2206.

Wogalter, M. S., Young, S. L., & Laughery, K. R., Eds. (2001). *Human factors perspectives on warnings, Volume 2: Selections from Human Factors and Ergonomics Society annual meetings 1993–2000*. Santa Monica, CA: Human Factors and Ergonomics Society.

Wolff, J. S., & Wogalter, M. S. (1998). Comprehension of pictorial symbols: Effects of context and test method. *Human Factors, 40*, 173–186.

Young, S. L. (1991). Increasing the noticeability of warnings: Effects of pictorial, color, signal icon, and border. In *Proceedings of the Human Factors Society 35th Annual Meeting* (pp. 580–584). Santa Monica, CA: Human Factors and Ergonomics Society.

Young, S. L., & Lovvoll, D. R. (1999). Intermediate processing stages: Methodological considerations for research on warnings. In M. Wogalter, D. DeJoy, & K. Laughery (Eds.), *Warnings and risk communication* (pp. 27–52). London: Taylor & Francis.

Young, S. L., & Wogalter, M. S. (1990). Comprehension and memory of instruction manual warnings: Conspicuous print and pictorial icons. *Human Factors, 32*, 637–649.

Zwaga, H. J. G., & Easterby, R. S. (1984). Developing effective symbols for public information. In R. S. Easterby & H. J. G. Zwaga (Eds.), *Information design: The design and evaluation of signs and printed material*. New York: Wiley.

Zeitlin, I. R. (1994). Failure to follow safety instructions: Faulty communication or risky decisions? *Human Factors, 36*, 172–181.

ABOUT THE AUTHORS

Kevin B. Bennett is a full professor in the Department of Psychology at Wright State University in Dayton, Ohio. He has more than 26 years of experience in human factors, including positions in industry, the military, and academia. Bennett's primary research interests are in display and interface design and how this technology can be used to provide graphical decision support for complex sociotechnical systems. He is a Fellow of the Human Factors and Ergonomics Society and was a co-recipient of the Jerome H. Ely Award for the most outstanding article in *Human Factors* in 1993. Bennett has served on the editorial board of *Human Factors* for more than a decade. He is director for Wright State University's human factors/industrial organizational Ph.D. program and teaches courses in human factors in system development, interface design, and display design.

C. Shawn Burke works at the Institute for Simulation and Training (IST) at the University of Central Florida as a research associate. Burke earned her doctorate in industrial and organizational psychology in May 2000 from George Mason University; her primary research interests include teams, team training, team performance, shared mental models, leadership, and team performance measurement. She applies her expertise in these areas to law enforcement agencies, fire services, medical teams, Navy helicopter crews, and private organizations. Prior to working at IST, Burke was employed as a Consortium Fellow at the U.S. Army Research Institute in Alexandria, Virginia, for six years. She has also presented at numerous conferences and is a member of the American Psychological Association and the Society for Industrial and Organizational Psychology.

John G. Casali, Ph.D., CPE, is the Grado Chaired Professor of Industrial and Systems Engineering (ISE) and director of the Auditory Systems Laboratory at Virginia Tech. He is president (2006) of the National Hearing Conservation Association and a Fellow of the Human Factors and Ergonomics Society (HFES) and the Institute of Industrial Engineers (IIE). Casali received the Paul M. Fitts Education Award (1997) and the Jack A. Kraft Innovator Award (1991) from HFES, and the David Baker Outstanding Researcher Award (1999) and the Alexander Holtzman Outstanding Educator Award (2005) from IIE. He served as head of Virginia Tech's ISE Department from 1995 to 2002, during which time the department attained its highest national ranking (5th), became endowed and named, and gained exemplary department status at the university. Casali has published more than

200 papers in the human factors literature, with emphasis on auditory displays, hearing protection, and driving safety. He has served as principal investigator on more than 60 projects funded to a total of more than $5.5 million by a variety of corporations and government and military agencies. Casali has served on numerous standards committees of ANSI, ISO, and the National Fire Protection Association and holds three U.S. patents.

Angela DiDomenico is a research scientist at the Liberty Mutual Research Institute for Safety, Hopkinton, Massachusetts. She performs research in the biomechanics laboratory to determine the factors that significantly cause or contribute to workplace injuries. DiDomenico's research interests include postural stability and fall prevention, occupational biomechanics, workload assessment, and measurement of human performance and motion. Her recent investigations involve measuring postural stability at elevations and determining anticipatory locomotor adjustments during goal-directed walking. She earned her doctorate in industrial and systems engineering (human factors) at Virginia Polytechnic Institute and State University in Blacksburg, Virginia. An associate ergonomics professional (A.E.P.), DiDomenico is an active member of the Human Factors and Ergonomics Society (HFES), the Gait and Clinical Movement Analysis Society, the American Society of Biomechanics, and the American Society of Safety Engineers. She serves as the newsletter editor for the HFES Industrial Ergonomics Technical Group.

Joseph Dumas is a nationally recognized expert on usability evaluation methods and author (with J. Redish) of *A Practical Guide to Usability Testing* and many research papers and analysis articles for practitioners. Dumas has spent more than 25 years consulting on usability engineering methods with major high-tech industry leaders. He has a Ph.D. in cognitive psychology from SUNY at Buffalo. He is a senior human factors specialist at the Design and Usability Center at Bentley College, where he consults with organizations on the development of products. He also teaches courses in Bentley's master's degree program in human factors in information design.

Samir N. Y. Gerges obtained his bachelor's degree in 1964, M.Sc. in 1970, and Ph.D. from the Institute of Sound and Vibration Research, Southampton University, UK, in 1974. After five years in the aeronautical industry, in 1978 he became professor of mechanical engineering at the Federal University of Santa Catarina, Brazil, where he teaches acoustics, noise control, and signal processing to undergraduate and postgraduate students. His interests include hearing protectors, experimental and numerical vibroacoustics modeling and analysis, noise vibration harshness (NVH), and aeronautical applications for Ford, GM, Embraer, and Fiat. A founding member of the Brazilian Acoustical Society (SOBRAC), Gerges was president from 1994 to 1997 and from 2000 to 2002, as well as editor of the *SOBRAC Journal*. He is also a founding member and current vice president of the Ibero-American Federation of Acoustics and a Fellow of the Acoustical Society of America. He recently served as president and organizer of Internoise 2005. He has published four books.

Krystyna Gielo-Perczak is a scientist at the Liberty Mutual Research Institute for Safety in Hopkinton, Massachusetts. She obtained an M.Sc. (Hon.) in aeronautics and mechanical engineering and a Ph.D. in mechanical engineering and biomechanics from the Department of Aeronautics and Mechanical Engineering at the Technical University of Warsaw. She received postdoctoral training in biomechanics in the Department of Mechanical Engineering at the University of Torino (Italy). She served as a faculty member and visiting lecturer at universities worldwide, including the University of Oregon, the University of Toronto, and Victoria University of Technology (Australia). She worked at the University of Waterloo as technical director of the Gait Laboratory. Her research interests include modeling and simulation of the musculoskeletal system, control theory, and a systems approach to designing safer workplaces. Gielo-Perczak's work incorporates biomechanics, medicine, and engineering to improve approaches of existing preventive strategies of musculoskeletal injuries to industrial workers. She organized the Computer Simulation Tutorial Workshop at the International Society of Biomechanics Congress. She is a member of the editorial board of *Theoretical Issues in Ergonomics Science* and a consulting editor for *Occupational Ergonomics*.

William R. Howse is chief of the Rotary Wing Aviation Research Unit of the U.S. Army Research Institute for the Behavioral and Social Sciences (ARI). He received B.A. (1972) and M.A. (1976) degrees in experimental psychology from California State University, Los Angeles, and a Ph.D. (1982) in audiology and speech pathology from Florida State University. He worked at the U.S. Army Aeromedical Research Laboratory investigating noise damage risk criteria and effects of hearing protection on auditory localization. Howse has worked at ARI since 1985. His experience includes work in psychophysics; educational program evaluation; design and evaluation of training systems; development of performance measures for individual, crew, and collective training systems; and development of personnel selection and classification systems. His activities involve the integration of automated and manual data collection processes and the development of artificially intelligent adaptive approaches to simulation-based training.

Julie A. Jacko is a professor of biomedical engineering in the Wallace H. Coulter Department of Biomedical Engineering at the Georgia Institute of Technology and the Emory University School of Medicine. She also holds a joint appointment as professor in the College of Computing at Georgia Tech. Her research focuses on human-computer interaction, human aspects of computing, universal access to electronic information technologies, and technological aspects of health care delivery. Her externally funded research has been supported by Intel Corporation, Microsoft Corporation, the National Science Foundation, the NIH Agency for Healthcare Research and Quality, the National Institute on Disability and Rehabilitation Research, and NASA. Jacko is the recipient of the National Science Foundation's Presidential Early Career Award for Scientists and Engineers for research involving computer users with visual impairments. This award is the highest honor bestowed on young scientists and engineers by the U.S. government. She routinely provides expert consultancy for organizations and corporations on systems usability and accessibility, emphasizing human aspects of interactive systems design. Jacko is serving an elected three-year term as president of the Association for Computing Machinery (ACM) Special Interest Group on Computer-Human Interaction (SIGCHI). She earned her B.S., M.S., and Ph.D. in industrial engineering from Purdue University.

Kenneth R. Laughery is an emeritus professor of psychology at Rice University in Houston, Texas. He earned a B.S. in metallurgical engineering and an M.S. and Ph.D. in psychology from Carnegie-Mellon University. He has worked at the State University of New York at Buffalo (1963–1972), the University of Houston (1972–1984), and Rice University (1984–2002). He is a Fellow of the American Psychological Association, a Fellow and past president of the Human Factors and Ergonomics Society, and past treasurer of the International Ergonomics Association. Laughery is the recipient of the Paul M. Fitts Education Award and the Arnold M. Small President's Distinguished Service Award from the Human Factors and Ergonomics Society. His interests and work focus on consumer product safety, industrial accident analysis, and risk perception and warnings. His extensive publications on these topics include three edited books and several handbook chapters. Over 25 years, he has been an expert witness on cases involving human factors and warnings issues; specific areas of emphasis are risk perception and warnings.

Kathlene Leonard recently completed her Ph.D. in industrial and systems engineering at the Georgia Institute of Technology. Her dissertation addressed the interaction needs of individuals with visual impairments for desktop and handheld computers. Her ongoing research interests emphasize the practical application of emergent theories concerning human aspects of personal and networked computing, universal access to information technologies, and usability. Leonard received her B.S. in industrial engineering from the University of Wisconsin-Madison and her Ph.D. in industrial and systems engineering from Georgia Tech. She was also the recipient of a 2004–2005 P.E.O. Scholar Award. She is the principal research analyst for Alucid Solution in Atlanta, Georgia.

Wayne S. Maynard is director of the Ergonomics and Tribology, Loss Prevention Department at the Liberty Mutual Research Institute for Safety, Hopkinton, Massachusetts. He is responsible for the technical product development of ergonomic consulting tools and resources and training for Liberty Mutual Group's Loss Prevention business units. He consults with industries of all types on complex ergonomics issues, slips and falls, product safety, and machine safety. He received a B.A. in zoology from the University of Maine at Orono and is a Certified Safety Professional, Certified Professional Ergonomist, and Associate in Loss Control Management. Maynard is a member of the American National Standards Institute (ANSI) B11 TR.1 *Ergonomic Guidelines for the Design, Installation and Use of Machine Tools*, former Accredited Standards Committee on Control of Cumulative Trauma Disorders Z365 (accredited by ANSI) and the standard *Management of Work-Related Musculoskeletal Disorders*, and member of ANSI/HFES TAG to ISO TC 159/SC3 *Anthropometry & Biomechanics*. He is a member of the American Society for Testing and

Materials (ASTM) F13 *Pedestrian/Walkway Safety and Footwear* and subcommittee F13.10 *Traction*, chair of ASTM F13.91 *Terminology,* and member of UL Standards Technical Panel (STP) 410 *Slip Resistance of Floor Surface Materials.*

Richard W. Pew holds a B.E.E. in electrical engineering from Cornell University (1956), an M.A. in psychology from Harvard University (1960), and a Ph.D. in psychology from The University of Michigan (1963). Since 1974, he has been at BBN Technologies, where he is a principal scientist; from 1976 to 1997, he served as manager of the Cognitive Sciences and Systems Department. He is working part time for BBN. Pew has 45 years of experience in human performance and experimental psychology as they relate to systems development. Throughout his career, he has been involved in the development and utilization of human performance models and in the conduct of experimental and field studies of human performance in applied settings. Before BBN, Pew spent 11 years on the faculty of the Psychology Department at Michigan. In 2001, the university created a Collegiate Chair in his name. Pew has been president of the Human Factors and Ergonomics Society and of Division 21 of the American Psychological Association. In 1999, he was awarded the HFES Arnold M. Small Distinguished Service Award for career-long contributions to the field and to the Society. He has authored more than 70 publications, consisting of book chapters, articles, and technical reports.

Eduardo Salas is trustee chair and professor of psychology at the University of Central Florida, where he also holds an appointment as program director for the Human Systems Integration Research Department at the Institute for Simulation and Training. Previously, for 15 years, he was a senior research psychologist and head of the Training Technology Development Branch of the Naval Air Warfare Center Training Systems Division. During this period, he served as a principal investigator for numerous research and development programs on teamwork, team training, advanced training technology, decision making under stress, and performance assessment. Salas has coauthored more than 300 journal articles and book chapters and has coedited 14 books. He is on or has been on the editorial boards of numerous journals, including *Journal of Applied Psychology*, *Personnel Psychology,* and *Journal of Organizational Behavior*. His expertise includes helping organizations foster teamwork, design and implement team training strategies, facilitate training effectiveness, manage decision making under stress, develop performance measurement tools, and design learning environments. He is working on designing tools and techniques to minimize human errors in aviation, military, and medical environments. He has consulted for a variety of manufacturing, industrial, and governmental organizations and pharmaceutical laboratories. Salas is a Fellow of the American Psychological Association (SIOP and Division 21) and the Human Factors and Ergonomics Society. He received his Ph.D. (1984) in industrial and organizational psychology from Old Dominion University.

About the Authors

Marilyn Salzman is a software engineering manager at Sun Microsystems, Inc. There she leads a team of user experience designers in the development of online tools for supporting customers throughout the customer life-cycle. Prior to joining Sun, she worked as an independent consultant and for companies such as US WEST Advanced Technologies and American Institutes for Research. Salzman's interests include designing, evaluating, and modeling end-to-end user experiences. Throughout her career, she has applied these approaches to the development of consumer products, productivity tools, educational virtual realities, wireless devices, Web sites, and e-commerce applications. She has published a variety of papers on designing and evaluating products. Salzman has a Ph.D. in applied cognitive psychology and human factors engineering from George Mason University and a B.S. in human factors engineering from Tufts University.

Philip J. Smith is codirector of the Institute for Ergonomics and professor in the Industrial and Systems Engineering Program at The Ohio State University. He teaches cognitive systems engineering, artificial intelligence, human-computer interaction, and the design of distributed work systems. His research focuses on the design of cooperative problem-solving systems to aid people in performing complex tasks such as information retrieval, planning, and diagnosis using contexts such as aviation, medicine, and military planning as testbeds. He is a Fellow of the Human Factors and Ergonomics Society and was a corecipient of the Jerome H. Ely *Human Factors* Article Award for the most outstanding article in the 1999 volume. In addition to research on the impact of system design on human performance, Smith has led efforts to design and implement a number of software systems that are in widespread use, including the Post Operations Evaluation Tool (POET), used by FAA traffic managers and airline dispatchers to evaluate performance in the aviation system, and the Transfusion Medicine Tutor (TMT), in use in more than 100 programs around the world to teach diagnostic problem solving to blood bankers.

Brian Stone teaches visual communication in the Department of Industrial, Interior and Visual Communication Design at The Ohio State University. A practicing designer who has garnered numerous awards for design excellence, usability, and user satisfaction, Stone is concerned with the creation of screen-based forms that enable interaction. These forms are integrated into products such as Web sites, interactive CD-ROMs, multimedia presentations, and kinetic typographic messages. He has written several articles on the subject and has presented his research at several international venues. He is also involved in Web usability and accessibility initiatives, working toward the delivery of communication that is usable, useful, and universal. Apple Computer has recognized him as one of the top educational technology leaders in the country with the Apple Distinguished Educator Award. Stone is also a recipient of the 2005 Order of Omega Faculty Recognition Award and the 2002 Alumni Award for Distinguished Teaching at The Ohio State University. He serves on the Executive Council for the Academy of Teaching at Ohio State and is the chair of graduate studies in the Department of Design.

Yvette J. Tenney received her B.A. and Ph.D. in cognitive psychology from Cornell University while a student of Eleanor Gibson and Ulric Neisser. At Cornell, she attended J. J. Gibson's weekly seminars and still owns several of his "purple perils." As a senior scientist at BBN Technologies, Tenney has been involved in human factors work, both theoretical and applied, for more than 20 years. She has worked with users in a diverse set of domains including intelligence analysis, computer security, telephony, finance, underwater exploration, harbor navigation, complex electronic troubleshooting, commercial aviation, command and control, tax auditing, environmental noise assessment, and statistical reasoning. Tenney has published articles on automation issues in *Human Factors* and the *International Journal of Aviation Psychology* and on practical problems of memory, including the effects of aging, in *Cognitive Science* and the *British Journal of Developmental Psychology*. She played a key role on the AMBR project, which compared and contrasted human performance models with human performance, and contributed to the book *Modeling Human Behavior with Integrated Cognitive Architectures: Comparison, Evaluation, and Validation*. Tenney holds one patent.

Dennis C. Wightman received a B.A. from Florida Technical University (1972) and M.A. (1977) and Ph.D. (1983) degrees from the University of South Florida. He worked in the Human Factors Laboratory of the Naval Training Equipment Center, Naval Air Warfare Center Training Systems Division and directed the Behavioral Research Program for the Visual Technology Research Simulator facility. Wightman joined the U.S. Army Research Institute for the Behavioral and Social Sciences in 1990 and became chief of the Rotary Wing Aviation Research Unit in 1997. He retired from federal service in 2005.

Katherine A. Wilson is a doctoral candidate in the applied experimental and human factors psychology program at the University of Central Florida (UCF). She earned a B.S. in aerospace studies from Embry-Riddle Aeronautical University in 1998 and an M.S. in modeling and simulation from UCF in 2002. Also in 2002, she was awarded the I/ITSEC Graduate Student Scholarship for her academic achievements and research conducted as a research assistant at the Institute for Simulation and Training. Wilson is the lead graduate student on a project funded by the Army Research Laboratory, with a focus on improving multicultural team adaptability. She is also working on several other projects that examine teams in the aviation and health care communities. Wilson has coauthored a number of articles and book chapters and has represented UCF at several national conferences.

Michael S. Wogalter is a professor of psychology at North Carolina State University in Raleigh and a faculty member in the ergonomics and experimental psychology Ph.D. graduate program. He held faculty appointments at the University of Richmond (1986–1989) and Rensselaer Polytechnic Institute (1989–1992). Wogalter received a B.A. in psychology from the University of Virginia (1978), an M.A. in human experimental psychology from the University of South Florida (1982), and a Ph.D. in human factors psychology from

Rice University (1986). Most of his research focuses on the factors that influence the effectiveness of warnings and the kinds of information that affect people's perceptions and behavior regarding potential hazards. He has authored more than 280 research publications, including 6 edited books and 3 edited journal issues, mostly on warnings and risk communication. He is on the editorial boards of several journals, including *Ergonomics*, *Applied Ergonomics*, *Theoretical Issues in Ergonomics Science*, and the *Journal of Safety Research*. Wogalter is a member of several scientific and professional organizations and is a Fellow of the Human Factors and Ergonomics Society and the International Ergonomics Association. In 2004, he was the recipient of the HFES A. R. Lauer Safety Award.

INDEX

Abrasive floor applications, 187
Absenteeism, noise influence on, 210–211
Abstract functions and priority measures, representing, 87–91
Accessibility, computer/technology
 approaches to, 148–152
 assistive technology in, 149–150
 design, 148–149
 future directions in, 160–162
 government research initiatives, 146–147
 growing populations needing, 142–143
 performance thresholds on GUI interaction and, 155–160
 in practice, 147–148
 professional organizations, 144–146
 universal access and, 150–152
 universal usability and, 152
Accessible and Usable Buildings and Facilities, 178
Accident data and CRM training, 60
Acidic cleaners, 188
ACM Special Interest Group on Accessible Computing, 145
Acoustic trauma, 207
Action equilibrium, 169
Active/electronic level-dependent HPDs, 229–230
Active noise reduction (ANR) devices, 223–224, 231–235
Actual work environments, 97–98
Acuity, visual, 156
Agents, cognitive, 76–77

Age-related macular degeneration (AMD), 142–143
Aging Technical Group, HFES, 145
Air traffic control, 13, 15t
 technology to enhance SA in, 18
 technology with the potential to impede SA in, 22
Alkaline cleaners, 188
Alliance for Technology Access (ATA), 145
Alternatives to representation aiding, 75
American Foundation for the Blind (AFB), 155
American National Standard Methods for Measuring the Intelligibility of Speech..., 220
American National Standards Institute (ANSI), 178, 205, 241
 warning systems and, 245–246, 249–252
American Society for Testing and Materials (ASTM), 173, 175, 178–179
Americans with Disabilities Act (ADA), 178, 179
Analysis
 communication, 50–51
 heuristic, 94
Analyzers, spectrum, 201–202
Anesthesiologists, 12, 13, 26
Ankle muscles, 169
Annoyance, noise as, 196, 212
Anticipated scenarios, 78–81, 90–91, 104
Anticipation techniques, 12–13
Anticipatory postural adjustment (APA), 169
Antifatigue mats, 185–186

Apple Computer, 147
Artifacts, 129
Assistive technology, 149–150, 160–162
Assistive Technology Act of 2004, 149
Asynchronous remote testing, 115
Attention and warnings, 249–250, 255–256
Attenuation performance testing for HPDs, 234
Audibility, signal. *See* Signal
Aviation. *See also* Crew resource management (CRM) training
 air traffic controllers, 13, 15t
 SA abilities required in, 26–27
 SAGAT method and, 10
 technology to enhance SA in, 18–20
 training programs to enhance SA in, 24–25
 vision and, 8
Aviation Safety Reporting System (ASRS), 44

Back-of-the-house mats, 185
Balance control and slips and falls, 168
Behavior
 influenced by warning systems, 244
 observation, 128, 129
 shaping constraints, 76
 skill- and rule-based, 78–79
Behavioral-based instruments in CRM
 communication analysis in, 50–51
 effectiveness of, 54–60
 Line/LOS Checklist, 48

281

Index

Line Operations Safety
    Audit (LOSA), 48–49
NOTECHS, 49–50
TARGETs, 47–48
team dimensional training
    tool (TDT), 50
Benchmark testing, 112–113,
    122
Beta field studies, 131
Between-subjects design for
    usability testing, 113
Biomechanical aspect of slips
    and falls, 166–170, 176–177
Blindness
    and low vision, 155–156
    mode-change, 5
Boundaries
    conditions, 95
    indeterminate, 4, 7
Braille, 150
Broadband noise, 214–215
Brungraber Mark II (PIAST),
    172, 178, 179

Carpeting and mats, 187
Cell phone users, 22–24
Center of mass (COM), 169, 176
Center of pressure (COP), 169
Ceramic tile, 182
Characteristics
    of situations, 5–6
    of usability tests, 111
Chemical etches and cleaners,
    187
Chess games, 3–4, 5
Cleaners, floor, 188
Cleaning, floor, 187–188
Cockpit Management Attitudes
    Questionnaire (CMAQ), 47
Cockpit resource management,
    36
Coefficient of friction (COF),
    170–174, 178
Cognitive-based instruments in
    CRM, 51
Cognitive processes, human, 75,
    78–79
    controlled, 96
Cognitive triad, the, 75–77
Cognitive walkthrough, 120–121
Coherence mapping, 77
Collages, 130
Color/contrast in warnings, 250
Combat Arms™ device, 226–227

Concomitant auditory injuries,
    209
Communication
    analysis, 50–51
    -Human Information
        Processing (C-HIP)
        model, 247–248
    in mixed cultures, 43–45
    visual, 93
Comparison testing, 112–113
Competencies, teamwork, 38–39
Complex work environments,
    96–97
Compliance, warnings,
    253–255, 256–258
Computer(s). *See also*
    Accessibility, computer/tech-
        nology
    -based simulation systems,
        46
    critical interaction scenarios,
        156–157
    graphical user interfaces
        (GUIs), 155–160, 156
    -human interaction, 93–94,
        141–142, 152–155
    software, 124, 125, 147
    surveys, 124–125
    warnings, 256–257
Computer System Usability
    Questionnaire (CSUQ), 125
Computer User Satisfaction
    Inventory (CUSI), 125
Concept
    mapping, 51
    testing, early, 112
Concrete flooring, 183
Conditions, boundary, 95
Configural displays, 80, 88–90,
    99
Conservation of mass, 87
Constraints, 76
    envelopes, 95
Consumer needs, 110
Consumer Product Safety
    Commission (CPSC), 241
Content mapping, 77
Contextual inquiry, 130
Correspondence, 77
Cost of compliance with warn-
    ings, 257–258
Crew resource management
    (CRM) training
    accident data and, 60

behavioral-based instruments,
    48–51
cockpit behavior and culture
    in, 43–45
cognitive-based instruments,
    51
communication analysis in,
    50–51
computer-based simulation
    systems, 46
concept mapping in, 51
content by community,
    62–63t
design and delivery, 52–53,
    64
effectiveness of, 54–60
evaluation and transfer,
    53–54, 65
future of, 66
implementation challenges,
    65
introduction of, 36
lessons learned from, 41–60
line/LOS Checklist in, 48
Line Operations Safety
    Audit (LOSA) in, 48–49
making a business case for,
    66
myths in, 64
national culture and, 42–45
need for more research in,
    64–65
NOTECHS in, 49–50
organizational input factors
    in, 40, 55
pathfinder, 51
personality traits and, 40,
    41–42, 55
reasons for, 35–36
regulatory factors in, 40–41
research, 52–54, 60–66
and research theoretical
    drivers, 37–41
RRLOE in, 45, 46–47
safety strategies other than,
    64
science of training and, 61
selecting pilots for, 41–42
self-report measures, 47–48
separate from technical
    skills training, 65
shared mental models in,
    39–40
simulations in, 45–47

Index

social-psychological constructs in, 40–41
standardization lacking in, 61–64
success of, 37
TARGETs in, 48
Team Dimensional Training tool (TDT), 50
team effectiveness and, 38
teamwork competencies and, 38–39
Critical band masking, 215–216
Critical event techniques, 13
Critical interaction scenarios, 156–157
Cultural factors
in CRM training, 42–45
in warning systems, 259–260
Customer support logs, 130
Custom Protect Ear db Blocker, 236

Danger Signals for Public and Work Areas, 216
Decibels, 199
Decision making
cognitive agents in, 76–77
recognition-primed, 8
warning compliance, 253–255
Definition of situation awareness, 2
Design
accessible, 148–149
conventional passive hearing protection device, 222–224
CRM training, 52–53, 64
facility, 181
floor, 181–183
inclusive, 144
interaction, 92–94
interface, 76
representational, 74
of signals for use in noise, 220–221
tread, 190
usability testing, 113
warning effectiveness and, 249–255
Detectors/sensing devices and warnings, 264–265
Diagnostic testing, 112
Diaries, 130

Dichotic sound transmission HPDs, 224
Digital active noise reduction HPDs, 232–233
Digits game, the, 82–84
Direct perception, 74
Disabilities, human functioning, 153–155
Discount field studies, 132
Display warnings, 262–264
Diversity. *See* Cultural factors
Domains
law-driven, 84–86
mapping, 76
Dosimeters, 201
Driving, 15t
cell phone use while, 22–24
dynamic warnings and, 263–264
rule-based behavior and, 78
technology to enhance SA in, 20–21
technology with the potential to impede SA in, 22–24
training programs to enhance SA in, 25
Dynamic warnings, 262

Earmuffs, 203–204, 230–231
Earplugs, 203–205, 225
Combat Arms™, 226–227, 228f
E-A-R Ultra 9000™ earmuffs, 230
Ecological interface design, 74
Ecological psychology approach to SA, 3–5
Effectiveness
of CRM training, 54–60
team, 38
warning, 249–258
Emergency medicine, 17
EnableMart, 150
Endsley, M. R., 18
definition of situation awareness, 2, 5
SAGAT method, 10–12
on team situation awareness, 7
on technology with the potential to impede SA, 22
English XL tester, 172, 178, 190
Enhanced Safety through SA Integration (ESSAI), 24–25

Enhancement of situation awareness
in air traffic control, 18
in aviation, 18–20, 24–25
in driving, 20–21, 25
in infantry, 25–26
in medicine, 21, 26
training programs for, 24–26
Environmental Protection Agency (EPA), 204, 241
Equal energy rule or trading relationship, 199
Equilibrium, reaction and action, 169
Ethnographic field studies, 131
Ethnographic interviews, 130–131
Evaluation
CRM training transfer and, 53–54, 65
of designs based on representation aiding, 98–102
heuristic, 119, 122
Evaluative field studies, 131
Evaluator effect, 116
Event-based approach to training (EBAT), 52
Event sampling, 129
Examples of situation awareness, 2
Expert
judgments, 13
reviews, 118–120
Explicitness of warnings, 254–255
Exploratory sequential data analysis (ESDA), 50–51
External memory aids, 84

Facility design and slip/fall prevention, 181
Familiarity, 256–257
Federal Caustic Poison Act (FCPA), 245
Field methods
contextual inquiry, 130
defined, 127
ethnographic, 130–131
strengths and weaknesses of, 132
tools, 129–130
types and uses of, 128
Field studies
beta, 131

discount, 132
ethnographic, 131
evaluative, 131
longitudinal panel, 131–132
Flight Management Attitudes Questionnaire (FMAQ), 43, 47
Floor
  cleaners, 188
  cleaning procedures, 187–188
  design and selection, 181–183
  housekeeping procedures, 188–189
  mats, 184–186
  surfaces and treatments, 186–187
  tile manufacturers, 173, 178
Focus groups, 127
Food and Drug Administration (FDA), 241
Footwear, slip-resistant, 190–191
Formative usability testing, 111
Form mapping, 77
Frameworks for understanding situation awareness, 2–5
Freedom Scientific, 150
Frequency, sound, 198
Friction, coefficient of (COF), 170–174, 178
Functional abilities and disabilities, human, 153–155
Functional blindness, 155

Game, digits, 82–84
Glazed tiles, 182
Glue-down mats, 185
Gold Violin, 150
Government research initiatives, accessibility, 146–147
Granite flooring, 182
Graphical user interface (GUI) interaction thresholds, 155–160, 156
Groupware, 15t, 16
Guidance systems, 7

Hazard
  control hierarchy, 246–247
  and injury surveillance, 180
  perceived, 255–256
Health Care Technical Group, HFES, 144–145

Hearing
  acoustic trauma and, 207
  concomitant auditory injuries, 209
  conservation and noise regulation, 202–205
  disorders, 154
  hyperacusis, 209
  loss influenced by noise, 206–209
  noise as a risk to, 197
  noise-induced threshold shift, 207–208
  protection devices, 203–205, 210, 221–236
  tinnitus, 209
Hearing protection devices (HPD)
  active noise, 231–235
  augmented, 224
  conventional passive, 222–224
  dichotic sound transmission, 224
  digital, 232–233
  level-dependent, 226–231
  masking and, 215–220
  OSHA requirements for, 203–205, 210
  uniform attenuation, 224–226
  variable-attenuation and verifiable-attenuation, 235–236
Hedonomics, 110
Heuristic analysis, 94
Heuristic evaluation, 119, 122
  value of, 123
Horizontal Dynamometer Pull-Meter, 173
Housekeeping, 188–189, 258–259
Human behavior representation, 28
Human body
  motion stability, 168–170
  musculoskeletal system, 166–167
  nervous system, 167–168
  sensorimotor system, 168
Human cognitive and perceptual processes, 75, 96
Human-computer interaction (HCI), 93–94, 141–142, 152–155

performance thresholds and clinical functions, 157–160
Human Factors: The Journal of Human Factors and Ergonomics Society, 1
Human factors/ergonomics, 93, 109
  in addressing noise, 197–198
  and computer and technology accessibility, 143–144
  professional organizations, 144–145
  relationship with other disciplines, 134
Hurricane Katrina, 29
Hyperacusis, 209

IBM, 125
Impact ratio, 119
Impairment
  hearing (*See* Hearing)
  visual (*See* Vision)
Impedance of situation awareness, in air traffic control, 22
Inclusive design, 144
Infantry, 15t
  training to enhance SA in, 25–26
Information processing
  approach to situation awareness, 3–4
  common ground with ecological psychology, 5–7
  modeling, 8
Injury
  and hazard surveillance, 180
  noise influence on work-related, 210–211
Inquiry, contextual, 130
Inspection methods
  acceptance of, 121–122
  expert review, 118–120
  reliability of, 122–123
  strengths and weaknesses of, 123–124
  validity of, 123
  walkthrough, 120–121
Intelligibility, speech, 220
Intensity level, sound, 199
Intensive care medicine, 16
Interaction
  design, 92–94

Index

human-computer, 93–94, 152–155
scenarios, identifying critical, 156–157
Interfaces, 76
Internet, the
 accessibility for special needs populations, 145
 broadband noise and, 214–215
 groupware and, 15t, 16
 remote testing and, 115–116
 user surveys, 125
 warnings, 260
Interviews, 126
 ethnographic, 130–131
 during field research, 128, 129

James Machine, 178
Job surveys, 175
Joint loading, 166
Journals
 photo, 130
 refereed, 146
Judgments, expert, 13

Kansei engineering and SEQUAM, 135
KLM Royal Dutch Airlines, 36
Knowledge-based behavior, 81
Knowledge-based processing, 82–84
Korean Airlines, 43

Laddering, 135
Law-driven domains, 84–86
Legal blindness, 155
Level-dependent HPDs, 226–231
Likert-type rating scales, 124
Line/LOS Checklist (LLC), 48
Line Operations Safety Audit (LOSA), 48–49
Line Oriented Flight Training (LOFT) simulator, 24–25, 41
Logs, customer support and product usage, 130
Longitudinal panel studies, 131–132
Long surveys, 124–125
Loose-lay matting, 185
Low vision, 155

Management systems, network, 100
Manual L-1: A Guide for the Preparation of Warning Labels..., 245
Manual on Uniform Traffic Control Devices..., 245
Manufacturing Chemists Association (MCA), 245
Mapping
 coherence, 77
 concept, 51
 content, 77
 domain, 76
 form, 77
 interfaces, 76
 semantic, 74
Marble flooring, 182
Masking and masked threshold in noise, 212–220
Mass, conservation of, 87
Mats, floor, 184–186, 187
Measuring situation awareness
 anticipation techniques, 12–13
 critical event techniques, 13
 physiological techniques, 13–14
 recall techniques, 9–12
 subjective techniques, 13
Medicine, 12, 13
 emergency, 17
 intensive care, 16
 representation aiding in, 79–80
 SA requirements in, 15t, 16
 technology to enhance SA in, 21
 training programs to enhance SA in, 26
Memory
 aids, external, 84
 load, 11–12
Mental models, shared, 39–40
Mental processes, controlled, 95
Message length, 252–253
Microbial cleaners, 188
Microphone-in-real-ear (MIRE) techniques, 234–235
Microsoft, 117, 147
Microworlds, 96–97
Mode-change blindness, 5

Modeling
 shared mental models and, 39–40
 situation awareness, 7–9, 28–29
 slip-and-fall systems, 176
 warning compliance, 257
Musculoskeletal system, 166–167

National culture and CRM training, 42–45
National Institute for Occupational Safety and Health (NIOSH), 197, 204
National Institute on Disability and Rehabilitation Research (NIDRR), 146–147
National Institutes of Health (NIH), 197
National Safety Council (NSC), 245
National Transportation Safety Board (NTSB), 36, 43
Nature of awareness, 6–7
Nervous system, 167–168
Network management systems, 100
Neutral cleaners, 188
Noise
 acoustic trauma, 207
 as annoyance, 196, 212
 broadband, 214–215
 critical band masking, 215–216
 defined, 195–196
 design of signals for use in, 220–221
 hearing protection devices and, 203–205
 induced threshold shift, 207–208
 influence on hearing loss, 206–209
 influence on human annoyance, 212
 influence on nonauditory health, 211
 influence on task performance, 209–210
 influence on work-related injury and absenteeism, 210–211
 masking, 212–220

OSHA limits, 202
parameters and measurement of sound and, 198–202
reduction devices, active (ANR), 223–224
reduction ratings (NRR), 204–205
regulation and hearing conservation, 202–205
as a risk to hearing, 197
role of human factors engineering in addressing, 197–198
signal audibility in, 212–220
upward spread of masking, 214
Noise-induced hearing loss (NIHL), 195, 197
Noise-induced permanent threshold shift (NIPTS), 207–208
Nonauditory health, noise influence on, 211
NOTECHS, 49–50
Noticing and encoding of warnings, 249–250

Observation, behavioral, 128, 129
Occupational Safety and Health Administration (OSHA), 175, 179, 200, 241
hearing protection devices requirements, 203, 205, 210
noise exposure limits, 202
Operational aspects of slips and falls, 174–175, 176–177
Organizational input factors
in CRM, 40, 55
management commitment and, 179–180
in slip-and-fall prevention, 174–175, 179–180
Overwarning, 252–253
Oxide reduction, 188

Panel studies, longitudinal, 131–132
Participants, number of testing, 114–115, 126
Part-tasks, 12–13
Passive level-dependent HPDs, 226–227, 228f
Pathfinder, 51

Peltor T7-SR™ earmuffs, 230
Perception
direct, 74
of hazards, 255–256
processes, human, 75
situation, 6–7
Perceptual cycles, 5
Performance
active noise reduction HPDs, 234
noise influence on task, 209–210
and process measurement, 47–48
thresholds, 157–160
Personality factors in CRM training, 40, 41–42, 55
Personal protective equipment (PPE) footwear, 190
Petrochemical systems with professional operators, 100–102
Photo journals, 130
Physical form and configuration, representing, 87
Physical interactivity and warnings, 253
Physicians, 12, 13
Physiological techniques, 13–14
Pictorials in warnings, 250–252, 253–254, 259–260, 261f
Polymerization, 188
Porcelain paver tile, 182
Postural control, 169
Posturo-kinetic capacity, 169
Power level, sound, 199
Pressure level, sound, 199
Private camera conversations, 135
Proceedings of the Human Factors and Ergonomics Society, 241
Process and performance measurement, team, 47–48
Process control systems, representation aids for, 84–91
Product usage logs, 130
Psychology, ecological, 3–5
Public safety, 241, 243–244
Purpose of warnings, 243–244

Quarry tile, 181–182
Questionnaire for User Interaction Satisfaction (QUIS), 125

Questionnaires, short, 124

Rand Corporation, 144
Rapid Iterative Test and Evaluation (RITE) Method, 117
Rapidly Reconfigurable Event-Set Based Line-Oriented Evaluations (RRLOE), 45, 46–47
Rasmussen, J.
on boundary conditions, 95
on complex work environments, 96–97
evaluation of actual work environments, 97
on knowledge-based behavior, 81
on modes of behavior, 77–78
on process control systems, 84
Reaction equilibrium, 169
Real-ear-attenuation-at-threshold (REAT) techniques, 234–235
Recall techniques, 9–12
Recessed mats, 185
Recognition-primed decision making, 8
Refereed journals, 146
Regulation, noise, 202–205
Regulatory factors in CRM training, 40–41
Reliability
of expert reviews, 119–120
of inspections, 122–123
of usability surveys, 125–126
of usability testing, 116–117
Reminders, warning, 244
Remote testing, 115–116
Reporting, employee, 175
Representation aiding
in actual work environments, 97–98
alternatives to, 75
cautions in using, 92
cognitive agents and, 76–77
and the cognitive triad, 75–77
in complex work environments, 96–97
configural displays in, 80, 88–90, 99
controlled cognitive tasks and, 96

Index

controlled task situation and, 96
display design and, 98–102, 103f
domain and, 76
empirical evaluations, 94–103
evaluations of designs based on, 98–102
goals of, 74–75
interaction design and, 92–94
interface and, 76
mappings, 75–77
mental processes and, 95
for network management systems, 100
in a pasteurization microworld, 99–100
in a petrochemical system with professional operators, 100–102
for process control systems, 84–91
rationale for preferring, 75
research, 104–106
skill- and rule-based behavior, 78–79
supporting anticipated scenarios, 77–81, 90–91, 104
supporting knowledge-based processing, 82–84
supporting problem solving in unanticipated scenarios, 81–92, 104–105
Representational design, 74
Research
active noise reduction HPDs, 233–234
CRM training, 52–54, 60–66
level-dependent HPDs, 230–231
representation aiding, 104–106
warnings, 245–246
Reviews, expert, 118–120
Robotics, 15t, 17–18
Role-play exercises, 45–46
Rubber tile, 182
Rule-based behavior, 78–79

SA Global Assessment Technique (SAGAT), 10–12, 14
Sample sizes, 114–115, 126
Science of training, 61
Selection of SA abilities for specific domains, in aviation, 26–27
Self-ratings, 13
Semantic mapping, 74
Sensorimotor system, 168
Shared mental models, 39–40
Signal
audibility in noise analysis of, 213–220
general concepts in, 212–213
design for use in noise, 220–221
words in warnings, 250
Signal-to-noise ratio, 212
Simulations, CRM skills, 45–47
Situational Judgment Test (SJT), 42
Situation Awareness Model for Person-in-the-Loop Evaluation (SAMPLE), 8–9
Situation awareness (SA)
abilities, techniques for selecting, 26–28
applied to many domains, 1
and characteristics of situations, 5–6
definition of, 2
ecological psychology approach to, 3–5
examples of, 2
frameworks for understanding, 2–5
implications of studying, 14, 15t
information processing approach to, 3–4, 5–7, 8
measuring, 9–14
modeling, 7–9, 28–29
nature of, 6–7
perceptual cycles, 5
requirements in different areas, 16–18
team, 7
technology to enhance, 18–21
training programs to enhance, 24–26
Situation perception, 6–7
Situation Test of Aircrew Response Styles (STARS), 42
Skill-based behavior, 78–79
Slip resistance, 171–174
footwear, 190–191
Slips and falls
biomechanical aspect of, 166–170, 176–177
connectivity considerations of system components in, 175–177
facility design and, 181
floor-cleaning procedures and, 187–188
floor design and selection and, 181–183
floor mats and, 184–186
floor surfaces and treatments and, 186–187
future directions in preventing, 191
housekeeping procedures and, 188–189
industry safety guidelines and standards, 177–179
injury and hazard surveillance and, 180
management commitment to reducing, 179–180
motion stability and, 168–170
multidimensional approach to reducing, 175–177
operational aspect of, 174–175, 176–177
prevention, 174–175
reporting, 175
slip-resistant footwear and, 190–191
stiction and, 170–174
system components of, 166–175
training and education in, 189–190
tribological aspect of, 170–174, 176–177
workers' compensation claims, 165
Soccer, 13
SA abilities required in, 27–28
Social-psychological constructs in CRM, 40–41
Software Usability Measurement Inventory (SUMI), 125
Software Usability Scale (SUS), 124
Sonomax SonoCustom™, 236

Sound level meters (SLMs), 200–201
Sound parameters and measurement, 198–202
Spatio-temporal signals, 78
Special needs populations, 142–143, 145
Spectrum analyzers, 201–202
Speech intelligibility in noise, 220
Speed of sound, 198–199
Sports, 13, 15t
 SA abilities required for, 27–28
Stability, biomechanical attribution of, 168–170
Standard for Slip Resistance of Floor Surface Materials, 179
Standard for the Provision of Slip Resistance on Walking-Working Surfaces, 178
Standard Practice for Validation and Calibration of Walkway Surface Tribometers..., 179
Standard Test Method for Using a Horizontal Pull Slipmeter (HPS), 178
Standard Test Method for Using a Portable Inclineable Articulated Strut Slip Tester (PIAST), 178
Standard Test Method for Using a Variable Incidence Tribometer (VIT), 178
Standardization of CRM training, 61–64
Stiction, 171–174
Subjective techniques, 13
Subpart D. Walking-Working Surfaces, 179
Subpart R-Steel Erection, 179
Summative usability testing, 111
Surface-tension adhesion, 171
Surveillance, injury and hazard, 180
Surveys, 124–126, 134–135
 slips and falls, 175
Synchronous remote testing, 115

Tactile SA System (TSAS), 20
Targeted Acceptable Responses to Generated Events or Tasks (TARGETs), 47–48
Tasks
 -based reviews, 119
 constraints, 98–99
 controlled cognitive, 96
 noise influence on performance of, 209–210
 situation, controlled, 96
Team dimensional training tool (TDT), 50
Teams
 effectiveness, 38
 process and performance measurement, 47–48
 recall techniques applied to, 12
 reviews, 119
 situation awareness, 7
 slip and fall prevention by, 174–175
 and teamwork competencies, 38–39
 training, 52–53
Teamwork competencies, 38–39
Technical skills training separate from CRM skills training, 65
Techniques for measuring situation awareness
 anticipation, 12–13
 critical event, 13
 physiological, 13–14
 recall, 9–12
 subjective, 13
Technology, assistive, 149–150
Technology and warnings detectors/sensing devices, 264–265
 in different contexts, 260
 displays, 262–264
 dynamic aspects of, 262
 potential barriers to, 266
Temporary threshold shift (TTS), 197
Terrazzo flooring, 182
Testing, usability
 benchmark, 112–113, 122
 comparison, 112–113
 defined, 111
 diagnostic, 112
 early concept, 112
 focus on fixing problems, 117
 formative, 111
 number of participants and, 114–115
 reliability of, 116–117
 remote, 115–116
 strengths and weaknesses of, 117
 think-aloud, 111, 114
Test Method for Determining the Static Coefficient of Friction of Ceramic Tile..., 178
Test Method for Static Coefficient of Friction of Polish-Coated Surfaces..., 178
Think-aloud testing, 111, 114
Tic-Tac-Toe representation, 82–84
Tiles
 ceramic, 182
 porcelain paver, 182
 quarry, 181–182
 rubber, 182
 vinyl composition, 172, 182
Time sampling, 129
Tinnitus, 209
Total blindness, 155
Training programs. *See also* Crew resource management (CRM) training
 to enhance SA, 24–26
 slip-and-fall prevention, 189–190
Trauma, acoustic, 207
Tread material, 190–191
Triad, cognitive, 75–77
Tribological aspect of slips and falls, 170–174, 176–177

Unanticipated scenarios, representations to support problem solving in, 81–92, 104–105
Underwriters Laboratories (UL), 179, 241
Uniform attenuation HPDs, 224–226
United Airlines, 36
Universal access, 150–152
Universal Access in the Information Society, 146
Universal access to computers and technology, 144
Universal Principles of Design, 149
Universal usability, 152

Index

Universal Walkway Tester (UWT), 172
Upward spread, noise, 214
Usability, universal, 152
Usability assessment
  characteristics, 111
  defined, 110
  diagnostic testing, 112
  early applications of, 109–110
  early concept testing, 112
  field methods, 127–132
  focus groups, 127
  focus on fixing problems, 117
  formative, 111
  general rules for using, 133
  human factors/ergonomics professionals and other disciplines using, 134
  inspection methods, 118–124
  interviews, 126
  number of participants in, 114–115
  reliability of, 116–117
  remote, 115–116
  strengths and weaknesses of, 117
  summative, 111
  surveys, 124–126, 134–135
  testing, 111–117
  thoughtful use of different methods in, 133
  variations and importance of thinking aloud in, 114
  walkthroughs, 120–121
Usability Magnitude Estimation (UME), 112–113
Users
  assessing product experience of, 134–135
  -based reviews, 119
  field research on, 129–130
  logs, 130
  special needs populations of, 142–143, 145
  universal access and, 150–152
  visually impaired computer, 142–143
  walkthroughs with, 121
  warnings tailored to, 265–266

Validity
  of inspections, 123
  of usability surveys, 125–126
Variable-attenuation and verifiable-attenuation HPDs, 235–236
Vibrations, sound, 199
Vinyl composition tile (VCT), 172, 182
Vision, 8
  acuity, 156
  assessment, 158–159t
  disabilities, 154–155
  driving and, 22–23
  HCI performance thresholds and, 157–160
  impairment of, 142–143, 155–156
  slips and falls and, 168
  and visual display design, 93–94
Visual communication, 93
Visual display design, 93

Walkthroughs, 120–121
Walkway surface slipmeters, 171–173
Warnings
  attention and, 249–250, 255–256
  audience diversity and, 259–260
  cleanser, 257, 258–259
  color/contrast in, 250
  compliance decisions and, 253–255, 256–258
  cost of compliance with, 257–258
  cultural factors in, 259–260
  design factors that influence effectiveness of, 249–255
  detectors/sensing devices, 264–265
  display, 262–264
  dynamic, 262
  effectiveness, 249–258
  explicitness of, 254–255
  familiarity and, 256–257
  future directions in, 259–266
  goals of, 241, 258–259
  hazard control hierarchy and, 246–247
  influencing behavior, 244
  location/placement of, 249–250
  medium used by, 242
  message length in, 252–253
  modeling and compliance with, 257
  nondesign factors that influence effectiveness of, 255–258
  over-, 252–253
  physical interactivity in, 253
  pictorials in, 250–252, 253–254, 259–260, 261f
  potential barriers to effective, 266
  providing information, 244
  public safety and, 241, 243–244
  purpose of, 243–244
  reminders and, 244
  research and applications history, 245–246
  signal words in, 250
  size of, 249
  systems approach to, 246–247
  tailored to the user, 265–266
  technology and, 260–266
  theoretical approaches to, 247–249
Wavelength, sound, 198–199
Waxes and slip-resistant applications to floors, 187
W3C Web Accessibility Initiative, 145
Web Site Analysis and Measurement Inventory (WAMMI), 125
Well and grate systems, 185
Wildcard effect, 113
Windows 95 software, 147
Within-subjects design for usability testing, 113
WOMBAT Situational Awareness and Stress Tolerance Test, 26–27
Words, signal, 250
Work environments
  actual, 97–98
  complex, 96–97
Worker's compensation claims, 165
World Health Organization, 196